Specifications and Criteria for Biochemical Compounds

THIRD EDITION

Prepared by the
Committee on Specifications and Criteria
for Biochemical Compounds
Division of Chemistry and
Chemical Technology
National Research Council

with support from the
National Institute of
General Medical Sciences

D1505735

NATIONAL ACADEMY OF SCIENCES
Washington, D.C. 1972

NOTICE: This publication is the result of a study undertaken under the aegis of the National Research Council of the National Academy of Sciences with the express approval of the Governing Board of the NRC. Such approval indicated that the Board considered the problem involved to be of national significance, requiring scientific and technical competence for its study, and particularly suited to the resources of the NRC. The institutional responsibilities of the NRC were then discharged in the following manner:

The members of the study committee were selected for their individual scholarly competence and judgment with due consideration for balance and breadth of interests. Responsibility for all aspects of this publication rests with these members, to whom sincere appreciation is hereby expressed.

Although publications of NRC study committees are not submitted for approval to the Academy membership nor to the Council, each publication is reviewed according to procedures established and monitored by the Academy's Report Review Committee. Distribution is approved, by the President, only after satisfactory completion of this review process.

ISBN 0-309-01917-6
Library of Congress Catalog Card Number 60-60094

Available from
Printing and Publishing Office
National Academy of Sciences
2101 Constitution Avenue, N.W.
Washington, D.C. 20418

Printed in the United States of America

Preface

This publication is the result of a program to improve the quality of chemicals available for biochemical research by establishing criteria, standards, or specifications useful for describing such chemicals, particularly with regard to purity. These pages represent an effort to satisfy a long-felt need of biochemists for more knowledge of the chemicals used in their investigations.

The program was initiated by Dr. Sam R. Hall* who, as a result of discussion with prominent biochemists, brought the matter to the attention of the Biochemistry Study Section of the National Institutes of Health at its meeting of January 1955. The Study Section recommended that a joint committee be established by the two largest organized groups of biochemists, the American Society of Biological Chemists and the Division of Biological Chemistry of the American Chemical Society. The governing bodies of these societies referred the problem to the then newly formed Committee on Biological Chemistry of the Division of Chemistry and Chemical Technology, National Academy of Sciences–National Research Council. Under the chairmanship of Prof. Herbert E. Carter, this Committee surveyed the need by submitting a questionnaire to the memberships of both societies. The response was 42 percent (considerably beyond normal expectations), and about 90 percent of the replies indicated a serious need for improvement in the quality and the standards of biochemicals.†

* Dr. Hall was at that time Executive Secretary of the Endocrinology Study Section of the National Institutes of Health. He played a leading role in the early development of this program and remained active in it until he left NIH in 1956 to join the staff of the American Cancer Society.
† *Science*, **123**, 54 (1956).

This response, together with many additional comments and communications, was regarded by the Committee on Biological Chemistry as a mandate for action.

Much of the earlier work on the project was carried out by the Committee on Biological Chemistry. However, the need for a continuing examination of problems in this area has been recognized by the formation of a Committee on Specifications and Criteria for Biochemical Compounds, a standing committee of the Division of Chemistry and Chemical Technology of the National Academy of Sciences–National Research Council. This committee has been responsible for preparing the third edition of this work. Many individuals were involved, and they relied heavily on the work of the numerous persons who contributed to the preparation of the earlier editions. A cumulative list of all the members of the former Committee on Biological Chemistry and the earlier subcommittees is to be found in the second edition, published in 1967. Following this preface is a list of present members of the Committee on Specifications and Criteria for Biochemical Compounds and its subcommittees. It should be realized that many other individuals helped immeasurably with the task of preparing this volume, and the Committee expresses its gratitude to these unnamed persons.

The principle guiding the Committee and its subcommittees has been to select those criteria and specifications that permit adequate characterization of quality for each compound included. For the sake of uniformity, the following definitions have been adopted as guidelines: *Criteria* state what a compound or material is, or what it does, or both; criteria describe any or all of the following: (a) chemical properties, (b) physical properties, (c) some kind of activity; criteria describe the properties of the purest and most highly active specimen reported to date. *Specifications* are based on criteria; they give acceptable ranges for the selected criteria and procedures for determining that these requirements have been met. Specifications may also describe properties that the material shall not possess, as well as requirements for preserving, handling, packaging, dating, and labeling.

Fine biochemicals—unlike inorganic chemicals—are supplied by a comparatively small number of specialized manufacturers who are, however, widely dispersed on this continent, as well as in Europe, Asia, and South America. The preparations may involve not only a variety of procedures, but also starting materials of various degrees of purity. Some classes of compounds, such as the sugars and amino acids, were relatively well characterized before this program was initiated. For those compounds, the criteria and specifications represent accurate descriptions of existing high-grade commercial preparations. Many other substances, however, such as the enzymes and coenzymes, cannot be described so rigorously. The subcommittees felt that it would be more appropriate to describe such substances in terms of criteria by which the user could form his own judgment as to their purity.

The Committee believes that the major burden of evaluating the acceptability of biochemicals for any specific use or purpose belongs, in the long run, to the users themselves. Although the representations of suppliers may generally be relied upon, users should be acquainted with the best available and practicable guide for recognizing the degree of purity required, and it should be the user's responsibility to determine what, if any, further purification may be needed. Many of the compounds for which criteria and specifications are given in this book are labile, and precautions should be taken to ensure their stability. Normally, recommendations of the manufacturer should be followed when storing these materials.

This edition contains criteria and specifications for 521 compounds. The first edition consisted of two volumes of loose-leaf pages, covering 225 compounds, and the second was a hard-cover volume containing data on 392 compounds.

Although there has been a progressive increase in the size of successive editions, the coverage is, of course, by no means complete. Each of the subcommittees has additional compounds under consideration for future editions, and other subcommittees will probably be appointed to consider categories of biochemicals not yet included. It is considered that the criteria and specifications here selected for the individual compounds constitute the best information available to date. In the future,

improved techniques will undoubtedly reveal better criteria, and improvements in manufacturing practice will put on the market products having a degree of purity not now commercially feasible.

The problem of setting criteria and specifications for radioactively labeled biochemicals is one that has long plagued the Committee, for it was realized that this Pandora's Box should be opened with care. Nevertheless, with so many labeled compounds available commercially, a start in this direction had to be made. As a first step, some general comments concerning the use of these materials are included in this volume, in a section prepared by Dr. Horace S. Isbell. For the performance of this task, the Committee is extremely grateful to Dr. Isbell, who has acquired an extensive knowledge of the problem through his many years of experience in the synthesis and use of isotopically labeled materials.

In accordance with unit prefixes and abbreviations recommended by the International Bureau of Weights and Measures,* nm has been used for nanometre (formerely, $m\mu$ for millimicron) and μm for micrometre (formerly, μ for micron). Where the name in common use for a compound differs from that in the Subject Index to *Chemical Abstracts,* the latter name is usually given also (parenthetically in boldface type) in the heading for a specification.

The Committee on Specifications and Criteria for Biochemical Compounds would greatly appreciate having any errors or omissions found in this publication brought to its attention. This request is earnestly directed to all purchasers and other interested persons. The Committee is also eager to receive criticisms and suggestions regarding either substantive text or publication format. As in many ventures with multiple authorship, there have been differences of opinion on selection of information and on the best manner to assemble and present the material. It is hoped that suggestions from the scientific public will result in improvement of future editions.

Another important aspect is evaluation by users of the adequacy of the data in actual situations. If some users find, for example, that the specifications and criteria are of limited value to them because of the manner of presentation or because certain additional information is not supplied, these facts will be of interest to the Committee, and communications will be welcome. In this connection, it should be pointed out that the biochemical characterizations were developed for general use in research. The criteria and specifications of the biochemical materials in this publication relate to purity standards and not to use of the materials for pur-

* "The International System of Units (SI)," National Bureau of Standards Special Publication 330 (January 1971).

poses other than biochemical research. Users with special requirements should discuss their needs directly with the manufacturers.

Certain commercial products and instruments are identified in this publication in order to specify the experimental procedure in sufficient detail. In no case does such identification imply recommendation or endorsement by the National Academy of Sciences–National Research Council, nor does it imply that the product or equipment so identified is necessarily the best available for the purpose.

Reports of errors, suggestions for revisions, and other comment on either substance or format may be addressed to:

Committee on Specifications and Criteria for
   Biochemical Compounds
Division of Chemistry and Chemical Technology
National Academy of Sciences–National Research
   Council
Washington, D.C. 20418

In conclusion, the Committee expresses appreciation for encouragement and support by the National Institutes of Health under Contract PH 43-64-44, Task Order No. 13, without which this project could not have been realized. The Committee owes a special debt of gratitude to Dr. R. Stuart Tipson, who edited the manuscript. Thanks are owing also to Dr. Waldo E. Cohn, Director of the NAS–NRC Office of Biochemical Nomenclature, for valuable advice regarding approved nomenclature.

COMMITTEE ON SPECIFICATIONS AND CRITERIA FOR BIOCHEMICAL COMPOUNDS

Donald L. MacDonald, *Chairman,* Oregon State University

Carbohydrates and Related Compounds

ARTHUR G. HOLSTEIN, *Chairman,* Pfanstiehl Laboratories, Inc.
CLAUDE T. BISHOP, National Research Council, Canada
ROBERT R. ENGLE, National Cancer Institute
HEWITT G. FLETCHER, JR., National Institute of Arthritis and Metabolic Diseases
LEON GOODMAN, University of Rhode Island
HORACE S. ISBELL, National Bureau of Standards (ret.)
REX MONTGOMERY, University of Iowa
EDWARD WELTON, Merck & Co., Inc.
MELVILLE L. WOLFROM, Ohio State University (deceased)

Carotenoids and Related Compounds

JOHN W. PORTER, *Chairman,* University of Wisconsin
C. R. BENEDICT, Texas A&M University
RICHARD E. DUGAN, Veterans' Administration Hospital, Madison, Wisconsin
T. W. GOODWIN, The University of Liverpool
O. ISLER, F. Hoffmann LaRoche & Co., Basel
SYNNØVE LIAAEN-JENSEN, Norwegian Institute of Technology
JAMES A. OLSON, Middlesex Hospital Medical School, London
F. W. QUACKENBUSH, Purdue University

Coenzymes and Related Compounds

JOHN LOWENSTEIN, *Chairman,* Brandeis University

Enzymes

RICHARD W. VON KORFF, *Chairman,* Maryland Psychiatric Research Center
MARY E. DEMPSEY, University of Minneapolis Medical School

Contents

Carbohydrates and Related Compounds (continued) *Code Number* *Page*

Carotenoids and Related Compounds

Nucleotides and Related Compounds (continued)

Specifications
and
Criteria
for
Biochemical
Compounds

Amino Acids and Related Compounds

GENERAL REMARKS

The Subcommittee on Amino Acids has selected for consideration those amino acids that, for the most part, are readily available on the commercial market. Of these compounds, some are racemic and some are the natural, optically active L forms. The latter compounds have been isolated from the hydrolyzates of various proteins or by resolution of the synthetic racemic amino acids.

The manufacturers of amino acids have performed an outstanding service in furnishing materials to the biochemical field in volume and at relatively low cost. Each year sees a lowering of costs, with consequent gain to the consumer. In general, the quality of the amino acids sold has been high and for this the manufacturers deserve added credit. What constitutes an adequate standard of purity, however is, in the long run, the responsibility of the investigator.

The Subcommittee initially selected for consideration only those amino acids that are generally recognized to be present in proteins. Subsequently, several new amino acids, derivatives, and intermediate metabolites were added. These have been retained, but no new additions have been made. Several of the amino acids are available in the D form, but these have not been listed separately because most of their properties are the same as those of their L counterparts, the primary differences being in their direction of rotation and their enzymic response. Occasionally, slight differences are found in their R_f values on paper chromatograms.

ANALYTICAL AND IDENTIFICATION PROCEDURES[1,2]

Melting Point

The melting points of the amino acids and their salts are generally not sharp, and are therefore of little value either as a means of identification or as a criterion of purity.

Loss of Weight on Drying

Unless otherwise specified, the general procedure for determining volatile contaminants (normally water) is the following, described in section 31.005 ("Vacuum Drying—Official Final Action") of the *Official Methods of Analysis of the Association of Official Analytical Chemists,* 11th ed., AOAC, Washington, D.C. (1970), p. 525.

Dry 2–5 g of prepared sample (ground, if necessary, and mixed to uniformity) in flat dish (Ni, Pt, or Al with tight-fit cover), 2 h at ≤ 70 °C (preferably 60 °C), under pressure ≤ 50 mmHg. Bleed oven with current of air (drying by passing through anhydrous $CaSO_4$, P_2O_5, or other efficient desiccant) during drying to remove H_2O vapor. Remove dish from oven, cover, cool in desiccator, and weigh. Redry 1 h and repeat process until change in weight between successive dryings at 1-h intervals is ≤ 2 mg.

Unless it is otherwise indicated, the pure amino acids should not lose more than 0.5% of their weight.

Ash (sulfated)

Weigh 5 g of the amino acid into a tared 50–100-ml platinum dish. Add 5 ml of 10% (by wt) sulfuric acid, heat gently with a low flame until the sample is well carbonized, and then ash in a muffle furnace at about 550 °C. Cool, add 2–3 ml of 10% sulfuric acid, evaporate on a steam bath, dry on a hot plate, and ignite at 800±25 °C for 15 min.

Heavy Metals (as Pb)

The sample is ashed at a low temperature, and the ash taken up with hydrochloric acid. The solution is filtered if turbid, buffered with sodium acetate, and treated with hydrogen sulfide. The turbidity caused by heavy-metal sulfides is compared with known standards.

Reagents

1. Standard lead solution: Dissolve 1.5985 g of lead nitrate, $Pb(NO_3)_2$, and 1 ml of concentrated nitric acid in water, and make up to 1 liter. Dilute 10 ml of this stock solution to 1 liter to obtain a solution containing 10 μg of lead per ml. The dilute solution is not stable, and should therefore be freshly prepared at frequent intervals.

2. Hydrogen sulfide solution: Prepare by allowing hydrogen sulfide to bubble through distilled water for 10 min.

3. Sodium acetate, 20 g made up to 100 ml with distilled water.

4. Hydrochloric acid, HCl, 6 M solution.

5. Acetic acid, glacial, C.P.

Procedure

1. To a 2-g sample in a small casserole, add 1 ml of a 20% solution of magnesium nitrate, dry, and ash. The temperature should not exceed 500 °C. It may be necessary to cool the sample, wet it with concentrated nitric acid, dry, and reignite it to complete the oxidation of the organic matter.

2. Add to the residue 10 ml of 6 M hydrochloric acid and 1 ml of nitric acid. Cover and digest on the steam bath for 15 to 20 min. Uncover, wash down the sides of the casserole and the cover glass, and evaporate to dryness. Add 2 ml of 6 M hydrochloric acid, warm gently, dilute, and filter through a small Whatman No. 40 filter paper into a 50-ml color-comparison tube. Add 2 ml of sodium acetate solution, and neutralize part of the acid with ammonium hydroxide, if necessary. The pH should be about 5 before the hydrogen sulfide is added.

3. Dilute the buffered solution to 40 ml, add 10 ml of hydrogen sulfide solution, mix, and compare with standards of the same volume prepared by adding graduated amounts of lead to solutions containing the quantities of reagents used in the test and treated exactly like the sample.

The limit of sensitivity of this test is about 20 μg of lead.

$$\frac{\text{equivalent lead } (\mu g)}{\text{mass of sample (g)}} \times 10^{-4} = \% \text{ heavy metals}$$

Ultimate Analyses

These are performed by standard procedures. The values found should agree with the calculated values within the following limits:

C, 1% relative; N, 3% relative. Sulfated ash should not exceed 0.1%.

Specific Rotation

The specific optical rotations are determined on samples previously dried to constant weight by the procedure given under "Loss of Weight on Drying" and made up to a 1–2% solution. The numerical value obtained will depend upon the solvent system employed, according to the rule of Clough, Lutz, and Jirgenson. This rule applies only to those α-amino acids that contain a single center of asymmetry, and may be given as follows: The addition of acid to an aqueous solution of an L-amino acid will cause the specific rotation to be shifted toward a more positive value; aqueous solutions of D-amino acids show a negative shift in direction of rotation on like treatment.

The above rule is not rigidly applicable to α-amino acids having two or more centers of optical asymmetry. It does become applicable, however, when based upon shifts in the partial molar rotation of the α-asymmetric center as determined by calculations developed by Winitz, Birnbaum, and Greenstein, *J. Am. Chem. Soc.*, **77**, 716 (1955).

$$[\alpha]_D^t = \frac{100\alpha}{\rho l},$$

where $[\alpha]_D^t$ is the specific rotation at t degrees Centigrade using the D-line of sodium;

α is the observed rotation in degrees,

ρ is the mass concentration of amino acid in grams per 100 ml of solution,

l is the length of the polarimeter tube in decimeters, and

t is the temperature.

The D-line of sodium is used except where indicated otherwise. Conditions of concentration, temperature, and solvent are those found in the literature. For several of the amino acids, solutions in water give the highest specific rotation, although this may be offset by low solubility (and hence a low observed rotation). When HCl is employed, a large excess is often necessary to insure the presence of the amino acid as a single species. Readings should be taken in a darkened room, preferably by two or more individuals. The method is not of the highest accuracy, the results being simply those agreed upon by the majority of observers. A difference of ±2% in such readings may well be tolerated.

Paper Chromatography[3, 4]

Although the paper-chromatographic methods described here will not give the quantities of impurities present, the latter can be approximated by noting the size and intensity of the spots in comparison with those given by known quantities of the contaminant. The sulfur-containing amino acids, and the iodotyrosines and iodothyronines may decompose during and after application to the paper. It is, therefore, advisable to avoid the use of heat and strong light during their application to the paper and to develop the chromatograms immediately after the spot has dried.

Spotting with 50 μg of the amino acid is recommended. The following solvents and color reagents are used for developing and staining:

Solvents

1. 88% w/v Phenol–H_2O = 100:20 v/v. Add 15 mg of 8-quinolinol per 100 ml. Run in the presence of 100 ml of 2 M NH_4OH.
2. 1-Butanol–Acetic Acid–H_2O = 450:50:125 v/v.
3a. 2-Butanol–3.3% aq NH_4OH = 150:60 v/v. Aqueous NH_4OH is prepared by diluting 65 ml of 14.8 M NH_4OH to 500 ml with H_2O.
3b. 2-Butanol–3% aq NH_4OH = 150:50 v/v (1:6 aq NH_4OH atmosphere).
4. 2-Methyl-1-butanol–Formic Acid 88% w/v–Water = 70:1:29 v/v.
5. 2-Methyl-1-butanol–2 M NH_4OH = 70:30 v/v.
6. tert-Butanol–Formic Acid, 88% w/v–Water = 70:1:29 v/v.
    Method A. Solvent 1 followed by Solvent 2.
    Method B. Solvent 1 followed by Solvent 3.
    Method C. Solvent 2 followed by Solvent 3.

Paper

Whatman No. 1 and/or Whatman No. 3.

Color Reagents

Ninhydrin for all amino acids: 0.25% w/v in acetone. Sakaguchi for arginine: 1-Naphthol–hypobromite reagent. Prepare a 0.01% w/v solution of 1-naphthol in ethanol containing 5% w/v urea. Add KOH to 5% w/v just before spraying. Air dry a few minutes, and spray lightly with a solution of 0.8 ml of Br_2 in 100 ml of 5% w/v KOH.

Nitroprusside for cystine and cysteine: Reagent 1: Sodium nitroprusside (1.5 g) is dissolved in 5 ml of 1 M H_2SO_4. Then 95 ml of methanol and 10 ml of 28% ammonia are added. The solution is filtered and stored in the refrigerator. Reagent 2: Two grams of NaCN is dissolved in 5 ml of water and diluted to 100 ml with methanol. Prepare fresh.

Pauly for histidine and tyrosine: Reagents: Sulfanilamide (1% w/v) in 10% v/v HCl; 5% w/v $NaNO_2$; half-saturated Na_2CO_3. Place 5 ml of sulfanilamide solution and 5 ml of $NaNO_2$ in a 100-ml separatory funnel. Shake for 1 min. Then add 50 ml of 1-butanol. Shake for 1 min and let stand for 4 min. Decant the butanol layer, and spray or dip the chromatogram. Dry the sheet in a current of air and then dip into the Na_2CO_3 solution. Imidazoles give a deep cherry-red color.

Ehrlich for tryptophan: A mixture of 1 g of p-dimethylaminobenzaldehyde, 90 ml of acetone, and 10 ml of concentrated HCl is freshly prepared.

Platinic iodide for cystine, cysteine, and methionine: Add, in the following order, 4 ml of 0.002 M $PtCl_6$, 0.25 ml of 1 M KI, 0.4 ml of 2 M HCl, and 76 ml of acetone. The dried chromatograms are dipped into this reagent. Cystine, cysteine, methionine, and some reducing substances also give a white spot on a red-purple background.

Ceric sulfate–sodium arsenite for iodoamino acids: See directions for iodo compounds below.

Periodate–Nessler for serine and threonine: The paper is sprayed with Nessler's reagent almost saturated with $NaIO_4$.

Isatin for proline and hydroxyproline: Spray with 0.40% w/v isatin in acetone, dry the chromatogram in air, and heat it for 10 min in an oven at 70–76 °C and 100% humidity. Proline and hydroxyproline give blue colors: cystine and tyrosine often also give blue colors. Glutamic and aspartic acids give pink spots that turn blue on standing. The other amino acids give pink spots that fade. Pipecolic acid also gives a blue color with isatin.

Qualitative Detection of Iodo Compounds on Paper Chromatograms

Reagents

A. 10% ceric sulfate, $Ce(HSO_4)_4$, w/v in 10% H_2SO_4 v/v.
B. 5% w/v aqueous sodium arsenite (Na_2AsO_3).
C. 0.05% w/v methylene blue.
D. Ammonia (sp. gr., 0.90).

Procedure

After development of the chromatogram, the paper is dried thoroughly in an atmosphere free of phenol. The paper is then sprayed lightly on both sides with a mixture of reagents A and B in the proportion of 2 and 3 volumes, respectively (use immediately after mixing). The paper is dried in an air stream (fan) for 5 min. It is then sprayed lightly with reagent C, dried for 5 min, and then subjected to NH_3 vapor. The completion of the neutralization can be detected by a change in the background color from pale purple to bright yellow. The chromatogram is again dried. The iodo compounds appear as bright blue spots against a yellow background. The minimum amount detectable is of the order of 0.1 to 0.2 μg of iodine.

Estimation of Carboxyl Groups (Van Slyke)

Principle

Amino acids are oxidized at acid pH to yield CO_2, which is determined by collection in barium hydroxide.

$$R \cdot CH(NH_2)COOH + \text{[ninhydrin]} \rightarrow R \cdot CHO + NH_3 + CO_2$$

Apparatus

Carbon dioxide can be determined either in the well-known Van Slyke-Neill manometric gas-analysis apparatus, or more conveniently in two 25-ml Erlenmeyer flasks without lips, one for the reaction and one for standard $Ba(OH)_2$. The flasks are connected by means of a ∪-tube. The ∪-tube has a small sidearm that permits the entire apparatus to be evacuated.

Reagents

1. Citrate buffers. pH 4.7: Grind together 17.65 g of trisodium citrate dihydrate and 8.40 g of citric acid monohydrate to a fine powder. pH 2.5: Grind together 2.06 g of trisodium citrate dihydrate and 19.15 g of citric acid monohydrate.

2. Standard $Ba(OH)_2$. 0.125 M: Saturated $Ba(OH)_2$ is adjusted to 0.15 M with CO_2-free water. Five volumes of this solution are mixed with 1 volume of 12% w/v $BaCl_2$ prepared in CO_2-free water. Three milliliters of the standard baryta should be titrated with M/7 HCl to pH 8, using the veronal (barbital) buffer color standard with 1 drop of 1% phenolphthalein as the indicator to obtain the blank value.

3. Veronal (barbital) buffer. pH 8: Dissolve 10.3 g of sodium veronal in 500 ml of water. Mix 7 ml of this solution with 4 ml of M/14 HCl; add phenolphthalein. This is the color to which the $Ba(OH)_2$ is titrated.

Method

1. Hydrolyzate. Adjust to faintly **acid** to bromophenol blue.

2. Removal of preformed CO_2. To 1–5 ml of solution in the 25-ml Erlenmeyer flask, add 50 to 100 mg of either pH 2.5 or 4.7 citrate buffer and 1 drop of alcohol. Boil off preformed CO_2. Stopper and cool to 15 °C. In the meantime, remove CO_2 from titration flask by a stream of CO_2-free air. Add 3 ml of 0.125 M $Ba(OH)_2$ to the titration flask.

3. Oxidation. Add 50 to 100 mg of ninhydrin to reaction flask. Connect to $Ba(OH)_2$ flask by means of ∪-tube and rubber tubing. Make connections glass to glass. Evacuate at the water pump, and then immerse entire apparatus, as far as the top of the ∪-tube, in boiling water for 7 min or longer.

4. Distillation of CO_2. Cool the $Ba(OH)_2$ flask in water, but keep the reaction vessel at 100 °C. Shake for 3 min to facilitate the distillation and absorption of all CO_2.

5. Titration. Titrate $Ba(OH)_2$ with M/7 HCl from a 5-ml burette, using 1 drop of 1% phenolphthalein as the indicator. Subtract reagent blank.

1 ml M/7 HCl = 1 mg of carboxyl nitrogen
= 0.857 mg of carboxyl carbon

Estimation of Amino Group (Van Slyke, Modified by Sobel)

Principle

Ammonia resulting from the oxidative deamination of amino acids with ninhydrin is aerated into boric acid and titrated.

Reagents

1. Octyl alcohol saturated with thymol.
2. Ninhydrin, solid.
3. Buffer: pH 2.5 citrate buffer [See "Estimation of Carboxyl Groups (Van Slyke)"]. Dissolve in water to 10% w/v.
4. H_2O_2: 30% w/v.
5. Saturated KOH: To a cylinder containing water, under mineral oil 1-in. thick, add solid KOH to saturation. Preserve under oil.
6. Indicator: Ten parts of 0.1% w/v bromocresol green plus 1 or 2 parts of 0.1% w/v methyl red in 95% ethanol.
7. Boric acid: Twenty grams of H_3BO_3 is diluted to 1 liter with water. Add 20 ml of indicator per liter.
8. Hydrochloric acid: Standard 0.0714 M HCl.

Method

Place 1 ml of solution containing 20 to 100 mg of N in an aeration tube and add 0.3 ml of buffer and 50 mg of ninhydrin. The pH should be 2.4 to 2.6. Mix and place in boiling water for 10 min. At the end of 2 min heating, shake to dissolve the ninhydrin. At the end of 10 min, add 3 drops of 30% w/v H_2O_2, shake, and heat 3 min longer.

Set up the tubes for aeration. Add 1 ml of saturated KOH, and aerate into the boric acid solution for 40 min using 1.5 ml of 2% boric acid to trap the NH_3. Titrate with HCl.

$$\text{Volume of 0.0714 M HCl (ml)} \times 1{,}000 = \text{mass of amino acid N (mg)}$$

Estimation of Amino Group (Titration)[5]

Principle

The amino group is titrated with a solution of perchloric acid in glacial acetic acid.

Reagents

1. Perchloric acid, approximately 0.2 M. Dissolve 17.3 ml of 70% w/v perchloric acid in 800 ml of glacial acetic acid, add 44 g of acetic anhydride, and dilute to 1 liter with glacial acetic acid. Allow to stand overnight, and then standardize an aliquot against 250.0 mg of pure dry Na_2CO_3 (see below). Do Not Heat!
2. Bromothymol blue indicator. 0.1% w/v in glacial acetic acid.
3. Glacial acetic acid, reagent grade.

Method

Weigh 1.000 g of amino acid into a 150-ml beaker and add 25 ml of glacial acetic acid and 10 drops of indicator. Titrate with 0.2 M perchloric acid, with constant (mechanical) stirring, to a faint pink color.

Calculations

$$\frac{\text{mass of } Na_2CO_3 \text{ (g)}}{0.0530 \times \text{volume of perchloric acid (ml)}} = \text{molar concentration of } HClO_4 \text{ (M)}$$

$$\frac{\text{volume of perchloric acid (ml)} \times \text{molar concentration (M)} \times \text{Mol. Wt. of amino acid} \times 100}{\text{sample mass (g)}} = \% \text{ of amino acid}$$

Comment

As the $HClO_4$–CH_3COOH mixture has a high coefficient of expansion, it should be standarized each time it is used.

Estimation of Carboxyl Group (Titration)[6]

Principle

The carboxyl group is titrated with NaOH in the presence of HCHO.

Reagents

1. Sodium hydroxide, 1 M, free of carbonate.
2. Formaldehyde: 35 to 37%, reagent grade.

Method

Dissolve 2 to 3 g of amino acid in 100 ml of HCHO and titrate with 1 M NaOH using a pH meter to follow the reaction. The endpoint is where addition of 0.1 ml of NaOH causes a marked increase in pH. Correct for the reagent blank by titrating 100 ml of HCHO to the same pH as the endpoint.

Calculation

$$\frac{\text{net volume of 1 M NaOH (ml)} \times \text{Mol. Wt. of amino acid} \times 100}{\text{sample mass (g)}} = \% \text{ of amino acid}$$

Infrared Spectroscopy

Infrared spectra can be used for the identification of amino acids, and are also useful in specific cases for the determination of small amounts of impurities that are difficult to detect by paper chromatography. Examples (with appropriate absorption bands of the impurity) are (a) DL-leucine (11.78 μm) in the epimeric mixture of L-isoleucine with D-*allo*isoleucine, (b) the epimeric mixture of L-isoleucine and D-*allo*isoleucine (12.48 μm) in DL-leucine, and (c) DL-*allo*threonine (9.95 or 11.99 μm) in DL-threonine. Minimum detectable amounts are about 0.5% in the leucine–isoleucine pairs and about 2% of *allo*threonine in threonine.

Ultraviolet Absorption

Tyrosine and tryptophan possess characteristic absorption in the ultraviolet region. The molar absorption coefficients for tyrosine in acid are 1,340 at 274.5 nm and 8,200 at 223.0 nm, both of which are maxima, and 170 at 245.0 nm, which represents a minimum in the curve for this region; the corresponding values in alkaline solution are 2,330 at 293.5 nm and 11,050 at 240.0 nm for the maxima and 1,000 at 269.5 nm for the minimum. For tryptophan in either acid or alkaline solution, the maxima are 4,550 and 5,550 at 287.5 nm and 278.0 nm, respectively, and the minimum is 1,930 at 242.0 nm.

Determination of Optical Purity with L- or D-Amino Acid Oxidase or Bacterial Decarboxylases

Most of the amino acids isolated from proteins can be obtained in the pure L form. L-Cystine, however, is a notable exception, for it is often not only contaminated by L-tyrosine (removable by the charcoal treatment in the recrystallization procedures), but also by D- and *meso*-cystine as a result of some racemization by the boiling HCl used in the hydrolysis of the original protein. Experience in several laboratories has shown that commercially available L-cystine contains 5 to 15% of D-cystine.

Contamination by the D isomer in samples of supposedly pure L isomers of cystine, hydroxyproline, proline, leucine, phenylalanine, and tyrosine, for example, can be determined by using D-amino acid oxidase as follows: To each of four Warburg vessels are added 1,000 μmol of the isomer to be tested, and to two of these flasks 1 μmol of the optical enantiomorph is also introduced. To these four flasks, as well as to two others which will serve as enzyme blanks, 1.5 to 2.5 ml of buffer solution is added. The buffers are made up of 0.1 M sodium pyrophosphate at pH 8.1. A saturated solution (300 to 400 μl) of a crude D-amino acid oxidase preparation from hog kidney is placed in the side arm of all six vessels. Crude D-amino acid oxidase prepara-

tions usually contain sufficient catalase to make addition of this enzyme unnecessary. After a 10- to 15-min equilibration period, the flasks are tipped, and the gas evolution or consumption is noted from time to time until the reaction has ended (15 to 120 min). Ten micromoles of an optically pure, completely susceptible amino acid consumes 112 μl of oxygen gas or evolves 224 μl of carbon dioxide gas, under standard conditions. This method is obviously applicable only where 1 μmol of added susceptible isomer is readily and quantitatively oxidized or decarboxylated in the presence of the 1,000-fold amount of the resistant enantiomorph.

The 1,000 μmol of the L isomer alone should consume less than 1 μmol of O, but the added 1 μmol of D isomer should be quantitatively oxidized, as shown by the increment in the value over the L isomer. All values are corrected for the enzyme blanks.

Tests for contamination of a D-amino acid with an L-amino acid may be similarly made by the use of snake venom L-amino acid oxidase or the appropriate, specific, bacterial L-amino acid decarboxylase, each at essentially the saturation level. With the L-amino acid oxidase, 0.1 M 2-amino-2-(hydroxymethyl)-1,3-propanediol (tris) buffer is used at pH 7.2. Twenty-five units of crystalline catalase are added to the main compartment of the Warburg vessels in determinations with L-amino acid oxidase. With bacterial decarboxylases, 0.1 M acetate buffer is employed at pH 4.9.

Recrystallization

Many of the amino acids are easy to purify, for they are readily and rapidly crystallizable either from water or from water–ethanol mixtures. For this reason, the Subcommittee has described, where possible, the appropriate solvent required for each of the amino acids. The general, but not invariable, procedure is to bring the solvent to the boiling temperature and to pour it over the solid amino acid in a large Erlenmeyer flask. Enough of the hot solvent is added to bring most of the amino acid into solution. The mixture is then placed on a hotplate and gently boiled. If the remainder of the solid does not dissolve, more of the hot solvent is added to the boiling mixture until practically all the solid has dissolved. At this point, acid-washed charcoal (Norit or Darco) is carefully added (if added too quickly, frothing will occur) in the proportion of about 1 g of charcoal for every 100 g of amino acid, and the mixture is boiled carefully for 2 to 5 min. At the end of this period, the mixture is filtered through two layers of filter paper (usually Whatman No. 4) in a hot-water-jacketed funnel. The hot filtrate may then be chilled without further treatment (for proline or tryptophan, for example), or a second hot solvent may be added, slowly and with careful stirring, to some specified volume (alanine or arginine, for example), or a reagent such as hot dilute ammonium hydroxide may be introduced (cystine) prior to chilling. In all cases, the amino acid is obtained in crystalline form within a few hours, filtered off with suction, washed successively with ethanol and ether, and dried. The recoveries are generally better than 80%. The recrystallization procedure, as described, should remove all traces of color from the commercial preparations (tryptophan may require two recrystallizations to accomplish this purpose), and practically all traces of salts, heavy metals, proteins, more soluble compounds, and adventitious impurities, and frequently, although not invariably, traces of the racemic form. The more thorough investigator may wish to perform two or more recrystallizations of a given amino acid preparation before he is completely satisfied.

Ion-Exchange Chromatography

The widely used chromatographic analysis of amino acids on sulfonated polystyrene resins, introduced by Moore and Stein in 1951, is an excellent tool for evaluating the purity of these compounds. This procedure will indicate the presence, for example, of *meso*-cystine in L-cystine or *allo*isoleucine in isoleucine. Readers desiring to use this technique are referred to the publications on these methods.[7,8]

Color of Solution

Pure amino acids yield colorless solutions. Lack of color and turbidity may be verified spectrophotometrically. A clear and colorless solution should show a transmission of 98% or more in a 10-cm cell at 430 nm; one which is substantially clear and colorless should show more than 95%.

REFERENCES

1. J. P. Greenstein and M. Winitz, *Chemistry of the Amino Acids,* John Wiley & Sons, Inc., New York (1961).
2. A. Meister, *Biochemistry of the Amino Acids,* Academic Press Inc., New York (1957).
3. R. J. Block and D. Bolling, *The Amino Acid Composition of Proteins and Foods,* 2nd ed., Charles C Thomas, Springfield, Ill. (1951).
4. R. J. Block, E. L. Durrum, and G. Zweig, *A Manual of Paper Chromatography and Paper Electrophoresis,* 2nd ed., Academic Press Inc., New York (1958).
5. L. J. Harris, "A Simple Process for Estimating Amino or Other Basic Groups in Amino Acids, etc.: The 'Glacial Acetic Acid' Method," *Biochem. J.,* **29**, 2820 (1935).
6. S. P. L. Sörensen, "Enzymstudien," *Biochem. Z.,* **7**, 45 (1908).
7. S. Moore, D. H. Spackman, and W. H. Stein, "Chromatography of Amino Acids on Sulfonated Polystyrene Resins. An Improved System," *Anal. Chem.,* **30**, 1185 (1958).
8. D. H. Spackman, W. H. Stein, and S. Moore, "Automatic Recording Apparatus for Use in the Chromatography of Amino Acids," *Anal. Chem.,* **30**, 1190 (1958).

AA-1
N-Acetyl-DL-glutamic Acid
Ac-DLGlu

$HOOC-CH_2CH_2\,CH(NHCOCH_3)COOH$

Formula: $C_7H_{11}NO_5$
Formula Wt.: 189.17
Calc. %: C, 44.44; H, 5.86; N, 7.41; O, 42.29

Source or Method of Preparation: Synthetic, via acetylation of DL-glutamic acid, or acetylation of L-glutamic acid followed by racemization of the *N*-acetyl-L-glutamic acid so derived.

Bibliography: M. Bergmann and L. Zervas, *Biochem. Z.*, **203**, 280 (1928); J. P. Greenstein and M. Winitz, *Chemistry of the Amino Acids*, John Wiley & Sons, Inc., New York (1961), p. 1948.

Specific Rotation: None.
Homogeneity: Paper-chromatographic evidence for purity should be presented from a minimum of two solvent systems that will permit the detection of small amounts of the most likely impurity: glutamic acid.
  Technique and results are to be presented by the supplier.
Volatile Matter: Not more than 0.3%.
Water-Insoluble Material: 5 g in 50 ml of water has a turbidity not greater than that given by 0.4 mg of $BaSO_4$ in 50 ml of water.
Ash (sulfated): Less than 0.1%.
Heavy Metals (as Pb): Not more than 20 ppm.
Likely Impurities: *N*-Acetyl-L-glutamic acid, L-glutamic acid, DL-glutamic acid.
Crystallization Medium: Minimum amount of boiling water.
Stability and Storage: Stable.
Melting Point: 185 °C.

AA-2
N-Acetyl-L-glutamic Acid
Ac-Glu

$HOOC-CH_2CH_2-\overset{\displaystyle NHCOCH_3}{\underset{\displaystyle H}{C}}-COOH$

Formula: $C_7H_{11}NO_5$
Formula Wt.: 189.17
Calc. %: C, 44.44; H, 5.86; N, 7.41; O, 42.29

Source or Method of Preparation: Synthetic, via acetylation of L-glutamic acid.

Bibliography: M. Bergmann and L. Zervas, *Biochem. Z.*, **203**, 280 (1928); J. P. Greenstein and M. Winitz, *Chemistry of the Amino Acids*, John Wiley & Sons, Inc., New York (1961), p. 1948; A. Meister, L. Levintow, R. B. Kingsley, and J. P. Greenstein, *J. Biol. Chem.*, **192**, 535 (1951).

Specific Rotation: −16.6° in water, $\rho = 1$ g/100 ml, $t = 25$ °C.
Homogeneity: Paper-chromatographic evidence for purity should be presented from a minimum of two solvent systems that will permit the detection of small amounts of the most likely impurity: L-glutamic acid.
  Technique and results are to be presented by the supplier.
  Alternatively, the presence of contaminating L-glutamic acid may be detected by the use of glutamic acid decarboxylase (*Clostridium welchii*).
Volatile Matter: Not more than 0.3%.
Water-Insoluble Material: 5 g in 50 ml of water has a turbidity not greater than that given by 0.4 mg of $BaSO_4$ in 50 ml of water.

Ash (sulfated): Less than 0.1%.
Heavy Metals (as Pb): Not more than 20 ppm.
Likely Impurities: L-Glutamic acid.
Crystallization Medium: Minimum amount of boiling water.
Stability and Storage: Stable.
Melting Point: 201 °C.

AA-3
N-Acetyl-DL-histidine*
Ac-DLHis

$CH_2CH(NHCOCH_3)COOH$

Formula: $C_8H_{11}N_3O_3$
Formula Wt.: 197.20
Calc. %: C, 48.72; H, 5.62; N, 21.31; O, 24.34

Source or Method of Preparation: Synthetic, via acetylation of DL-histidine, or simultaneous acetylation of L-histidine and racemization.

Bibliography: M. Bergmann and L. Zervas, *Biochem. Z.*, **203**, 280 (1928); J. P. Greenstein and M. Winitz, *Chemistry of the Amino Acids*, John Wiley & Sons, Inc., New York (1961), p. 1989.

Specific Rotation: None.
Homogeneity: Determined by paper chromatography.
  One dimensional: Solvents 2 and 3. Two dimensional: Method C.
  Color reagent: Pauly; indicator dyes (*N*-acetyl-DL-histidine should not give a positive ninhydrin reaction).
  R_f 0.14, Solvent 2, descending.
  R_f 0.06, Solvent 3, descending.
  R_{BCP} 0.21, Solvent 2, descending. R_{BCP} refers to the distance moved by the compound calculated as the fraction of the distance moved by bromocresol purple, which is applied as a 0.1% w/v solution in ethanol.
Volatile Matter: Not more than 0.3%.
Water-Insoluble Material: 5 g in 50 ml of water has a turbidity not greater than that given by 0.4 mg of $BaSO_4$ in 50 ml of water.
Ash (sulfated): Less than 0.1%.
Heavy Metals (as Pb): Not more than 20 ppm.
Likely Impurities: DL-Histidine, *N*-acetyl-L-histidine.
Crystallization Medium: Water, then acetone to 80%.
Stability and Storage: Stable.
Melting Point: 148 °C for monohydrate.

* Available as monohydrate.

AA-4
N-Acetyl-L-histidine*
Ac-His

$CH_2-\overset{\displaystyle NHCOCH_3}{\underset{\displaystyle H}{C}}-COOH$

Formula: $C_8H_{11}N_3O_3$
Formula Wt.: 197.20
Calc. %: C, 48.72; H, 5.62; N, 21.31; O, 24.34

Source or Method of Preparation: Synthetic, via acetylation of L-histidine.

Bibliography: M. Bergmann and L. Zervas, *Biochem. Z.*, **203**, 280 (1928); J. P. Greenstein and M. Winitz, *Chemistry of the Amino Acids*, John Wiley & Sons, Inc., New York (1961), p. 1738.

Specific Rotation: +46.2° in water, $\rho = 1$ g/100 ml, $t = 25$ °C.
Homogeneity: Determined by paper chromatography.
 One dimensional: Solvents 2 and 3. Two dimensional: Method C
 Color reagent: Pauly; indicator dyes (*N*-acetyl-L-histidine should not give a positive ninhydrin reaction).
  R_f 0.14, Solvent 2, descending.
  R_f 0.06, Solvent 3, descending.
  R_{BCP} 0.21, Solvent 2, descending. R_{BCP} refers to the distance moved by the compound calculated as the fraction of the distance moved by bromocresol purple, which is applied as a 0.1% w/v solution in ethanol.
 Alternatively, the presence of contaminating L-histidine may be detected by the use of snake venom L-amino acid oxidase (*Crotalus adamanteus*).
Volatile Matter: Not more than 0.3%.
Water-Insoluble Material: 5 g in 50 ml of water has a turbidity not greater than that given by 0.4 mg of $BaSO_4$ in 50 ml of water.
Ash (sulfated): Less than 0.1%.
Heavy Metals (as Pb): Not more than 20 ppm.
Likely Impurities: L-Histidine.
Crystallization Medium: Water, then acetone to 80%.
Stability and Storage: Stable.
Melting Point: 169 °C for monohydrate.

 * Available as monohydrate.

AA-5
N-Acetyl-DL-tryptophan
Ac-DLTrp

Formula: $C_{13}H_{14}N_2O_3$
Formula Wt.: 246.27
Calc. %: C, 63.40; H, 5.73; N, 11.38; O, 19.49

Source or Method of Preparation: Synthetic.

Bibliography: J. P. Greenstein and M. Winitz, *Chemistry of the Amino Acids*, John Wiley & Sons, Inc., New York (1961), pp. 2330–2336.

Specific Rotation: None.
Homogeneity: Determined by paper chromatography.
 One dimensional: Solvent 5.
 Color reagent: Ehrlich.
  R_f 0.46, Solvent 5, ascending.
Volatile Matter: Not more than 0.3%.
Water-Insoluble Material: 5 g in 50 ml of water has a turbidity not greater than that given by 0.4 mg of $BaSO_4$ in 50 ml of water
Ash (sulfated): Less than 0.1%.
Heavy Metals (as Pb): Less than 20 ppm.
Likely Impurities: DL-Tryptophan.
Crystallization Medium: Ethanol, then water to excess.
Stability and Storage: Stable.
Melting Point: 206 °C.

AA-6
N-Acetyl-L-tryptophan
Ac-Trp

Formula: $C_{13}H_{14}N_2O_3$
Formula Wt.: 246.27
Calc. %: C, 63.40; H, 5.73; N, 11.38; O, 19.49

Source or Method of Preparation: Synthetic, via acetylation of L-tryptophan or resolution of *N*-acetyl-DL-tryptophan.

Bibliography: J. L. Warnell and C. P. Berg, *J. Am. Chem. Soc.*, **76**, 1708 (1954); V. du Vigneaud and R. R. Sealock, *J. Biol. Chem.*, **96**, 511 (1932); C. P. Berg, *J. Biol. Chem.*, **100**, 79 (1933); J. P. Greenstein and M. Winitz, *Chemistry of the Amino Acids*, John Wiley & Sons, Inc., New York (1961), p. 2726.

Specific Rotation: +30.1° in water containing 1 equivalent of NaOH, $\rho = 1$ g/100 ml, $t = 25$ °C.
Homogeneity: Determined by paper chromatography.
 One dimensional: Solvent 5.
 Color reagent: Ehrlich.
  R_f 0.46, Solvent 5, ascending.
Volatile Matter: Not more than 0.3%.
Water-Insoluble Material: 5 g in 50 ml of water has a turbidity not greater than that given by 0.4 mg of $BaSO_4$ in 50 ml of water
Ash (sulfated): Less than 0.1%.
Heavy Metals (as Pb): Less than 20 ppm.
Likely Impurities: L-Tryptophan, *N*-acetyl-D-tryptophan.
Crystallization Medium: Water.
Stability and Storage: Stable.
Melting Point: 188 °C.

AA-7
DL-Alanine
DLAla

CH₃CH(NH₂)COOH

Formula: $C_3H_7NO_2$
Formula Wt.: 89.09
Calc. %: C, 40.44; H, 7.92; N, 15.72; O, 35.92

Source or Method of Preparation: Synthetic.
Specific Rotation: None.
Homogeneity: Determined by paper chromatography.
 One dimensional: Solvent 2. Two dimensional: Methods A and B.
 Color reagent: Ninhydrin.
  R_f 0.45, Solvent 2, ascending.
  R_{BCP} 0.32, Solvent 2, descending. R_{BCP} refers to the distance moved by the amino acid calculated as the fraction of the distance moved by bromocresol purple, which is applied as a 0.1% w/v solution in ethanol.
Volatile Matter: Not more than 0.3%.
Water-Insoluble Material: 5 g in 50 ml of water has a turbidity not greater than that given by 0.4 mg of $BaSO_4$ in 50 ml of water.
Ash (sulfated): Less than 0.1%.
Heavy Metals (as Pb): Not more than 20 ppm.
Likely Impurities: 2, 2'-Iminodipropionic acid.
Crystallization Medium: Water, then ethanol to 80% v/v, or water-ethanol (1:4 v/v).
Stability and Storage: Stable.

AA-8
L-Alanine
Ala

$$CH_3-\underset{\underset{H}{|}}{\overset{\overset{NH_2}{|}}{C}}-COOH$$

Formula: $C_3H_7NO_2$
Formula Wt.: 89.09
Calc. %: C, 40.44; H, 7.92; N, 15.72; O, 35.92

Source or Method of Preparation: Resolution of synthetic, or from natural sources.

Specific Rotation: $+14.7°$ in 1.0 M HCl, $\rho = 5.8$ g/100 ml, $t = 15\,°C$.
Homogeneity: Determined by paper chromatography.
  One dimensional: Solvent 2. Two dimensional: Methods A and B.
  Color reagent: Ninhydrin.
    R_f 0.45, Solvent 2, ascending.
    R_{BCP} 0.32, Solvent 2, descending. R_{BCP} refers to the distance moved by the amino acid calculated as the fraction of the distance moved by bromocresol purple, which is applied as a 0.1% w/v solution in ethanol.
Volatile Matter: Not more than 0.3%.
Water-Insoluble Material: 5 g in 50 ml of water has a turbidity not greater than that given by 0.4 mg of $BaSO_4$ in 50 ml of water.
Ash (sulfated): Less than 0.1%.
Heavy Metals (as Pb): Not more than 20 ppm.
Likely Impurities: 2, 2'-Iminodipropionic acid, vitamin B_6.
Crystallization Medium: Water, then ethanol to 80% v/v, or water–ethanol (1:4 v/v).
Stability and Storage: Stable.

AA-9
L-Anserine Nitrate

$$CH_2-\underset{\underset{H}{|}}{\overset{\overset{NHCOCH_2CH_2NH_2 \cdot HNO_3}{|}}{C}}-COOH$$

Formula: $C_{10}H_{17}N_5O_6$
Formula Wt.: 303.28
Calc. %: C, 39.60; H, 5.65; N, 23.10; O, 31.65

Source or Method of Preparation: Synthetic.

Bibliography: O. K. Behrens and V. du Vigneaud, *J. Biol. Chem.*, **120**, 517 (1937); J. P. Greenstein and M. Winitz, *Chemistry of the Amino Acids*, John Wiley & Sons, Inc., New York (1961), pp. 74–75, 1521–1522.

Specific Rotation: $+12.2°$ in water, calculated as free anserine, $\rho = 5$ g/100 ml, $t = 30\,°C$.
Homogeneity: Determined by paper chromatography.
  One dimensional: Solvents 4 and 6.
  Color reagent: Ninhydrin.
    R_f 0.29, Solvent 4, ascending.
    R_f 0.34, Solvent 6, ascending.
Volatile Matter: Not more than 0.3%.
Water-Insoluble Material: Gives a clear colorless solution in water.
Ash (sulfated): Less than 0.1%.
Heavy Metals (as Pb): Less than 20 ppm.
Likely Impurities: 1-Methylimidazole-5-alanine, histidine.
Crystallization Medium: Dilute methanol.
Stability and Storage: Stable.
Melting Point: 225 °C.

AA-10
L-Arginine Hydrochloride*
Arg·HCl

$$HN=\underset{HN-CH_2CH_2CH_2-\underset{\underset{H}{|}}{\overset{\overset{NH_2}{|}}{C}}-COOH}{\overset{\overset{NH_2 \cdot HCl}{|}}{C}}$$

Formula: $C_6H_{15}ClN_4O_2$
Formula Wt.: 210.67
Calc. %: C, 34.21; H, 7.18; Cl, 16.83; N, 26.60; O, 15.19

Source or Method of Preparation: Gelatin, blood meal, seed globulins.

Specific Rotation: $+26.9°$ in 6 M HCl, calculated as free L-arginine, $\rho = 2$ g/100 ml, $t = 25\,°C$.
Homogeneity: Determined by paper chromatography.
  One dimensional: Solvent 2. Two dimensional: Methods A and B.
  Color reagent: Ninhydrin.
    R_f 0.26, Solvent 2, ascending.
    R_{BCP} 0.09, Solvent 2, descending. R_{BCP} refers to the distance moved by the amino acid calculated as the fraction of the distance moved by bromocresol purple, which is applied as a 0.1% w/v solution in ethanol.
Volatile Matter: Not more than 0.5%.
Water-Insoluble Material: 5 g in 50 ml of dilute HCl has a turbidity not greater than that given by 0.4 mg of $BaSO_4$ in 50 ml of water.
Ash (sulfated): Less than 0.1%.
Heavy Metals (as Pb): Not more than 20 ppm.
Likely Impurities: Ornithine.
Crystallization Medium: Water pH 5 to 7, then ethanol to 80% v/v.
Stability and Storage: Stable if kept dry.

* Also available as free L-arginine: Formula Wt.: 174.21; N = 32.16%.

AA-11
L-Argininosuccinic Acid
Arg(Suc)

$$HOOC-\underset{\underset{NH}{|}}{\overset{\overset{H}{|}}{C}}-CH_2-COOH$$
$$HN=\underset{HN-(CH_2)_3-\underset{\underset{H}{|}}{\overset{\overset{NH_2}{|}}{C}}-COOH}{C}$$

Formula: $C_{10}H_{18}N_4O_6$
Formula Wt.: 290.28
Calc. %: C, 41.37; H, 6.25; N, 19.30; O, 33.07

Source or Method of Preparation: Biosynthetic. Isolated as the amorphous alkaline barium salt.

Bibliography: S. Ratner, W. P. Anslow, Jr., and B. Petrack, *J. Biol. Chem.*, **204**, 115 (1953).

Specific Rotation: $+16.4°$ in H_2O, $\rho = 2.90$ g/100 ml, $t = 24\,°C$; $+5.2°$ in 0.5 M HCl, $\rho = 2.90$ g/100 ml, $t = 24\,°C$; $+26.6°$ in 0.5 M NaOH, $\rho = 2.90$ g/100 ml, $t = 24\,°C$.
Homogeneity: Determined by paper chromatography.
  One dimensional: Solvent 1 and Solvent Py [pyridine–H_2O (70:30) on paper washed with acetic acid]. Two dimensional: Solvent Py, followed by Solvent 1.
  Color reagent: Ninhydrin.
    R_f 0.38, Solvent 1, ascending.
    R_f 0.22, Solvent Py, ascending.

Volatile Matter: Cannot be heated without change.
Water-Insoluble Material: 5 g in 50 ml of water has a turbidity not greater than that given by 0.4 mg of $BaSO_4$ in 50 ml of water.
Ash (sulfated): Less than 0.1%.
Heavy Metals (as Pb): Not more than 20 ppm.
Likely Impurities: Fumaric acid. Readily undergoes ring closure to the anhydride at neutral or acid pH.[1]
Crystallization Medium: Water, then ethanol to 60%.
Stability and Storage: Stable as solid barium salt when stored at 0–5 °C under anhydrous conditions.
Additional Information Desirable: Colorimetric estimation.[2]

References
1. S. Ratner, B. Petrack, and O. Rochovansky, *J. Biol. Chem.*, **204**, 95 (1953).
2. S. Ratner, H. Morell, and E. Carvalho, *Arch. Biochem. Biophys.*, **91**, 280 (1960).

AA-12
L-Argininosuccinic Anhydride

Formula: $C_{10}H_{16}N_4O_5$
Formula Wt.: 272.26
Calc. %: C, 44.11; H, 5.92; N, 20.58; O, 29.38

II   III

Source or Method of Preparation: From solutions of argininosuccinic acid allowed to stand at neutral pH (anhydride II) or acid pH (anhydride III).

Bibliography: S. Ratner, B. Petrack, and O. Rochovansky, *J. Biol. Chem.*, **204**, 95 (1953); R. G. Westall, *Biochem. J.*, **77**, 135 (1960).

Specific Rotation of II: $-10.0°$ in H_2O, $\rho = 2.72$ g/100 ml, $t = 23$ °C; $+8.9°$ in 0.5 M HCl, $\rho = 2.72$ g/100 ml, $t = 23$ °C; $+5.2°$ in 0.5 M NaOH, $\rho = 2.72$ g/100 ml, $t = 23$ °C.
Homogeneity: Determined by paper chromatography.
  One dimensional: Solvent 1 and Solvent Py (pyridine: H_2O, 70:30 on paper washed with acetic acid). Two dimensional: Solvent Py, followed by Solvent 1.
  Color reagent: Ninhydrin.
    R_f 0.39 Anhydride II, 0.49 Anhydride III, Solvent 1, ascending.
    R_f 0.38 Anhydride II, 0.25 Anhydride III, Solvent Py, ascending.
Volatile Matter: Cannot be heated without change.
Water-Insoluble Material: 5 g in 50 ml of water has a turbidity not greater than that given by 0.4 mg of $BaSO_4$ in 50 ml of water.
Ash (sulfated): Less than 0.1%.
Heavy Metals (as Pb): Not more than 20 ppm.
Likely Impurities: In solution, various mixtures of anhydride II, anhydride III, and argininosuccinic acid are developed, depending on pH and temperature.

Crystallization Medium: Water, then ethanol to 70%, v/v.
Stability and Storage: Stable.
Additional Information Desirable: Anhydride III, but not anhydride II, may be estimated colorimetrically.[1]

Reference
1. S. Ratner, H. Morell, and E. Carvalho, *Arch. Biochem. Biophys.*, **91**, 280 (1960).

AA-13
L-Asparagine Monohydrate*
Asn·H_2O or Asp(NH_2)·H_2O

Formula: $C_4H_{10}N_2O_4$
Formula Wt.: 150.14
Calc. %: C, 32.00; H, 6.71; N, 18.66; O, 42.63

Source or Method of Preparation: Lupine seedlings.

Specific Rotation: $+32.6°$ in 0.1 M HCl, calculated as asparagine, $\rho = 1$ g/100 ml, $t = 20$ °C.
Homogeneity: Determined by paper chromatography.
  Two dimensional: Methods A and B.
  Color reagent: Ninhydrin.
    R_f 0.53, Solvent 1, descending.
    R_{BCP} 0.11, Solvent 2, descending. R_{BCP} refers to the distance moved by the amino acid calculated as the fraction of the distance moved by bromocresol purple, which is applied as a 0.1% w/v solution in ethanol.
Assay: Amide nitrogen (2 M HCl, reflux for 2 h): 9.4%.
Volatile Matter: Not more than 12%.
Water-Insoluble Material: 5 g in 50 ml of dilute HCl has a turbidity not greater than that given by 0.4 mg of $BaSO_4$ in 50 ml of water.
Ash (sulfated): Less than 0.1%.
Heavy Metals (as Pb): Not more than 20 ppm.
Likely Impurities: Aspartic acid, tyrosine.
Crystallization Medium: Water, or water and ethanol.
Stability and Storage: Slowly effloresces in dry air.

* Also available as anhydrous L-asparagine: Formula Wt. 132.12; N = 21.21%.

AA-14
DL-Aspartic Acid
DLAsp

$$HOOC-CH_2CH(NH_2)COOH$$

Formula: $C_4H_7NO_4$
Formula Wt.: 133.10
Calc. %: C, 36.09; H, 5.30; N, 10.52; O, 48.08

Source or Method of Preparation: Synthetic.

Specific Rotation: None.
Homogeneity: Determined by paper chromatography.
  One dimensional: Solvent 1. Two dimensional: Methods A and B.

Color reagent: Ninhydrin.
R_f 0.25, Solvent 1, ascending.
R_f 0.22, Solvent 2, descending.
R_{BCP} 0.11, Solvent 2, descending. R_{BCP} refers to the distance moved by the amino acid calculated as the fraction of the distance moved by bromocresol purple, which is applied as a 0.1% w/v solution in ethanol.
Volatile Matter: Not more than 0.3%.
Water-Insoluble Material: 5 g in 50 ml of dilute HCl has a turbidity not greater than that given by 0.4 mg of $BaSO_4$ in 50 ml of water.
Ash (sulfated): Less than 0.1%.
Heavy Metals (as Pb): Not more than 20 ppm.
Crystallization Medium: Water, then ethanol to 80% v/v.
Stability and Storage: Stable.

AA-15
L-Aspartic Acid
Asp

$$HOOC-CH_2-\underset{\underset{H}{|}}{\overset{\overset{NH_2}{|}}{C}}-COOH$$

Formula: $C_4H_7NO_4$
Formula Wt.: 133.10
Calc. %: C, 36.09; H, 5.30; N, 10.52; O, 48.08

Source or Method of Preparation: Natural sources, or hydrolysis of L-asparagine.

Specific Rotation: $+25.4°$ in 5 M HCl, $\rho = 2$ g/100 ml, $t = 25$ °C.
Homogeneity: Determined by paper chromatography.
One dimensional: Solvent 1. Two dimensional: Methods A and B.
Color reagent: Ninhydrin.
R_f 0.25, Solvent 1, ascending.
R_f 0.22, Solvent 1, descending.
R_{BCP} 0.11, Solvent 2, descending. R_{BCP} refers to the distance moved by the amino acid calculated as the fraction of the distance moved by bromocresol purple, which is applied as a 0.1% w/v solution in ethanol.
Volatile Matter: Not more than 0.3%.
Water-Insoluble Material: 5 g in 50 ml of dilute HCl has a turbidity not greater than that given by 0.4 mg of $BaSO_4$ in 50 ml of water.
Ash (sulfated): Less than 0.1%.
Heavy Metals (as Pb): Not more than 20 ppm.
Likely Impurities: Glutamic acid, NH_4Cl, cystine, asparagine.
Crystallization Medium: Water, then ethanol to 80%, v/v.
Stability and Storage: Stable.

AA-16
L-Carnosine

$$\underset{\underset{\underset{\underset{HC\!-\!-\!N}{\overset{||}{}\quad\overset{||}{}}}{C\quad CH}}{\overset{\overset{N}{|}}{\overset{|}{H}}}}{\overset{\overset{NHCOCH_2CH_2NH_2}{|}}{CH_2-\underset{H}{\overset{|}{C}}-COOH}}$$

Formula: $C_9H_{14}N_4O_3$
Formula Wt.: 226.24
Calc. %: C, 47.78; H, 6.24; N, 24.77; O, 21.22

Source or Method of Preparation: Synthetic, from histidine.

Bibliography: R. H. Sifferd and V. du Vigneaud, *J. Biol. Chem.*, **108**, 753 (1933); N. C. Davis and E. Smith, *Biochem. Prep.*, **4**, 38 (1955); J. P. Greenstein and M. Winitz, *Chemistry of the Amino Acids*, John Wiley & Sons, Inc., New York (1961), pp. 74, 1521.

Specific Rotation: $+20.5°$ in water, $\rho = 2$ g/100 ml, $t = 25$ °C.
Homogeneity: Determined by paper chromatography.
One dimensional: Solvents 4 and 5.
Color reagent: Pauly.
R_f 0.24, Solvent 4, ascending.
R_f 0.32, Solvent 5, ascending.
Volatile Matter: Not more than 0.3%.
Water-Insoluble Material: Gives a clear colorless solution in water.
Ash (sulfated): Less than 0.1%.
Heavy Metals (as Pb): Less than 20 ppm.
Likely Impurities: Histidine, β-alanine.
Crystallization Medium: Water, then ethanol to excess.
Stability and Storage: Stable.

AA-17
L-Citrulline

$$NH_2CONHCH_2CH_2CH_2-\underset{\underset{H}{|}}{\overset{\overset{NH_2}{|}}{C}}-COOH$$

Formula: $C_6H_{13}N_3O_3$
Formula Wt.: 175.19
Calc. %: C, 41.13; H, 7.48; N, 23.99; O, 27.40

Source or Method of Preparation: Synthetic, via action of cyanate or urea on copper complex of L-ornithine.

Bibliography: L. H. Smith, *J. Am. Chem. Soc.*, **77**, 6691 (1955); P. B. Hamilton and R. A. Anderson, *Biochem. Prep.*, **3**, 100 (1953); J. P. Greenstein and M. Winitz, *Chemistry of the Amino Acids*, John Wiley & Sons, Inc., New York (1961), pp. 2492–2494.

Specific Rotation: $+4.0°$ in water, $\rho = 2$ g/100 ml, $t = 25$ °C; $+24.2°$ in 5 M HCl, $\rho = 2$ g/100 ml, $t = 25$ °C; $+17.5°$ in glacial acetic acid, $\rho = 2$ g/100 ml, $t = 25$ °C.
Homogeneity: Determined by paper chromatography.
One dimensional: Solvents 1 and 5.
Color reagent: Ninhydrin.
R_f 0.46, Solvent 1, ascending.
R_f 0.46, Solvent 5, ascending.
Volatile Matter: Not more than 0.3%.
Water-Insoluble Material: Gives a clear colorless solution in water.
Heavy Metals (as Pb): Not more than 20 ppm.
Likely Impurities: Ornithine, arginine.
Crystallization Medium: Water, then 5 volumes of ethanol.
Stability and Storage: Stable.

AA-18
Creatine Monohydrate

$$\underset{HN=\overset{|}{C}NH_2}{CH_3N\!CH_2COOH \cdot H_2O}$$

Formula: $C_4H_{11}N_3O_3$
Formula Wt.: 149.15
Calc. %: C, 32.21; H, 7.43; N, 28.18; O, 32.18

Source or Method of Preparation: Synthetic.

Specific Rotation: None.
Homogeneity: Determined by paper chromatography.
One dimensional: Solvents 1 and 3. Two dimensional: Method B.
Color reagent: Heat at 100 °C for 1 h to convert into creatinine,

then spray with freshly prepared mixture of either 10% w/v sodium hydroxide (10 ml) and saturated aqueous picric acid (50 ml), or 5% w/v sodium nitroprusside (2 ml), 3% w/v hydrogen peroxide (5 ml), and 10% w/v sodium hydroxide (1 ml) diluted to 15 ml with water.

R_f 0.90, Solvent 1, descending.

R_{BCP} 1.08, Solvent 1, descending. R_{BCP} refers to the distance moved by the amino acid calculated as the fraction of the distance moved by bromocresol purple, which is applied as a 0.1% w/v solution in ethanol.

Volatile Matter: Loses water of hydration at 105 °C.
Water-Insoluble Material: Gives a clear colorless solution in water.
Ash (sulfated): Not more than 0.1%.
Heavy Metals (as Pb): Not more than 50 ppm.
Likely Impurities: Creatinine and other guanidino compounds.
Crystallization Medium: Water.
Stability and Storage: Stable.
Additional Information: Gives a negative color reaction with picric acid and a positive color reaction with 1-naphthol and 2,3-butanedione.

AA-19
Creatinine

$$HN=C\begin{array}{c}NH-CO\\ \\ H_3CN---CH_2\end{array}$$

Formula: $C_4H_7N_3O$
Formula Wt.: 113.12
Calc. %: C, 42.47; H, 6.24; N, 37.15; O, 14.14

Source or Method of Preparation: Synthetic.

Specific Rotation: None.
Homogeneity: Determined by paper chromatography.
  One dimensional: Solvents 1 and 2. Two dimensional: Method A.
  Color reagent: Spray with freshly prepared mixture of either 10% w/v sodium hydroxide (10 ml) and saturated aqueous picric acid (50 ml) or 5% w/v sodium nitroprusside (2 ml), 3% w/v hydrogen peroxide (5 ml), and 10% w/v sodium hydroxide (1 ml) diluted to 15 ml with water.
  R_f 0.97, Solvent 1, descending.
  R_{BCP} 0.56, Solvent 3, descending. R_{BCP} refers to the distance moved by the amino acid calculated as the fraction of the distance moved by bromocresol purple, which is applied as a 0.1% w/v solution in ethanol.
Volatile Matter: Not more than 0.3%.
Water-Insoluble Material: Gives a clear colorless solution in water
Ash (sulfated): Not more than 0.05%.
Heavy Metals (as Pb): Not more than 30 ppm.
Likely Impurities: Creatine, ammonium chloride.
Crystallization Medium: Dilute hydrochloric acid followed by neutralization with ammonium hydroxide.
  Water, then excess of acetone.
Stability and Storage: Stable when stored under anhydrous conditions.
Additional Information: Gives a negative color reaction with 1-naphthol and 2,3-butanedione, and a positive color reaction with picric acid.

AA-20
L-Cysteic Acid*
Cys(O₃H)

$$HO_3S-CH_2-\overset{\overset{\displaystyle NH_2}{|}}{\underset{\underset{\displaystyle H}{|}}{C}}-COOH$$

Formula: $C_3H_7NO_5S$
Formula Wt.: 169.16
Calc. %: C, 21.30; H, 4.17; N, 8.28; O, 47.29; S, 18.96

Source or Method of Preparation: Synthetic, via oxidation of cystine with bromine.

Bibliography: E. Freidmann, *Beitr. Chem. Physiol. Pathol.*, **3**, 1 (1903); H. T. Clarke and J. M. Inouye, *J. Biol. Chem.*, **94**, 541 (1931–32); J. P. Greenstein and M. Winitz, *Chemistry of the Amino Acids*, John Wiley & Sons, Inc., New York (1961), p. 1908.

Specific Rotation: +8.7° in water, calculated as monohydrate, $\rho = 7$ g/100 ml, $t = 20$ °C.
Homogeneity: Determined by paper chromatography.
  One dimensional: Solvents 1 and 2. Two dimensional: Method A.
  Color reagent: Ninhydrin.
    R_f 0.04, Solvent 1, descending.
    R_f 0.01, Solvent 2, descending.
    R_{BCP} 0.05, Solvent 1, descending. R_{BCP} refers to the distance moved by the amino acid calculated as the fraction of the distance moved by bromocresol purple, which is applied as a 0.1% w/v solution in ethanol.
Volatile Matter: Not more than 0.3%.
Water-Insoluble Material: 5 g in 50 ml of water has a turbidity not greater than that given by 0.4 mg of $BaSO_4$ in 50 ml of water.
Ash (sulfated): Less than 0.1%.
Heavy Metals (as Pb): Not more than 20 ppm.
Likely Impurities: Cystine, oxides of cysteine.
Crystallization Medium: Water, then 2 volumes of ethanol.
Stability and Storage: Stable.

* Available as the monohydrate.

AA-21
L-Cysteine Hydrochloride Monohydrate*
Cys·HCl·H₂O

$$HSCH_2-\overset{\overset{\displaystyle NH_2 \cdot HCl}{|}}{\underset{\underset{\displaystyle H}{|}}{C}}-COOH \cdot H_2O$$

Formula: $C_3H_{10}ClNO_3S$
Formula Wt.: 175.64
Calc. %: C, 20.51; H, 5.74; Cl, 20.19; N, 7.98; O, 27.33; S, 18.26.

Source or Method of Preparation: Reduction of cystine.

Specific Rotation: +6.53° in 5 M HCl, calculated as cysteine, $\rho = 2$ g/100 ml, $t = 25$ °C.
Homogeneity: Determined by paper chromatography.[1]
  One dimensional: Solvent 2.
  Two dimensional: Method A, after condensing with *N*-ethylmaleimide.
  Color reagent: Ninhydrin.
  Specific reagent: Nitroprusside.
    R_f 0.08, Solvent 2, ascending.

Volatile Matter: Cannot be dried by AOAC method without decomposition.

Water-Insoluble Material: 5 g in 50 ml of dilute HCl has a turbidity not greater than that given by 0.4 mg of $BaSO_4$ in 50 ml of water.

Ash (sulfated): Less than 0.1%.

Heavy Metals (as Pb): Not more than 20 ppm.

Likely Impurities: Cystine, tyrosine, H_2S.

Crystallization Medium: Methanol, then ether (peroxide-free).

Stability and Storage: Decomposes and oxidizes slowly; hygroscopic.

Reference

1. R. J. Block, E. L. Durrum, and G. Zweig, *Paper Chromatography and Paper Electrophoresis*, Academic Press Inc., New York (1958).

* Also available as anhydrous L-cysteine: Formula Wt. 121.16; N = 11.56%.

AA-22

L-Cystine

Cys Cys

$$SCH_2-\underset{H}{\overset{NH_2}{C}}-COOH$$
$$SCH_2-\underset{H}{\overset{NH_2}{C}}-COOH$$

Formula: $C_6H_{12}N_2O_4S_2$

Formula Wt: 240.30

Calc. %: C, 29.99; H, 5.03; N, 11.66; O, 26.63; S, 26.68

Source or Method of Preparation: Hair and wool wastes

Specific Rotation: $-212°$ in 1 M HCl, $\rho = 1$ g/100 ml, $t = 25$ °C.

Homogeneity: Determined by paper chromatography.

One dimensional: Solvent 2.

Color reagent: Ninhydrin.

Specific reagents: Nitroprusside—NaCN.

R_f 0.12, Solvent 2, ascending.

R_{BCP} 0.03, Solvent 2, descending. R_{BCP} refers to the distance moved by the amino acid calculated as the fraction of the distance moved by bromocresol purple, which is applied as a 0.1% w/v solution in ethanol.

Volatile Matter: Not more than 0.3%.

Water-Insoluble Material: 5 g in 50 ml of dilute HCl has a turbidity not greater than that given by 0.4 mg of $BaSO_4$ in 50 ml of water.

Ash (sulfated): Less than 0.1%.

Heavy Metals (as Pb): Not more than 20 ppm.

Likely Impurities: D-Cystine, *meso*-cystine, tyrosine. L-Cystine and *meso*-cystine can be distinguished by infrared spectroscopy* or by column chromatography[1]† and separated by recrystallization of the hydrochlorides.[2]

Crystallization Medium: 1.5 M HCl, then NH_4OH to neutrality.

Stability and Storage: Stable.

References

1. N. Wright, *J. Biol. Chem.*, **120**, 641 (1937).
2. H. S. Loring and V. du Vigneaud, *J. Biol. Chem.*, **102**, 287 (1933).

* See "Infrared Spectroscopy," p. 5.
† See "Ion-Exchange Chromatography," p. 7.

AA-23

3-(3,4-Dihydroxyphenyl)-DL-alanine
DLPhe(OH)$_2$ or DLDopa

Formula: $C_9H_{11}NO_4$

Formula Wt.: 197.19

Calc. %: C, 54.82; H, 5.62; N, 7.10; O, 32.46

Source or Method of Preparation: Synthetic.

Bibliography: J. P. Greenstein and M. Winitz, *Chemistry of the Amino Acids*, John Wiley & Sons, Inc., New York (1961), pp. 2713–2718.

Specific Rotation: None.

Homogeneity: Determined by paper chromatography.

One dimensional: Solvent 2.

Color reagent: Ninhydrin.

Specific reagent: Freshly prepared 0.5% w/v sodium 1,2-naphthoquinone-4-sulfonate in 0.2 M borate buffer, pH 8.9.

R_f 0.16, Solvent 2, descending.

R_{BCP} 0.21, Solvent 2, descending. R_{BCP} refers to the distance moved by the compound calculated as the fraction of the distance moved by bromocresol purple, which is applied as a 0.1% w/v solution in ethanol.

Volatile Matter: Not more than 0.3%.

Water-Insoluble Material: 5 g in 50 ml of dilute HCl has a turbidity not greater than that given by 0.4 mg of $BaSO_4$ in 50 ml of water.

Ash (sulfated): Less than 0.2%.

Heavy Metals (as Pb): Not more than 20 ppm.

Likely Impurities: Vanillin, hippuric acid, 3-methoxytyrosine.

Crystallization Medium: Dilute hydrochloric acid, then dilute ammonium hydroxide to pH 5.

Stability and Storage: Unstable in aqueous alkali.

AA-24

3-(3,4-Dihydroxyphenyl)-L-alanine
Phe(OH)$_2$ or Dopa

Formula: $C_9H_{11}NO_4$

Formula Wt.: 197.19

Calc. %: C, 54.82; H, 5.62; N, 7.10; O, 32.46

Source or Method of Preparation: Synthetic, from L-tyrosine, resolution of the racemate, or natural sources, e.g., velvet beans.

Bibliography: T. Torquati, *Arch. Farmacol. Sper.*, **15**, 308 (1913); M. Guggenheim, *Z. Physiol. Chem.*, **88**, 276 (1913); E. Waser and M. Lewandowski, *Helv. Chim. Acta*, **4**, 657 (1921); C. R. Harington and S. S. Randall, *Biochem. J.*, **25**, 1029 (1931); K. Vogler and H. Baumgartner, *Helv. Chim. Acta*, **35**, 1776 (1952); J. P. Greenstein and M. Winitz, *Chemistry of the Amino Acids*, John Wiley & Sons, Inc., New York (1961), pp. 68–69, 2714–2718.

Specific Rotation: $-12.0°$ in 1 M HCl, $\rho = 2$ g/100 ml, $t = 25$ °C.

Homogeneity: Determined by paper chromatography.

One dimensional: Solvent 2.

Color reagent: Ninhydrin.

Specific reagent: Freshly prepared 0.5% w/v sodium 1,2-naphthoquinone-4-sulfonate in 0.2 M borate buffer at pH 8.9.

R_f 0.16, Solvent 1, descending.

R_{BCP} 0.21, Solvent 1, descending. R_{BCP} refers to the distance moved by the compound calculated as the fraction of the distance moved by bromocresol purple, which is applied as a 0.1% w/v solution in ethanol.

Volatile Matter: No more than 0.3%.

Water-Insoluble Material: 5 g in 50 ml of dilute HCl has a turbidity not greater than that given by 0.4 mg of $BaSO_4$ in 50 ml of water.

Ash (sulfated): Less than 0.2%.

Heavy Metals (as Pb): Not more than 20 ppm.

Likely Impurities: Vanillin, hippuric acid, 3-methoxytyrosine, tyrosine, 3-aminotyrosine.

Crystallization Medium: Dilute hydrochloric acid, then dilute ammonium hydroxide to pH 5.

Stability and Storage: Unstable in aqueous alkali.

AA-25
3,5-Diiodo-L-tyrosine
Tyr(I₂)

Formula: $C_9H_9I_2NO_3$
Formula Wt.: 432.99
Calc. %: C, 24.96; H, 2.10; I, 58.62; N, 3.24; O, 11.09

Source or Method of Preparation: Iodination of tyrosine.

Specific Rotation: +1.5° in 1 M HCl, ρ = 5 g/100 ml, t = 25 °C.

Homogeneity: Determined by paper chromatography.
One dimensional: Solvent 3b. Two dimensional: Method C.
Color reagent: Ninhydrin.
Specific reagents: $Ce(HSO_4)_4$ + $NaAsO_2$.
R_f 0.29, Solvent 3b, descending.
R_f 0.16, Solvent 3, descending.
R_{BCP} 0.87, Solvent 2, descending. R_{BCP} refers to the distance moved by the amino acid calculated as the fraction of the distance moved by bromocresol purple, which is applied as a 0.1% w/v solution in ethanol.

Volatile Matter: No more than 0.5%.

Water-Insoluble Material: 5 g in 50 ml of dilute NaOH has a turbidity not greater than that given by 0.4 mg of $BaSO_4$ in 50 ml of water.

Ash (sulfated): Less than 0.1%.

Heavy Metals (as Pb): Not more than 20 ppm.

Likely Impurities: Tyrosine, 3-iodotyrosine, iodide, iodine.

Crystallization Medium: Cold dilute NH_4OH, then acetic acid to pH 6.

Stability and Storage: Decomposes slowly to liberate iodine and iodide.

AA-26

L-Ethionine [L-2-Amino-4-(ethylthio)butyric Acid]

Formula: $C_6H_{13}NO_2S$
Formula Wt.: 163.24
Calc. %: C, 44.15; H, 8.03; N, 8.58; O, 19.60; S, 19.64

Source or Method of Preparation: Synthetic, from L-methionine, or resolution of the racemate.

Bibliography: S. M. Birnbaum, L. Levintow, R. B. Kingsley, and J. P. Greenstein, *J. Biol. Chem.*, **194**, 455 (1952); H. M. Dyer, *J. Biol. Chem.*, **124**, 519 (1938); J. P. Greenstein and M. Winitz, *Chemistry of the Amino Acids*, John Wiley & Sons, Inc., New York (1961), pp. 2658–2659.

Specific Rotation: +23.7° in 5 M HCl, ρ = 2 g/100 ml, t = 25 °C.

Homogeneity: Determined by paper chromatography.
One dimensional: Solvents 1 and 3. Two dimensional: Method B.
Color reagent: Ninhydrin.
Specific reagent: Platinic iodide.
R_f 0.85, Solvent 1, descending.
R_f 0.16, Solvent 3, descending.
R_{BCP} 1.13, Solvent 1, descending. R_{BCP} refers to the distance moved by the compound calculated as the fraction of the distance moved by bromocresol purple, which is applied as a 0.1% w/v solution in ethanol.

Volatile Matter: Not more than 0.3%.

Water-Insoluble Material: 5 g in 50 ml of water has a turbidity not greater than that given by 0.4 mg of $BaSO_4$ in 50 ml of water.

Ash (sulfated): Less than 0.1%.

Heavy Metals (as Pb): Not more than 20 ppm.

Likely Impurities: N-acetyl-D-ethionine, N-acetyl-L-ethionine, L-methionine, D-ethionine.

Crystallization Medium: Water, then 4 volumes of ethanol.

Stability and Storage: Stable.

AA-27
L-Glutamic Acid
Glu

Formula: $C_5H_9NO_4$
Formula Wt.: 147.13
Calc. %: C, 40.81; H, 6.17; N, 9.52; O, 43.50

Source or Method of Preparation: Wheat gluten, beet sugar filtrate.

Specific Rotation: +31.5° in 5 M HCl, ρ = 2 g/100 ml, t = 20 °C.

Homogeneity: Determined by paper chromatography.
One dimensional: Solvent 1. Two dimensional: Methods A and B.
Color reagent: Ninhydrin.
R_f 0.33, Solvent 1, ascending.
R_f 0.33, Solvent 1, descending.
R_{BCP} 0.21, Solvent 2, descending. R_{BCP} refers to the distance moved by the amino acid calculated as the fraction of the distance moved by bromocresol purple, which is applied as a 0.1% w/v solution in ethanol.

Volatile Matter: Not more than 0.3%.

Water-Insoluble Material: 5 g in 50 ml of dilute HCl has a turbidity not greater than that given by 0.4 mg of $BaSO_4$ in 50 ml of water.

Ash (sulfated): Less than 0.1%.

Heavy Metals (as Pb): Not more than 20 ppm.

Likely Impurities: Aspartic acid, cystine.

Crystallization Medium: Water, then ethanol to 80% v/v.

Stability and Storage: Stable.

AA-28
L-Glutamine
Gln or Glu(NH$_2$)

$$H_2NCOCH_2CH_2-\underset{\underset{H}{|}}{\overset{\overset{NH_2}{|}}{C}}-COOH$$

Formula: $C_5H_{10}N_2O_3$
Formula Wt.: 146.15
Calc. %: C, 41.09; H, 6.85; N, 19.17; O, 32.82

Source or Method of Preparation: Synthetic, or beet extract.

Specific Rotation: +7.0° in water, $\rho = 2$ g/100 ml, $t = 25$ °C; +31.8° in 1 M HCl, $\rho = 2$ g/100 ml, $t = 25$ °C.
Homogeneity: Determined by paper chromatography.
  One dimensional: None. Two dimensional: Methods A and B.
  Color reagent: Ninhydrin.
    R_f 0.65, Solvent 1, descending.
    R_{BCP} 0.13, Solvent 2, descending. R_{BCP} refers to the distance moved by the amino acid calculated as the fraction of the distance moved by bromocresol purple, which is applied as a 0.1% w/v solution in ethanol.
Assays: Amide nitrogen (2 M HCl, refluxing for 2 h, or water at 100 °C for 3 h): 9.6%.
Volatile Matter: Not more than 0.5%.
Water-Insoluble Material: 5 g in 50 ml of dilute HCl has a turbidity not greater than that given by 0.4 mg of $BaSO_4$ in 50 ml of water.
Ash (sulfated): Less than 0.1%.
Heavy Metals (as Pb): Not more than 20 ppm.
Likely Impurities: Glutamic acid, ammonium salt of pyroglutamic acid,* tyrosine, asparagine, isoglutamine, arginine, nickel.
Crystallization Medium: Water, then ethanol to 80% v/v.
Stability and Storage: Stable; converted into ammonium salt of pyroglutamic acid in hot water.

* Pyroglutamic acid can be made visible in two-dimensional chromatography by spraying with an acid–base indicator.

AA-29
Glycine
Gly

$$H_2NCH_2COOH$$

Formula: $C_2H_5NO_2$
Formula Wt.: 75.07
Calc. %: C, 32.00; H, 6.72; N, 18.66; O, 42.63

Source or Method of Preparation: Synthetic.

Specific Rotation: None.

Homogeneity: Determined by paper chromatography.
  One dimensional: Solvent 1. Two dimensional: Methods A and B.
  Color reagent: Ninhydrin.
    R_f 0.48, Solvent 1, ascending.
    R_f 0.56, Solvent 1, descending.
    R_{BCP} 0.15, Solvent 2, descending. R_{BCP} refers to the distance moved by the amino acid calculated as the fraction of the distance moved by bromocresol purple, which is applied as a 0.1% w/v solution in ethanol.
Volatile Matter: No more than 0.2%.

Water-Insoluble Material: 5 g in 50 ml of water has a turbidity not greater than that given by 0.4 mg of $BaSO_4$ in 50 ml of water.
Ash (sulfated): Less than 0.1%.
Heavy Metals (as Pb): Not more than 20 ppm.
Likely Impurities: Ammonium salt, iminodiacetic acid and its ammonium salt, nitrilotriacetic acid, NH_4Cl.
Crystallization Medium: Water, then ethanol to 80% v/v.
Stability and Storage: Stable.

AA-30
L-Histidine*
His

Formula: $C_6H_9N_3O_2$
Formula Wt.: 155.16
Calc. %: C, 46.44; H, 5.85; N, 27.09; O, 20.62

Source or Method of Preparation: Blood meal, hemoglobin.

Specific Rotation: +13.0° in 6 M HCl, $\rho = 1$ g/100 ml, $t = 25$ °C.
Homogeneity: Determined by paper chromatography.
  One dimensional: Solvent 2. Two dimensional: Methods A and B.
  Color reagent: Ninhydrin.
  Specific reagent: Pauly.
    R_f 0.80, Solvent 1, descending.
    R_f 0.22, Solvent 2, ascending.
    R_{BCP} 0.07, Solvent 2, descending. R_{BCP} refers to the distance moved by the amino acid calculated as the fraction of the distance moved by bromocresol purple, which is applied as a 0.1% w/v solution in ethanol.
Volatile Matter: Not more than 0.5% for histidine · HCl.
Water-Insoluble Material: 5 g in 50 ml of dilute HCl has a turbidity not greater than that given by 0.4 mg of $BaSO_4$ in 50 ml of water.
Ash (sulfated): Less than 0.1%.
Heavy Metals (as Pb): Not more than 20 ppm.
Likely Impurities: Arginine.
Crystallization Medium: Water, then ethanol to 80% v/v.
Stability and Storage: Stable.

* Also available as the hydrochloride.

AA-31
L-Histidine Monohydrochloride Monohydrate
His · HCl · H$_2$O

Formula: $C_6H_{12}ClN_3O_3$
Formula Wt.: 209.63
Calc. %: C, 34.38; H, 5.77; Cl, 16.91; N, 20.05; O, 22.90

Source or Method of Preparation: Blood meal, hemoglobin.

Specific Rotation: +13.0° in 6 M HCl, calculated as free L-histidine, $\rho = 1$ g/100 ml, $t = 25$ °C.

Homogeneity: Determined by paper chromatography.

One dimensional: Solvent 2. Two dimensional: Methods A and B.

Color reagent: Ninhydrin.

Specific reagent: Pauly.

R_f 0.80, Solvent 1, descending.

R_f 0.22, Solvent 2, ascending.

R_{BCP} 0.07, Solvent 2, descending. R_{BCP} refers to the distance moved by the amino acid calculated as the fraction of the distance moved by bromocresol purple, which is applied as a 0.1% w/v solution in ethanol.

Volatile Matter: Not more than 0.5% for histidine·HCl.

Water-Insoluble Material: 5 g in 50 ml of dilute HCl has a turbidity not greater than that given by 0.4 mg of $BaSO_4$ in 50 ml of water.

Ash (sulfated): Less than 0.1%.

Heavy Metals (as Pb): Not more than 20 ppm.

Likely Impurities: Arginine.

Crystallization Medium: Water, then ethanol to 80% v/v.

Stability and Storage: Stable.

AA-32

L-Homoserine (2-Amino-4-hydroxybutyric Acid)
Hse

Formula: $C_4H_9NO_3$

Formula Wt.: 119.12

Calc. %: C, 40.33; H, 7.61; N, 11.76; O, 40.30

$$HOCH_2CH_2-\overset{\overset{\displaystyle NH_2}{|}}{\underset{\underset{\displaystyle H}{|}}{C}}-COOH$$

Source or Method of Preparation: Resolution of synthetic racemate.

Bibliography: S. M. Birnbaum and J. P. Greenstein, *Arch. Biochem. Biophys.*, **42**, 212 (1953); M. D. Armstrong, *J. Am. Chem. Soc.*, **71**, 3399 (1949); J. P. Greenstein and M. Winitz, *Chemistry of the Amino Acids*, John Wiley & Sons, Inc., New York (1961), pp. 2612–2616.

Specific Rotation: −8.8° in water, $\rho = 5$ g/100 ml, $t = 26$ °C; +18.3° in 2 M HCl, $\rho = 2$ g/100 ml, $t = 26$ °C

Homogeneity: Determined by paper chromatography.

One dimensional: Solvents 2 and 3. Two dimensional: Method C.

Color reagent: Ninhydrin.

R_f 0.60, Solvent 1, descending.

R_f 0.03, Solvent 3, descending.

R_{BCP} 0.78, Solvent 1, descending. R_{BCP} refers to the distance moved by the compound calculated as the fraction of the distance moved by bromocresol purple, which is applied as a 0.1% w/v solution in ethanol.

Volatile Matter: Not more than 0.3%.

Water-Insoluble Material: Gives a clear colorless solution in water.

Ash (sulfated): Less than 0.1%.

Heavy Metals (as Pb): Not more than 20 ppm.

Likely Impurities: *N*-(Chloroacetyl)-L-homoserine, *N*-(chloroacetyl)-D-homoserine, D-homoserine, homoserine lactone, homoserine anhydride (diketopiperazine compound).

Crystallization Medium: Water, then 9 volumes of ethanol.

Stability and Storage: Stable in solid state (anhydrous conditions). Stable in basic and dilute aqueous (neutral) solutions. Cyclizes to lactone in strongly acidic solutions. Cyclizes to diketopiperazine compound in concentrated aqueous and slightly acidic solutions.

AA-33

erythro-3-Hydroxy-DL-aspartic Acid

$$HOOC-CH(OH)CH(NH_2)COOH$$

Formula: $C_4H_7NO_5$

Formula Wt.: 149.10

Calc. %: C, 32.21; H, 4.73; N, 9.40; O, 53.65

Source or Method of Preparation: Synthetic, by amination of 3-chloromalic acid.

Bibliography: H. D. Dakin, *J. Biol. Chem.*, **48**, 273 (1921); J. P. Greenstein and M. Winitz, *Chemistry of the Amino Acids*, John Wiley & Sons, Inc., New York (1961), pp. 214–215, 2416–2418.

Specific Rotation: None.

Homogeneity: Determined by paper chromatography.

One dimensional: Solvents 1 and 2.

Color reagent: Ninhydrin.

R_f 0.06, Solvent 1, ascending.

R_f 0.07, Solvent 2, descending.

R_{BCP} 0.12, Solvent 2, descending. R_{BCP} refers to the distance moved by the compound calculated as the fraction of the distance moved by bromocresol purple, which is applied as a 0.1% w/v solution in ethanol.

Volatile Matter: Not more than 0.3%.

Water-Insoluble Material: 5 g in 50 ml of dilute HCl has a turbidity not greater than that given by 0.4 mg of $BaSO_4$ in 50 ml of water.

Ash (sulfated): Less than 0.1%.

Heavy Metals (as Pb): Not more than 20 ppm.

Likely Impurities: 3-Chloromalic acid, ammonium chloride, *threo*-3-hydroxyaspartic acid.

Crystallization Medium: Water.

Stability and Storage: Stable.

AA-34

5-Hydroxy-L-lysine Monohydrochloride
Lys(OH)·HCl

Formula: $C_6H_{15}ClN_2O_3$

Formula Wt.: 198.65

Calc. %: C, 36.27; H, 7.61; Cl, 17.85; N, 14.10; O, 24.16

$$HCl \cdot NH_2CH_2-\overset{\overset{\displaystyle OH}{|}}{\underset{\underset{\displaystyle H}{|}}{C}}-CH_2CH_2-\overset{\overset{\displaystyle NH_2}{|}}{\underset{\underset{\displaystyle H}{|}}{C}}-COOH$$

Source or Method of Preparation: Natural sources (gelatin), or resolution of racemate.

Bibliography: P. B. Hamilton and R. A. Anderson, *Biochem. Prep.*, **8**, 55 (1961); W. S. Fones, *Biochem. Prep.*, **8**, 62 (1961); W. S. Fones, *J. Am. Chem. Soc.*, **75**, 4865 (1953); J. P. Greenstein and M. Winitz, *Chemistry of the Amino Acids*, John Wiley & Sons, Inc., New York (1961), pp. 1996–2016.

Specific Rotation: +17.8° in 6 M HCl, calculated as free 5-hydroxy-L-lysine, $\rho = 2$ g/100 ml, $t = 25$ °C. Epimerization lowers the rotation of the product obtained from gelatin hydrolyzates.

Homogeneity: Determined by paper chromatography.

One dimensional: Solvents 1 and 2.

Color reagent: Ninhydrin.
R_f 0.67, Solvent 1, ascending.
R_f 0.06, Solvent 2, ascending.
R_{BCP} 0.15, Solvent 3a, descending. R_{BCP} refers to the distance moved by the compound calculated as the fraction of the distance moved by bromocresol purple, which is applied as a 0.1% w/v solution in ethanol.
Volatile Matter: No more than 0 5%.
Water-Insoluble Material: 5 g in 50 ml of water has a turbidity not greater than that given by 0.4 mg of $BaSO_4$ in 50 ml of water.
Ash (sulfated): Less than 0.1%.
Heavy Metals (as Pb): Not more than 20 ppm.
Likely Impurities: 5-*allo*-Hydroxy-D-lysine, 5-*allo*-hydroxy-L-lysine, histidine, lysine, ornithine.
Crystallization Medium: Water, then ethanol to 60% v/v, with further additions of ethanol to 90% as crystallization proceeds.
Stability and Storage: Stable.

AA-35
Hydroxy-L-proline
Pro(OH)

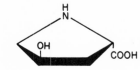

Formula: $C_5H_9NO_3$
Formula Wt.: 131.13
Calc. %: C, 45.79; H, 6.92; N, 10.68; O, 36.61

Source or Method of Preparation: Collagen, gelatin.

Specific Rotation: −76.7° in H_2O, ρ = 20 g/100 ml, t = 25 °C.
Homogeneity: Determined by paper chromatography.
One dimensional: Solvent 2. Two dimensional: Methods A and B.
Color reagents: Ninhydrin and isatin.
Specific reagents: Isatin followed by Ehrlich.
R_f 0.74, Solvent 1, descending.
R_f 0.20, Solvent 2, ascending.
R_{BCP} 0.18, Solvent 2, descending. R_{BCP} refers to the distance moved by the amino acid calculated as the fraction of the distance moved by bromocresol purple, which is applied as a 0.1% w/v solution in ethanol.
Assays: Amino group (ninhydrin according to Van Slyke): 0.0%.
Volatile Matter: Not more than 0.3%.
Water-Insoluble Material: 5 g in 50 ml of water has a turbidity not greater than that given by 0.4 mg of $BaSO_4$ in 50 ml of water.
Ash (sulfated): Less than 0.1%.
Heavy Metals (as Pb): Not more than 20 ppm.
Likely Impurities: Proline.
Crystallization Medium: Ethanol–methanol (1:1 v/v).
Stability and Storage: Stable.

AA-36
5-Hydroxy-DL-tryptophan Monohydrate
DLTrp(OH)·H₂O

Formula: $C_{11}H_{14}N_2O_4$
Formula Wt.: 238.25

Calc. %: C, 55.45; H, 5.92; N, 11.76; O, 26.86

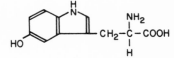

Source or Method of Preparation: Synthetic.

Bibliography: A. Ek and B. Witkop, *J. Am. Chem. Soc.*, **76**, 5579 (1954); W. R. Boehme, *J. Am. Chem. Soc.*, **75**, 2502 (1953); G. Frangatos and F. L. Chubb, *Can. J. Chem.*, **37**, 1374 (1959); J. P. Greenstein and M. Winitz, *Chemistry of the Amino Acids*, John Wiley & Sons, Inc., New York (1961), pp. 2734–2737.

Specific Rotation: None.
Homogeneity: Determined by paper chromatography.
One dimensional: Solvents 1, 2, 4, and 5.
Color reagent: Ninhydrin.
Specific reagent: Ehrlich.
R_f 0.58, Solvent 1, ascending.
R_f 0.17, Solvent 2, ascending.
R_f 0.35, Solvent 4, ascending.
R_f 0.19, Solvent 5, ascending.
Water-Insoluble Material: Gives a clear colorless solution in water.
Ash (sulfated): Less than 0.1%.
Heavy Metals (as Pb): Less than 20 ppm.
Likely Impurities: 5-Benzyloxy-DL-tryptophan.
Crystallization Medium: Water, then ethanol, in a nitrogen atmosphere.
Stability and Storage: Decomposes on contact with air. Store in an atmosphere of nitrogen.

AA-37
5-Hydroxy-L-tryptophan
Trp(OH)

Formula: $C_{11}H_{12}N_2O_3$
Formula Wt.: 220.23
Calc. %: C, 59.99; H, 5.49; N, 12.72; O, 21.80

Source or Method of Preparation: Resolution of racemate.

Bibliography: A. J. Morris and M. D. Armstrong, *J. Org. Chem.*, **22**, 306 (1957); J. P. Greenstein and M. Winitz, *Chemistry of the Amino Acids*, John Wiley & Sons, Inc., New York (1961), pp. 2732–2737.

Specific Rotation: −32.5° in water, ρ = 1 g/100 ml, t = 22 °C; +16.0° in 4 M HCl ρ = 1 g/100 ml t = 22 °C
Homogeneity: Determined by paper chromatography.
One dimensional: Solvents 1 and 2.
Color reagent: Ehrlich.
R_f 0.52, Solvent 1, ascending.
R_f 0.27, Solvent 2, ascending.
R_{BCP} 0.57, Solvent 3a, descending. R_{BCP} refers to the distance moved by the compound calculated as the fraction of the distance moved by bromocresol purple, which is applied as a 0.1% w/v solution in ethanol.
Volatile Matter: Not more than 0.3%.
Water-Insoluble Material: 5 g in 50 ml of dilute sodium hydroxide has a turbidity not greater than that given by 0.5 mg of $BaSO_4$ in 50 ml of water.
Ash (sulfated): Less than 0.1%.

Heavy Metals (as Pb): Not more than 20 ppm.
Likely Impurities: 5-Hydroxy-D-tryptophan, 5-benzyloxy-D-, DL-, or L-tryptophan.
Crystallization Medium: Water, then ethanol, in a nitrogen atmosphere.
Stability and Storage: Decomposes on contact with air. Store in an atmosphere of nitrogen.

AA-38
3-Iodo-L-tyrosine
Tyr(I)

Formula: $C_9H_{10}INO_3$
Formula Wt.: 307.09
Calc. %: C, 35.19; H, 3.28; I, 41.32; N, 4.56; O, 15.62

Source or Method of Preparation: Iodination of tyrosine.

Specific Rotation: $-4.4°$ in 1 M HCl, $\rho = 2$ g/100 ml, $t = 25$ °C
Homogeneity: Determined by paper chromatography,
   One dimensional: Solvent 3b. Two dimensional: Method C.
   Color reagent: Ninhydrin.
   Specific reagents: $Ce(HSO_4)_4$—$NaAsO_2$.
   R_f 0.38, Solvent 3, descending.
   R_f 0.59, Solvent 3b, descending.
   R_{BCP} 0.70, Solvent 2, descending. R_{BCP} refers to the distance moved by the amino acid calculated as the fraction of the distance moved by bromocresol purple, which is applied as a 0.1% w/v solution in ethanol
Volatile Matter: Not more than 0.3%.
Water-Insoluble Material: 5 g in 50 ml of dilute NaOH has a turbidity not greater than that given by 0.4 mg of $BaSO_4$ in 50 ml of water.
Ash (sulfated): Less than 0.1%.
Heavy Metals (as Pb): Not more than 20 ppm.
Likely Impurities: Tyrosine, diiodotyrosine, iodide, iodine.
Crystallization Medium: Dilute NH_4OH at room temperature, then acetic acid to pH 6.
Stability and Storage: Liberates iodine; keep cold.

AA-39
L-Isoleucine
Ile

Formula: $C_6H_{13}NO_2$
Formula Wt.: 131.18
Calc. %: C, 54.94; H, 9.99; N, 10.68; O, 24.40

Source or Method of Preparation: Natural sources.

Specific Rotation: $+40.6°$ in 6.1 M HCl, $\rho = 5.1$ g/100 ml, $t = 20$ °C.
Homogeneity: Determined by paper chromatography.
   One dimensional: Solvents 1 and 3b. Two dimensional: Methods A and B.

Color reagent: Ninhydrin.
   R_f 0.87, Solvent 1, descending.
   R_f 0.82, Solvent 2, ascending.
   R_{BCP} 0.76, Solvent 2, descending. R_{BCP} refers to the distance moved by the amino acid calculated as the fraction of the distance moved by bromocresol purple, which is applied as a 0.1% w/v solution in ethanol (This amino acid is poorly separated from leucine.)
Volatile Matter: Not more than 0.3%.
Water-Insoluble Material: 5 g in 50 ml of dilute HCl has a turbidity not greater than that given by 0.4 mg of $BaSO_4$ in 50 ml of water.
Ash (sulfated): Less than 0.1%.
Heavy Metals (as Pb): Not more than 20 ppm.
Likely Impurities: Leucine, tyrosine, methionine, *allo*isoleucine, D-isoleucine.
Crystallization Medium: Water, then ethanol to 80% v/v.
Stability and Storage: Stable.

AA-40
L-Isoleucine + D-*Allo*isoleucine*
Ile D*allo*Ile

Formula: $C_6H_{13}NO_2$
Formula Wt.: 131.18
Calc. %: C, 54.94; H, 9.99; N, 10.68; O, 24.40

Source or Method of Preparation: Synthetic.

Specific Rotation: $\pm3°$ in 5 M HCl, $\rho = 5$ g/100 ml, $t = 20$ °C.

Homogeneity: Determined by paper chromatography.
   One dimensional: Solvents 1 and 3b. Two dimensional: Methods A and B.
   Color reagent: Ninhydrin.
   R_f 0.87, Solvent 1, descending.
   R_f 0.82, Solvent 2, ascending.
   R_{BCP} 0.76, Solvent 2, descending. R_{BCP} refers to the distance moved by the amino acid calculated as the fraction of the distance moved by bromocresol purple, which is applied as a 0.1% w/v solution in ethanol. (This amino acid is poorly separated from leucine.)
Volatile Matter: Not more than 0.3%.
Water-Insoluble Material: 5 g in 50 ml of dilute HCl has a turbidity not greater than that given by 0.4 mg of $BaSO_4$ in 50 ml of water.
Ash (sulfated): Less than 0.1%.
Heavy Metals (as Pb): Not more than 20 ppm.
Likely Impurities: Leucine, D-isoleucine, L-*allo*isoleucine.
Crystallization Medium: Water, then ethanol to 80% v/v.
Stability and Storage: Stable.

* Some commercial samples of labeled DL-isoleucine are the epimeric mixture of L-isoleucine and D-*allo*isoleucine. The detection of *allo*isoleucine is best carried out by infrared spectroscopy (p. 5), or by column chromatography on ion-exchange resins (Refs. 7, 8, p. 7). If a pure sample of L-isoleucine is available, further assays can be done, using microbiological methods.

AA-41
L-Kynurenine Sulfate Monohydrate

Formula: $C_{10}H_{16}N_2O_8S$
Formula Wt.: 324.31
Calc. %: C, 37.03; H, 4.97; N, 8.64; O, 39.47; S, 9.89

Source or Method of Preparation: Synthetic, from L-tryptophan.

Bibliography: J. L. Warnell and C. P. Berg, *J. Am. Chem. Soc.*, **76**, 1708 (1954); J. P. Greenstein and M. Winitz, *Chemistry of the Amino Acids*, John Wiley & Sons, Inc., New York (1961), pp. 2723–2727.

Specific Rotation: +9.6° in water, $\rho = 1$ g/100 ml, $t = 25$ °C.
Homogeneity: Determined by paper chromatography.
One dimensional: Solvents 1 and 5.
Color reagent: Ninhydrin.
R_f 0.78, Solvent 1, ascending.
R_f 0.71, Solvent 5, ascending.
Water-Insoluble Material: Gives a clear colorless solution in water.
Ash (sulfated): Not more than 0.2%.
Heavy Metals (as Pb): Less than 20 ppm.
Likely Impurities: L-Tryptophan, *N*-acetyl-L-tryptophan.
Crystallization Medium: Water, then excess of ethanol.
Stability and Storage: Stable.
Melting Point: 178 °C.

AA-42
DL-Leucine
DLLeu

Formula: $C_6H_{13}NO_2$
Formula Wt.: 131.18
Calc. %: C, 54.94; H, 9.99; N, 10.68; O, 24.40

Source or Method of Preparation: Synthetic.

Specific Rotation: None.
Homogeneity: Determined by paper chromatography.
One dimensional: Solvents 2 and 3b. Two dimensional: Methods A and C.
Color reagent: Ninhydrin.
R_f 0.88, Solvent 1, descending.
R_f 0.85, Solvent 2, ascending.
R_f 0.82, Solvent 2, descending.
Volatile Matter: Not more than 0.3%.
Water-Insoluble Material: 5 g in 50 ml of dilute HCl has a turbidity not greater than that given by 0.4 mg of $BaSO_4$ in 50 ml of water.
Ash (sulfated): Less than 0.1%.
Heavy Metals (as Pb): Not more than 20 ppm.
Likely Impurities: L-Isoleucine, D-alloisoleucine.
Crystallization Medium: Water, then ethanol to 80% v/v.
Stability and Storage: Stable.

AA-43
L-Leucine
Leu

Formula: $C_6H_{13}NO_2$
Formula Wt.: 131.18
Calc. %: C, 54.94; H, 9.99; N, 10.68; O, 24.40

Source or Method of Preparation: Casein, wheat gluten, hemoglobin.

Specific Rotation: +15.6° in 5 M HCl, $\rho = 2$ g/100 ml, $t = 25$ °C.

Homogeneity: Determined by paper chromatography.
One dimensional: Solvents 2 and 3b Two dimensional: Methods A and C.
Color reagent: Ninhydrin.
R_f 0.88, Solvent 1, descending.
R_f 0.85, Solvent 2, ascending.
R_f 0.82, Solvent 2, descending.
Volatile Matter: Not more than 0.3%.
Water-Insoluble Material: 5 g in 50 ml of dilute HCl has a turbidity not greater than that given by 0.4 mg of $BaSO_4$ in 50 ml of water.
Ash (sulfated): Less than 0.1%.
Heavy Metals (as Pb): Not more than 20 ppm.
Likely Impurities: Isoleucine, valine, methionine.
Crystallization Medium: Water, then ethanol to 80% v/v.
Stability and Storage: Stable.

AA-44
L-Lysine Monohydrochloride*
Lys·HCl

Formula: $C_6H_{15}ClN_2O_2$
Formula Wt.: 182.65
Calc. %: C, 39.45; H, 8.28; Cl, 19.41; N, 15.34; O, 17.52

Source or Method of Preparation: Natural sources, or resolution of synthetic.

Specific Rotation: +25.9° in 5 M HCl, calculated as free L-lysine, $\rho = 2$ g/100 ml, $t = 25$ °C.
Homogeneity: Determined by paper chromatography.
One dimensional: Solvent 2. Two dimensional: Methods A and B.
Color reagent: Ninhydrin.
R_f 0.92, Solvent 1, descending.
R_f 0.18, Solvent 2, ascending.
R_{BCP} 0.05, Solvent 2, descending. R_{BCP} refers to the distance moved by the amino acid calculated as the fraction of the distance moved by bromocresol purple, which is applied as a 0.1% w/v solution in ethanol.
Volatile Matter: Not more than 0.5%.
Water-Insoluble Material: 5 g in 50 ml of water has a turbidity not greater than that given by 0.4 mg of $BaSO_4$ in 50 ml of water.
Ash (sulfated): Less than 0.1%.
Heavy Metals (as Pb): Not more than 20 ppm.

Likely Impurities: Arginine, D-lysine, 2,6-diaminoheptanedioic acid, glutamic acid.

Crystallization Medium: For lysine·HCl, water at pH 4 to 6, then ethanol to 80% v/v; for lysine·2HCl, methanol, then ether, in the presence of excess HCl.

Stability and Storage: Stable at less than 60% relative humidity; above this, the dihydrate is formed.

* Also available as the dihydrochloride.

AA-45
DL-Methionine
DLMet

$CH_3SCH_2CH_2CH(NH_2)COOH$

Formula: $C_5H_{11}NO_2S$
Formula Wt.: 149.21
Calc. %: C, 40.25; H, 7.43; N, 9.39; O, 21.45; S, 21.49

Source or Method of Preparation: Synthetic.

Specific Rotation: None.
Homogeneity: Determined by paper chromatography.
    One dimensional: Solvent 2. Two dimensional: Methods A and B.
    Color reagent: Ninhydrin.
    Specific reagent: Platinic iodide.
    R_f 0.83, Solvent 1, descending.
    R_f 0.50, Solvent 2, ascending.
    R_{BCP} 0.53, Solvent 2, descending. R_{BCP} refers to the distance moved by the amino acid calculated as the fraction of the distance moved by bromocresol purple, which is applied as a 0.1% w/v solution in ethanol.
Volatile Matter: Not more than 0.3%
Water-Insoluble Material: 5 g in 50 ml of dilute HCl has a turbidity not greater than that given by 0.4 mg of $BaSO_4$ in 50 ml of water.
Ash (sulfated): Less than 0.1%.
Heavy Metals (as Pb): Not more than 20 ppm.
Likely Impurities: Methionine sulfoxide, methionine sulfone, methanethiol, ethionine.
Crystallization Medium: Water, then ethanol to 80% v/v.
Stability and Storage: Protect from light. Develops an odor on long storage.

AA-46
L-Methionine
Met

$$CH_3SCH_2CH_2 - \overset{\overset{\displaystyle NH_2}{|}}{\underset{\underset{\displaystyle H}{|}}{C}} - COOH$$

Formula: $C_5H_{11}NO_2S$
Formula Wt.: 149.21
Calc. %: C, 40.25; H, 7.43; N, 9.39; O, 21.45; S, 21.49

Source or Method of Preparation: Natural sources, or resolution of synthetic.

Specific Rotation: +21.2° in 0.2 M HCl, $\rho = 0.8$ g/100 ml, $t = 25$ °C.
Homogeneity: Determined by paper chromatography.
    One dimensional: Solvent 2. Two dimensional: Methods A and B.
    Color reagent: Ninhydrin.

Specific reagent: Platinic iodide.
    R_f 0.83, Solvent 1, descending.
    R_f 0.50, Solvent 2, ascending.
    R_{BCP} 0.53, Solvent 2, descending. R_{BCP} refers to the distance moved by the amino acid calculated as the fraction of the distance moved by bromocresol purple, which is applied as a 0.1% w/v solution in ethanol.
Volatile Matter: Not more than 0.3%.
Water-Insoluble Material: 5 g in 50 ml of dilute HCl has a turbidity not greater than that given by 0.4 mg of $BaSO_4$ in 50 ml of water.
Ash (sulfated): Less than 0.1%.
Heavy Metals (as Pb): Not more than 20 ppm.
Likely Impurities: Methionine sulfoxide, methionine sulfone, methanethiol, D-methionine.
Crystallization Medium: Water, then ethanol to 80% v/v.
Stability and Storage: Protect from light. Develops an odor on long storage.

AA-47
DL-Methionine Sulfoxide
[2-Amino-4-(methylsulfinyl)butyric Acid]
DLMet(O)

$$CH_3\overset{\overset{\displaystyle }{}}{\underset{\underset{\displaystyle O}{\|}}{S}}CH_2CH_2CH(NH_2)COOH$$

Formula: $C_5H_{11}NO_3S$
Formula Wt.: 165.21
Calc. %: C, 36.35; H, 6.71; N, 8.48; O, 29.05; S, 19.41

Source or Method of Preparation: Synthetic, via oxidation of DL-methionine.

Bibliography: A. Lepp and M. S. Dunn, *Biochem. Prep.*, **4**, 80 (1955); J. A. Roper and H. McIlwain, *Biochem. J.*, **42**, 485 (1948); J. P. Greenstein and M. Winitz, *Chemistry of the Amino Acids*, John Wiley & Sons, Inc., New York (1961), pp. 2145–2148.

Specific Rotation: None.
Homogeneity: Determined by paper chromatography.
    One dimensional: Solvents 1 and 3. Two dimensional: Method B.
    Color reagent: Ninhydrin.
    Specific reagent: Platinic iodide.
    R_f 0.72, Solvent 1, descending.
    R_f 0.03, Solvent 2, descending.
    R_{BCP} 0.93, Solvent 1, descending. R_{BCP} refers to the distance moved by the compound calculated as the fraction of the distance moved by bromocresol purple, which is applied as a 0.1% w/v solution in ethanol.
Volatile Matter: Not more than 0.3%.
Water-Insoluble Material: 5 g in 50 ml of water has a turbidity not greater than that given by 0.4 mg of $BaSO_4$ in 50 ml of water.
Ash (sulfated): Less than 0.1%.
Heavy Metals (as Pb): Not more than 20 ppm.
Likely Impurities: DL-Methionine sulfone, DL-methionine.
Crystallization Medium: Water, then ethanol to excess.
Stability and Storage: Stable.

AA-48

S-Methyl-L-cysteine [3-(Methylthio)alanine]
Cys(Me)

Formula: $C_4H_9NO_2S$
Formula Wt.: 135.19
Calc. %: C, 35.54; H, 6.71;
    N, 10.36; O, 23.67; S, 23.71

$$CH_3SCH_2 - \overset{\overset{\displaystyle NH_2}{|}}{\underset{\underset{\displaystyle H}{|}}{C}} - COOH$$

Source or Method of Preparation: Synthetic, from L-cystine.

Bibliography: V. du Vigneaud, H. S. Loring, and H. A. Craft, *J. Biol. Chem.*, **105**, 481 (1934); W. H. Horner and E. J. Kuchinskas, *J. Biol. Chem.*, **234**, 2935 (1959); C. J. Morris and J. F. Thompson, *J. Am. Chem. Soc.*, **78**, 1605 (1956); J. P. Greenstein and M. Winitz, *Chemistry of the Amino Acids*, John Wiley & Sons, Inc., New York (1961), pp. 1902–1903.

Specific Rotation: $-32.0°$ in water, $\rho = 1$ g/100 ml, $t = 26$ °C.
Homogeneity: Determined by paper chromatography.
   One dimensional: Solvents 1, 2, and 3. Two dimensional: Method B.
   Color reagent: Ninhydrin.
     R_f 0.79, Solvent 1, descending.
     R_f 0.13, Solvent 2, descending.
     R_f 0.06, Solvent 3, descending.
     R_{BCP} 1.57, Solvent 2, descending. R_{BCP} refers to the distance moved by the compound calculated as the fraction of the distance moved by bromocresol purple, which is applied as a 0.1% w/v solution in ethanol.
Volatile Matter: Not more than 0.3%.
Water-Insoluble Material: Gives a clear colorless solution in water.
Ash (sulfated): Less than 0.1%.
Heavy Metals (as Pb): Less than 20 ppm.
Likely Impurities: Cysteine, cystine, S-methyl-DL-cysteine.
Crystallization Medium: Water, then 4 volumes of ethanol.
Stability and Storage: Stable.

AA-49

S-Methyl-L-methionine Chloride
Met(Me)

Formula: $C_6H_{14}ClNO_2S$
Formula Wt.: 199.70
Calc. %: C, 36.08; H, 7.07;
    Cl, 17.76; N, 7.02; O, 16.02;
    S, 16.06

$$\left[CH_3 \overset{+}{\underset{\underset{\displaystyle CH_3}{|}}{S}} CH_2CH_2 - \overset{\overset{\displaystyle NH_2}{|}}{\underset{\underset{\displaystyle H}{|}}{C}} - COOH \right] Cl^-$$

Source or Method of Preparation: Synthetic, from L-methionine.

Bibliography: A. Stevens and W. Sakami, *J. Biol. Chem.*, **234**, 2063 (1959).

Specific Rotation: $+33°$ in 0.2 M HCl, $\rho = 1$ g/100 ml, $t = 23$ °C.
Homogeneity: Determined by paper chromatography.
   One dimensional: Solvents 1 and 2.
   Color reagent: Ninhydrin.
   Specific reagent: Azide–iodine (weak).
     R_f 0.89, Solvent 1, ascending.
     R_f 0.11, Solvent 2, ascending.
     R_{BCP} 0.21, Solvent 3a, descending. R_{BCP} refers to the distance moved by the compound calculated as the fraction of the distance moved by bromocresol purple, which is applied as a 0.1% w/v solution in ethanol.

Volatile Matter: Cannot be dried by AOAC method without decomposition.
Water-Insoluble Material: 5 g in 50 ml of dilute HCl has a turbidity not greater than that given by 0.4 mg of $BaSO_4$ in 50 ml of water.
Ash (sulfated): Less than 0.1%.
Heavy Metals (as Pb): Not more than 20 ppm.
Likely Impurities: Methionine, methionine sulfoxide, methionine sulfone.
Crystallization Medium: Water, then ethanol to 95% v/v.
Stability and Storage: Undergoes slow decomposition. Store in a cool dry place. Protect from light.

AA-50

L-Ornithine Monohydrochloride*
Orn·HCl

Formula: $C_5H_{13}ClN_2O_2$
Formula Wt.: 168.62
Calc. %: C, 35.61; H, 7.77; Cl, 21.03; N, 16.62; O, 18.98

$$HCl \cdot NH_2CH_2CH_2CH_2 - \overset{\overset{\displaystyle NH_2}{|}}{\underset{\underset{\displaystyle H}{|}}{C}} - COOH$$

Source or Method of Preparation: Synthetic, via enzymic (arginase) or alkaline degradation of arginine.

Bibliography: A. Hunter, *Biochem. J.*, **33**, 27 (1939); D. E. Rivard and H. E. Carter, *J. Am. Chem. Soc.*, **77**, 1260 (1955); D. E. Rivard, *Biochem. Prep.*, **3**, 97 (1953); J. P. Greenstein and M. Winitz, *Chemistry of the Amino Acids*, John Wiley & Sons, Inc., New York (1961), pp. 2477–2488.

Specific Rotation: $+28.3°$ in 5 M HCl, $\rho = 2$ g/100 ml calculated as free base, $t = 25$ °C.
Homogeneity: Determined by paper chromatography.
   One dimensional: Solvents 1 and 2. Two dimensional: Method A.
   Color reagent: Ninhydrin.
     R_f 0.87, Solvent 1, descending.
     R_f 0.01, Solvent 2, descending.
     R_{BCP} 1.12, Solvent 1, descending. R_{BCP} refers to the distance moved by the compound calculated as the fraction of the distance moved by bromocresol purple, which is applied as a 0.1% w/v solution in ethanol.
Volatile Matter: Not more than 0.3%.
Water-Insoluble Material: 5 g in 50 ml of water has a turbidity not greater than that given by 0.4 mg of $BaSO_4$ in 50 ml of water.
Ash (sulfated): Less than 0.1%.
Heavy Metals (as Pb): Not more than 20 ppm.
Likely Impurities: Citrulline, arginine, D-ornithine.
Crystallization Medium: Water, then 4 volumes of ethanol.
Stability and Storage: Stable.

* Also available as the dihydrochloride.

AA-51

DL-Phenylalanine
DLPhe

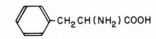

Formula: $C_9H_{11}NO_2$
Formula Wt.: 165.19
Calc. %: C, 65.44; H, 6.71; N, 8.48; O, 19.37

Source or Method of Preparation: Synthetic, from benzaldehyde.

Specific Rotation: None.
Homogeneity: Determined by paper chromatography.
 One dimensional: Solvent 2. Two dimensional: Method A.
 Color reagent: Ninhydrin.
  R_f 0.90, Solvent 1, descending.
  R_f 0.80, Solvent 2, ascending.
  R_{BCP} 0.69, Solvent 2, descending. R_{BCP} refers to the distance moved by the amino acid calculated as the fraction of the distance moved by bromocresol purple, which is applied as a 0.1% w/v solution in ethanol.
Volatile Matter: Not more than 0.3%.
Water-Insoluble Material: 5 g in 50 ml of dilute HCl has a turbidity not greater than that given by 0.4 mg of $BaSO_4$ in 50 ml of water.
Ash (sulfated): Less than 0.1%.
Heavy Metals (as Pb): Not more than 20 ppm.
Crystallization Medium: Water, then ethanol to 80% v/v.
Stability and Storage: Stable.

AA-52
L-Phenylalanine
Phe

Formula: $C_9H_{11}NO_2$
Formula Wt.: 165.19
Calc. %: C, 65.44; H, 6.71; N, 8.48; O, 19.37

Source or Method of Preparation: Resolution of synthetic, or from natural sources.

Specific Rotation: −34.0° in H_2O, $\rho = 2$ g/100 ml, $t = 25$ °C.
Homogeneity: Determined by paper chromatography.
 One dimensional: Solvent 2. Two dimensional: Method A.
 Color reagent: Ninhydrin.
  R_f 0.90, Solvent 1, descending.
  R_f 0.80, Solvent 2, ascending.
  R_{BCP} 0.69, Solvent 2, descending. R_{BCP} refers to the distance moved by the amino acid calculated as the fraction of the distance moved by bromocresol purple, which is applied as a 0.1% w/v solution in ethanol.
Volatile Matter: Not more than 0.3%.
Water-Insoluble Material: 5 g in 50 ml of dilute HCl has a turbidity not greater than that given by 0.4 mg of $BaSO_4$ in 50 ml of water.
Ash (sulfated): Less than 0.1%.
Heavy Metals (as Pb): Not more than 20 ppm.
Likely Impurities: Leucines, valine, methionine, tyrosine, D-phenylalanine.
Crystallization Medium: Water, then ethanol to 80% v/v.
Stability and Storage: Stable.

AA-53
L-Proline
Pro

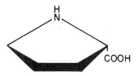

Formula: $C_5H_9NO_2$
Formula Wt.: 115.13
Calc. %: C, 52.16, H, 7.88; N, 12.17; O, 27.79

Source or Method of Preparation: Collagen, wheat gluten.

Specific Rotation: −85.1° in H_2O, $\rho = 2$ g/100 ml, $t = 25$ °C.
Homogeneity: Determined by paper chromatography.
 One dimensional: Solvent 2. Two dimensional: Method A.
 Color reagents: Ninhydrin and isatin.
  R_f 0.30, Solvent 2, ascending.
  R_{BCP} 0.35, Solvent 2, descending. R_{BCP} refers to the distance moved by the amino acid calculated as the fraction of the distance moved by bromocresol purple, which is applied as a 0.1% w/v solution in ethanol.
Assays: Amino group (ninhydrin according to Van Slyke): 0.0%.
Volatile Matter: Not more than 0.5%
Water-Insoluble Material: 5 g in 50 ml of water has a turbidity not greater than that given by 0.4 mg of $BaSO_4$ in 50 ml of water.
Ash (sulfated): Less than 0.1%.
Heavy Metals (as Pb): Not more than 20 ppm.
Likely Impurities: Hydroxyproline.
Crystallization Medium: Absolute ethanol.
Stability and Storage: Hygroscopic; keep in a desiccator.

AA-54
DL-Serine
DLSer

$$HOCH_2CH(NH_2)COOH$$

Formula: $C_3H_7NO_3$
Formula Wt.: 105.09
Calc. %: C, 34.28; H, 6.72; N, 13.33; O, 45.67

Source or Method of Preparation: Synthetic.

Specific Rotation: None.
Homogeneity: Determined by paper chromatography.
 One dimensional: Solvent 1. Two dimensional: Methods A and B.
 Color reagent: Ninhydrin.
 Specific reagents: Nessler and HIO_4.
  R_f 0.43, Solvent 1, ascending.
  R_f 0.47, Solvent 1, descending.
  R_{BCP} 0.14, Solvent 2, descending. R_{BCP} refers to the distance moved by the amino acid calculated as the fraction of the distance moved by bromocresol purple, which is applied as a 0.1% w/v solution in ethanol.
Volatile Matter: Dry to constant weight at 105 °C at atmospheric pressure: 0.3%.
Water-Insoluble Material: 5 g in 50 ml of dilute HCl has a turbidity not greater than that given by 0.4 mg of $BaSO_4$ in 50 ml of water.
Ash (sulfated): Less than 0.1%.
Heavy Metals (as Pb): Not more than 20 ppm.
Likely Impurities: Glycine.
Crystallization Medium: Water, then ethanol to 80% v/v.
Stability and Storage: Store in a desiccator, after drying.

AA-55
L-Serine
Ser

$$HOCH_2-\underset{\underset{H}{|}}{\overset{\overset{NH_2}{|}}{C}}-COOH$$

Formula: $C_3H_7NO_3$
Formula Wt.: 105.09
Calc. %: C, 34.28; H, 6.72; N, 13.33; O, 45.67

Source or Method of Preparation: Resolution of synthetic.

Specific Rotation: $+14.5°$ in 1.0 M HCl, $\rho = 9.3$ g/100 ml, $t = 25$ °C.
Homogeneity: Determined by paper chromatography.
  One dimensional: Solvent 1. Two dimensional: Methods A and B
  Color reagent: Ninhydrin.
  Specific reagents: Nessler and HIO_4.
    R_f 0.43, Solvent 1, ascending.
    R_f 0.47, Solvent 1, descending.
    R_{BCP} 0.14, Solvent 2, descending. R_{BCP} refers to the distance moved by the amino acid calculated as the fraction of the distance moved by bromocresol purple, which is applied as a 0.1% w/v solution in ethanol.
Volatile Matter: Dry to constant weight at 105 °C at atmospheric pressure: 0.3%.
Water-Insoluble Material: 5 g in 50 ml of dilute HCl has a turbidity not greater than that given by 0.4 mg of $BaSO_4$ in 50 ml of water.
Ash (sulfated): Less than 0.1%.
Heavy Metals (as Pb): Not more than 20 ppm.
Likely Impurities: Glycine, D-serine.
Crystallization Medium: Water, then ethanol to 80% v/v.
Stability and Storage: Store in a desiccator, after drying.

AA-56
DL-Threonine
DLThr

$$CH_3CH(OH)-CH(NH_2)COOH$$

Formula: $C_4H_9NO_3$
Formula Wt.: 119.12
Calc. %: C, 40.33; H, 7.62; N, 11.76; O, 40.30

Source or Method of Preparation: Synthetic.

Specific Rotation: None.
Homogeneity: Determined by paper chromatography.
  One dimensional: Solvent 1. Two dimensional: Methods A and B.
  Color reagent: Ninhydrin.
  Specific reagents: Nessler and HIO_4.
    R_f 0.55, Solvent 1, ascending.
    R_f 0.61, Solvent 1, descending.
    R_{BCP} 0.24, Solvent 2, descending. R_{BCP} refers to the distance moved by the amino acid calculated as the fraction of the distance moved by bromocresol purple, which is applied as a 0.1% w/v solution in ethanol.
Volatile Matter: Dry to constant weight at 105 °C at atmospheric pressure: 0.3%.
Water-Insoluble Material: 5 g in 50 ml of water has a turbidity not greater than that given by 0.4 mg of $BaSO_4$ in 50 ml of water.

Ash (sulfated): Less than 0.1%.
Heavy Metals (as Pb): Not more than 20 ppm.
Likely Impurities: Allothreonine,* glycine.
Crystallization Medium: Water, then ethanol to 80% v/v.
Stability and Storage: Store in a desiccator, after drying.

  * See "Infrared Spectroscopy," p. 5.

AA-57
L-Threonine
Thr

$$CH_3-\underset{\underset{HO}{|}}{\overset{\overset{H}{|}}{C}}-\underset{\underset{H}{|}}{\overset{\overset{NH_2}{|}}{C}}-COOH$$

Formula: $C_4H_9NO_3$
Formula Wt.: 119.12
Calc. %: C, 40.33; H, 7.62; N, 11.76; O, 40.30

Source or Method of Preparation: Resolution of synthetic.

Specific Rotation: $-28.4°$ in water, $\rho = 1.0$ g/100 ml, $t = 26$ °C.
Homogeneity: Determined by paper chromatography.
  One dimensional: Solvent 1. Two dimensional: Methods A and B.
  Color reagent: Ninhydrin.
  Specific reagents: Nessler and HIO_4.
    R_f 0.55, Solvent 1, ascending.
    R_f 0.61, Solvent 1, descending.
    R_{BCP} 0.24, Solvent 2, descending. R_{BCP} refers to the distance moved by the amino acid calculated as the fraction of the distance moved by bromocresol purple, which is applied as a 0.1% w/v solution in ethanol.
Volatile Matter: Dry to constant weight at 105 °C at atmospheric pressure: 0.3%.
Water-Insoluble Material: 5 g in 50 ml of water has a turbidity not greater than that given by 0.4 mg of $BaSO_4$ in 50 ml of water.
Ash (sulfated): Less than 0.1%.
Heavy Metals (as Pb): Not more than 20 ppm.
Likely Impurities: *Allo*threonine,* glycine, D-threonine.
Crystallization Medium: Water, then ethanol to 80% v/v.
Stability and Storage: Store in a desiccator, after drying.

  * See "Infrared Spectroscopy," p. 5.

AA-58
L-Thyroxine

Formula: $C_{15}H_{11}I_4NO_4$
Formula Wt.: 776.88
Calc. %: C, 23.19; H, 1.43; I, 65.34; N, 1.80; O, 8.24

Source or Method of Preparation: Natural sources, or resolution of synthetic.

Specific Rotation: $+26°$ in ethanol–1 M aq. HCl (2:1), $\rho = 1$–2 g/100 ml, $t = 22$ °C.
Homogeneity: Determined by paper chromatography.
  One dimensional: Solvent 3b. Two dimensional: Method C.

Color reagent: Ninhydrin.
Specific reagent: $Ce(HSO_4)_4$—$NaAsO_2$.
   R_{BCP} 1.14, Solvent 3b, descending.
   R_f 0.60, Solvent 3, descending.
   R_{BCP} 1.20, Solvent 2, descending. R_{BCP} refers to the distance moved by the amino acid calculated as the fraction of the distance moved by bromocresol purple, which is applied as a 0.1% w/v solution in ethanol.
Water-Insoluble Material: 5 g in 50 ml of dilute NaOH has a turbidity not greater than that given by 0.4 mg of $BaSO_4$ in 50 ml of water.
Ash (sulfated): Less than 0.1%.
Heavy Metals (as Pb): Not more than 20 ppm.
Likely Impurities: Tyrosine, iodotyrosines, and other iodothyronines,* iodide,* iodine.*
Crystallization Medium: Dilute NH_4OH at room temperature, then acetic acid to pH 6.
Stability and Storage: Decomposes slowly in the cold.

* These impurities are often present in thyroxine-^{131}I and may account for one half of the total radioactivity.

AA-59
3,3′,5-Triiodo-L-thyronine

Formula: $C_{15}H_{12}I_3NO_4$
Formula Wt.: 650.98
Calc. %: C, 27.67; H, 1.85; I, 58.49; N, 2.15; O, 9.83

Source or Method of Preparation: Synthetic.

Specific Rotation: +21.5° in ethanol–1 M aq. HCl (2:1), $\rho = 2$ g/100 ml, $t = 29.5$ °C.
Homogeneity: Determined by paper chromatography.
   One dimensional: Solvent 3b. Two dimensional: Method C.
   Color reagent: Ninhydrin.
   Specific reagent: $Ce(HSO_4)_4$–$NaAsO_2$.
   R_{BCP} 1.46, Solvent 3b, descending.
   R_{BCP} 1.13, Solvent 2, descending. R_{BCP} refers to the distance moved by the amino acid calculated as the fraction of the distance moved by bromocresol purple, which is applied as a 0.1% w/v solution in ethanol.
Water-Insoluble Material: 5 g in 50 ml of dilute NaOH has a turbidity not greater than that given by 0.4 mg of $BaSO_4$ in 50 ml of water.
Ash (sulfated): Less than 0.1%.
Heavy Metals (as Pb): Not more than 20 ppm.
Likely Impurities: Cf. thyroxine.
Crystallization Medium: Dilute NH_4OH at room temperature, then acetic acid to pH 6.
Stability and Storage: Decomposes slowly in the cold.

AA-60
DL-Tryptophan
DLTrp

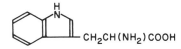

Formula: $C_{11}H_{12}N_2O_2$
Formula Wt.: 204.23
Calc. %: C, 64.69; H, 5.93; N, 13.72; O, 15.67

Source or Method of Preparation: Synthetic.

Specific Rotation: None.
Homogeneity: Determined by paper chromatography.
   One dimensional: Solvent 3b. Two dimensional: Methods A and B.
   Color reagent: Ninhydrin.
   Specific reagent: Ehrlich.*
   R_f 0.60, Solvent 3, ascending.
Water-Insoluble Material: 5 g in 50 ml of dilute NaOH has a turbidity not greater than that given by 0.4 mg of $BaSO_4$ in 50 ml of water.
Ash (sulfated): Less than 0.1%.
Heavy Metals (as Pb): Not more than 20 ppm.
Crystallization Medium: 65% v/v ethanol.
Stability and Storage: Store in a dark place; darkens on prolonged exposure to light.

* Impurities in the solvents may interfere with this reaction.

AA-61
L-Tryptophan
Trp

Formula: $C_{11}H_{12}N_2O_2$
Formula Wt.: 204.23
Calc. %: C, 64.69; H, 5.93; N, 13.72; O, 15.67

Source or Method of Preparation: Resolution of synthetic.

Specific Rotation: −33.3° in H_2O, $\rho = 1$ g/100 ml, $t = 25$ °C.
Homogeneity: Determined by paper chromatography.
   One dimensional: Solvent 3b. Two dimensional: Solvents A and B.
   Color reagent: Ninhydrin.
   Specific reagent: Ehrlich.*
   R_f 0.60, Solvent 3, ascending.
Volatile Matter: Not more than 0.3%.
Water-Insoluble Material: 5 g in 50 ml of dilute NaOH has a turbidity not greater than that given by 0.4 mg of $BaSO_4$ in 50 ml of water.
Ash (sulfated): Less than 0.1%.
Heavy Metals (as Pb): Not more than 20 ppm.
Likely Impurities: D-Tryptophan.
Crystallization Medium: 65% v/v ethanol.
Stability and Storage: Store in a dark place; darkens on prolonged exposure to light.

* Impurities in the solvents may interfere with this reaction.

AA-62
DL-Tyrosine
DLTyr

$HO-C_6H_4-CH_2CH(NH_2)COOH$

Formula: $C_9H_{11}NO_3$
Formula Wt.: 181.19
Calc. %: C, 59.66; H, 6.12; N, 7.73; O, 26.49

Source or Method of Preparation: Synthetic.

Specific Rotation: None.
Homogeneity: Determined by paper chromatography.
   One dimensional: Solvent 2. Two dimensional: Methods A and B.
   Color reagent: Ninhydrin.
   Specific reagent: Pauly.
      R_f 0.69, Solvent 1, descending.
      R_f 0.60, Solvent 2, ascending.
      R_{BCP} 0.41, Solvent 2, descending. R_{BCP} refers to the distance moved by the amino acid calculated as the fraction of the distance moved by bromocresol purple, which is applied as a 0.1% w/v solution in ethanol.
Volatile Matter: Not more than 0.3%.
Water-Insoluble Material: 5 g in 50 ml of dilute HCl has a turbidity not greater than that given by 0.4 mg of $BaSO_4$ in 50 ml of water.
Ash (sulfated): Less than 0.1%.
Heavy Metals (as Pb): Not more than 20 ppm.
Likely Impurities: o-Tyrosine (on paper chromatograms developed with Solvent 1).[1]
Crystallization Medium: Dilute NH_4OH, then acetic acid to pH 5.
Stability and Storage: Stable.

Reference

1. J. P. Lambooy, *J. Am. Chem. Soc.*, **78**, 771 (1956).

AA-63
L-Tyrosine
Tyr

$HO-C_6H_4-CH_2-\overset{NH_2}{\underset{H}{C}}-COOH$

Formula: $C_9H_{11}NO_3$
Formula Wt.: 181.19
Calc. %: C, 59.66; H, 6.12; N, 7.73; O, 26.49

Source or Method of Preparation: Casein, hair.

Specific Rotation: $-10.0°$ in 5 M HCl, $\rho = 2$ g/100 ml, $t = 25$ °C.
Homogeneity: Determined by paper chromatography.
   One dimensional: Solvent 2. Two dimensional: Methods A and B.
   Color reagent: Ninhydrin.
   Specific reagent: Pauly.
      R_f 0.69, Solvent 1, descending.
      R_f 0.60, Solvent 2, ascending.
      R_{BCP} 0.41, Solvent 2, descending. R_{BCP} refers to the distance moved by the amino acid calculated as the fraction of the distance moved by bromocresol purple, which is applied as a 0.1% w/v solution in ethanol.
Volatile Matter: Not more than 0.3%.
Water-Insoluble Material: 5 g in 50 ml of dilute HCl has a turbidity not greater than that given by 0.4 mg of $BaSO_4$ in 50 ml of water.
Ash (sulfated): Less than 0.1%.
Heavy Metals (as Pb): Not more than 20 ppm.
Likely Impurities: Ammonium salt, L-cystine.
Crystallization Medium: Dilute NH_4OH, then acetic acid to pH 5.
Stability and Storage: Stable.

AA-64
DL-Valine
DLVal

$(CH_3)_2CHCH(NH_2)COOH$

Formula: $C_5H_{11}NO_2$
Formula Wt.: 117.15
Calc. %: C, 51.26; H, 9.47; N, 11.96; O, 27.32

Source or Method of Preparation: Synthetic.

Specific Rotation: None.
Homogeneity: Determined by paper chromatography.
   One dimensional: Solvents 2 and 3b. Two dimensional: Methods A and B.
   Color reagent: Ninhydrin.
      R_f 0.81, Solvent 1, descending.
      R_{BCP} 0.57, Solvent 2, descending. R_{BCP} refers to the distance moved by the amino acid calculated as the fraction of the distance moved by bromocresol purple, which is applied as a 0.1% solution in ethanol.
Volatile Matter: Not more than 0.3%.
Water-Insoluble Material: 5 g in 50 ml of dilute HCl has a turbidity not greater than that given by 0.4 mg of $BaSO_4$ in 50 ml of water.
Ash (sulfated): Less than 0.1%.
Heavy Metals (as Pb): Not more than 20 ppm.
Crystallization Medium: Water, then ethanol to 80% v/v.
Stability and Storage: Stable.

AA-65
L-Valine
Val

$\overset{H_3C}{\underset{H_3C}{}}HC-\overset{NH_2}{\underset{H}{C}}-COOH$

Formula: $C_5H_{11}NO_2$
Formula Wt.: 117.15
Calc. %: C, 51.26; H, 9.47; N, 11.96; O, 27.32

Source or Method of Preparation: Resolution of synthetic.

Specific Rotation: $+26.7°$ in 6.0 M HCl, $\rho = 3.4$ g/100 ml, $t = 20$ °C.
Homogeneity: Determined by paper chromatography.
   One dimensional: Solvents 2 and 3b. Two dimensional: Methods A and B.
   Color reagent: Ninhydrin.
      R_f 0.81, Solvent 1, descending.
      R_{BCP} 0.57, Solvent 2, descending. R_{BCP} refers to the distance moved by the amino acid calculated as the fraction of the distance moved by bromocresol purple, which is applied as a 0.1% w/v solution in ethanol.
Volatile Matter: Not more than 0.3%.
Water-Insoluble Material: 5 g in 50 ml of dilute HCl has a turbidity not greater than that given by 0.4 mg of $BaSO_4$ in 50 ml of water.
Ash (sulfated): Less than 0.1%.
Heavy Metals (as Pb): Not more than 20 ppm.
Likely Impurities: D-Valine.
Crystallization Medium: Water, then ethanol to 80% v/v.
Stability and Storage: Stable.

Carbohydrates and Related Compounds

GENERAL REMARKS

The majority of commercially available carbohydrates* meet exceedingly high standards of purity, comparable with those for U.S.P. dextrose. Optical rotatory power is an almost ideal property for characterization of purity, particularly when the specific rotation is relatively large and when the wavelength of the light used is known and controlled. However, even specific-rotation measurements may fail to detect small amounts of contaminating carbohydrates. Therefore, paper, thin-layer, and gas–liquid chromatography—qualitative procedures of great delicacy—are here adopted as routine methods for ascertaining homogeneity.

When kept tightly sealed in amber-glass containers and stored away from strong light in a cool place, the carbohydrates are usually stable for an indefinitely long period of time. Exceptional cases are noted.

NOTES ON ANALYTICAL PROCEDURES

Specific Rotation

Unless otherwise noted, specific rotations are determined in water at 20 °C, using the sodium D-line, on

* In revising the specifications for some compounds, the opportunity has been taken to include the modern systematic name. This is given first, followed in some cases by one or more less-desirable names in parentheses. Some compounds have approved trivial names ["Rules of Carbohydrate Nomenclature," *J. Org. Chem.*, **28,** 281 (1963)]; in such cases, only that name is given.

samples dried to constant weight *in vacuo* at 60 °C (see "Loss of Weight on Drying," p. 28). The mass concentration, ρ, is expressed in grams of substance per 100 ml of solution. For substances that mutarotate, only the equilibrium rotation is given. To hasten mutarotation, one drop of a solution that has been made by diluting concentrated (28%) aqueous ammonia with an equal volume of water may be added in making up a 25-ml solution of the sugar. The specific rotation, $[\alpha]_D^{20}$, is calculated according to the formula

$$[\alpha]_D^{20} = 100\alpha/l\rho,$$

where α is the observed rotation in circular degrees and l is the length of the polarimeter tube in decimeters. For hydrated compounds, the calculation is based on the weight of the hydrated form, the weight loss on drying being used to adjust the specific rotation to correspond with the formula weight of the anhydrous compound.

Melting Point

At the time the Subcommittee on Carbohydrates was established, a review was made of available methods by which purity of a carbohydrate could be measured. Samples of several sugars were submitted to various laboratories. Excellent agreement was obtained in all instances except for melting points. The decision was then made that melting points be omitted except for certain sugar alcohols, lactones, glycosides, and a few sugars for which the values could be rigorously defended. Based on this decision, the melting points used are limited, and are usually given only as criteria, not as specifications.

Any of a large variety of apparatus may be used, the temperature being raised at a rate of 1 to 2 °C per minute during the actual determination. The values given are corrected for stem exposure, if any, and represent the range from the first appearance of liquid to the disappearance of the last crystal.

Loss of Weight on Drying

Unless otherwise specified under individual specifications, the general procedure for determining volatile contaminants (normally water) is the following one, described in Section 31.005 ("Vacuum Drying—Official Final Action") of the *Official Methods of Analysis of the Association of Official Analytical Chemists,* 11th ed., AOAC, Washington, D.C. (1970), p. 525.

Dry 2–5 g prepared sample (ground, if necessary, and mixed to uniformity) in flat dish (Ni, Pt, or Al with tight-fit cover), 2 h at ≤ 70 °C (preferably 60 °C) under pressure ≤ 50 mmHg. Bleed oven with current of air (drying by passing through anhydrous $CaSO_4$, P_2O_5, or other efficient desiccant) during drying to remove H_2O vapor. Remove dish from oven, cover, cool in desiccator, and weigh. Redry 1 hr and repeat process until change in weight between successive dryings at 1-h intervals is ≤ 2 mg.

Ash Determination

The method described in Section 30.014 ("Sulfated Ash—Official Final Action") of the *Official Methods of Analysis of the Association of Official Analytical Chemists,* 11th ed., AOAC, Washington, D.C. (1970), p. 526 is used.

Weigh 5 g sample into 50–100 ml Pt dish, add 5 ml 10% (by wt) H_2SO_4, ignite until sample is well carbonized, and then ash in muffle at about 550 °C. Cool, add 2–3 ml 10% H_2SO_4, evaporate on steam bath, dry on hotplate, and again ignite at 550 °C to constant weight. Express result as % sulfated ash.

Paper Partition Chromatography: General Procedure[1–3]

Paper

Whatman No. 1 filter paper, in sheets 18.25 in. × 22.5 in., is used, the "grain" of the paper (indicated by an arrow on the package) being in the direction of the longer dimension. Chromatography, whether ascending or descending, is conducted across the grain of the paper, the vertical dimension always being 18.25 in., and the horizontal dimension differing according to the number of samples to be tested or the capacity of the container used. Where descending chromatography is to

be used and it is desired to allow the liquid front to run off the paper, the lower end of the paper is serrated with a pair of dressmaker's pinking shears (coarse teeth) in order to obtain uniform runoff.

Chromatography

Either ascending or descending chromatography may be used, although the latter is often preferable since the process may be continued beyond the stage when the advancing front reaches the end of the paper. A 100-mg sample of the carbohydrate to be tested is dissolved in water to make a volume of 1.0 ml and a 2-μl aliquot is applied on a line that is lightly ruled about 6 cm from, and parallel to, the edge of the paper whence the solvent front will start. The application is made portionwise, so that the spot is not more than 3 mm in diameter. Evaporation may be hastened through the use of a gentle stream of warm air (the ordinary household hair dryer is useful for this purpose). Other spots may be placed on the line at about 4-cm intervals. The paper containing the materials is exposed for 4 h, at the temperature to be used for chromatography, to an atmosphere saturated with the solvent system to be used. The edge of the paper nearer the spots is now dipped into the liquid phase in a suitably enclosed chamber (either a glass cylinder or a chromatography chest, which is protected from drafts or other influences that might result in uneven temperatures in the chamber) and kept at a fairly constant temperature (±3 °C) for the length of time specified or until the liquid front has reached approximately 2.5 cm from the end of the paper. The paper is then dried in a gentle stream of air (hood or hair dryer), and the spots are developed.

Solvent Systems

(Reagent-grade materials are used without further purification.)

(a) System 1. 1-Butanol–pyridine–water (3.0:2.0: 1.5 v/v).

(b) System 2. 1-Propanol–water (0.9:2.8 v/v).

(c) Special solvent systems as described under individual specifications.

Spraying Reagents

1. Aniline hydrogen phthalate reagent:[4] 1.66 g of phthalic acid, 0.93 g of freshly distilled aniline, and 100 ml of 1-butanol saturated with water. The reagent is stable for a considerable period of time when stored in a brown-glass bottle in the refrigerator; it should be discarded after darkening becomes marked. The reagent is sprayed evenly upon the paper, but not in sufficient

quantity to run. The paper is then dried at 105–110 °C for 5 to 10 min; respraying and redrying intensifies the spots. Weak spots may be verified by viewing the paper under ultraviolet light.*

2. Periodate–permanganate reagent:[5] 2% (w/v) aqueous $NaIO_4$ and freshly prepared 1% (w/v) $KMnO_4$ in 2% (w/v) aqueous Na_2CO_3. The two solutions are mixed, 4:1 by volume, just before use, and are sprayed lightly and evenly on the paper. When the spots have appeared, excess reagent is removed by washing the paper in water. Weak spots may be intensified by spraying with a solution made as follows: 1 g of benzidine, 8 g of trichloroacetic acid, 20 ml of glacial acetic acid, 12 ml of water, and 150 ml of absolute alcohol.[6]

3. Ammoniacal silver nitrate reagent:[7] 5 g of $AgNO_3$, 95 ml of water, and 6 ml of concentrated ammonia. To avoid the danger of formation of explosive silver residues, the reagent should be made up just before use and be discarded promptly thereafter. The paper is sprayed with the reagent and left overnight in the dark to develop. The background coloring is then removed with "Kodak Liquid X-ray Fixer," and the paper is washed with water and dried.

Thin-Layer Chromatography: General Procedure[8–11]

Layers

Standard layers, 0.25 mm thick, of silica gel G are used (silica gel containing about 14% of plaster of Paris, E. Merck, Darmstadt, Germany), prepared with a commercial apparatus or in one of many other ways. An aqueous slurry of silica gel G consisting of two parts of water to one part of adsorbent is shaken vigorously for 1 min and spread in an even film over clean glass plates. The layers are allowed to dry for 30 min and are then activated by placing them in an oven at 110 °C for 1 h. The layers are stored over calcium chloride or other solid desiccant.

Chromatography

Ascending chromatography is used. A 10-mg sample of the carbohydrate to be tested is dissolved in methanol,

* Radiation of approximately 254 nm wavelength, such as is provided by the model V-41 "Mineralight" of Ultra-Violet Products, Inc., South Pasadena, California, is satisfactory.

the solution is diluted to 1.0 ml, and successive volumes of 2, 4, and 8 μl are applied to three spots that are 2 cm from the bottom of the layer and at least 1 cm apart. The application is made portionwise, with a micropipet, in such a way that the spot is not more than 4 mm in diameter. Evaporation may be hastened by carrying out the operation on a warm (50–60 °C) metal block or by using a gentle stream of warm air (hair dryer). The edge nearer the spots is then dipped into the prescribed solvent mixture, so that about 1 cm of the layer lies under the surface. The development is carried out at room temperature in a chamber just large enough to hold the layer. The sides of the chamber are at least half lined with filter paper previously wetted with the chosen solvent. Development is allowed to continue until the solvent front has moved 15 cm from the origin spots of the sample. The layer is then removed from the chamber, dried under a current of warm air, well sprayed with concentrated H_2SO_4, and kept in an oven at 110 °C for 10 min.

Solvent Systems

The solvents are reagent grade, used without further purification, and are prescribed for each compound.

Gas–Liquid Chromatography: General Procedure[12, 13]

Liquid Phases

The homogeneity of most carbohydrates can be verified by gas–liquid chromatography of their trimethylsilyl ethers. Two types of liquid phase are used. A nonpolar phase (Apiezon M or SE-30) is useful for establishing the absence of carbohydrate impurities of lower or higher molecular weight in the sample, and a polar phase [poly(ethylene glycol succinate) (EGS) or neopentyl glycol sebacate polyester (NPGS)] is useful in establishing the absence of carbohydrates having molecular weights similar to that of the sample.[14, 15]

Sample Preparation

The carbohydrate (about 10 mg) is treated with hexamethyldisilazane (0.2 ml), chlorotrimethylsilane (0.1 ml), and pyridine (1 ml) and the mixture is shaken vigorously at room temperature for 30 min. Samples of the reaction mixture are examined directly on the gas chromatograph.

Columns

Columns 4 to 6 ft long having 0.2-in. i.d., or analytical columns of ⅛-in. o.d., are packed with 10% (w/w) of poly(ethylene glycol succinate), neopentyl glycol sebacate polyester, Apiezon M, or SE-30 as the liquid phase on 80–100 mesh Chromosorb W.

Chromatography

Gas chromatographs are operated with gas flow-rates of 75 to 150 ml/min for the 0.2-in. i.d. columns, or 25–30 ml/min for the ⅛-in. o.d. columns. Nonpolar columns are temperature-programmed at linear rates from 125 to 250 °C, to cover the analysis of compounds ranging from derivatives of tetroses to those of oligosaccharides. Polar columns are operated at fixed temperatures in the range of 130 to 240 °C, in order to establish the absence of carbohydrates having molecular weights similar to those of the samples under investigation.

Heavy Metals

Dissolve 2 g of the compound in 20 ml of water and add 0.5 ml of 1 M HCl and 10 ml of a freshly prepared saturated aqueous solution of hydrogen sulfide. Any darkening produced should not be more than that in a blank to which 0.02 mg of Cu has been added.

Iron

Dissolve 1 g of the compound in 10 ml of water. Add 1 ml of concentrated HCl, about 30 mg of ammonium persulfate, and 15 ml of butanolic potassium thiocyanate. Shake vigorously and allow to separate. Any red color in the clear butanol layer is to be not darker than that in a blank to which 0.005 mg of Fe has been added.

Reagent: Dissolve 10 g of KSCN in 10 ml of water. Warm this solution to 25–30 °C, add sufficient butanol to make 100 ml, and shake vigorously until clear.

Arsenic

The method used is that given under "Dextrose" in *Reagent Chemicals,* American Chemical Society, Washington, D.C. (1955), pp. 4 and 151, and in various textbooks and collections of official methods of analysis. The ACS description is given here.

The arsenic in a sample (5 g) is determined by the Gutzeit method. If the amount of stain produced is less than that given by 2 μg of arsenic, the sample contains As at less than 0.5 ppm.

The widely used Gutzeit method for arsenic, which is prescribed for testing reagents, depends on the measurement or comparison of stains produced by the action of evolved arsine on strips of paper that have been impregnated with mercuric bromide from an alcoholic solution. Details of the method are given in collections of official methods of analysis and in textbooks. Because of the nature of the test it is important to have the greatest possible uniformity in the preparation of the stains from samples and from measured amounts of arsenic.

Apparatus

A wide-mouthed bottle of about 60-ml capacity serves for the generator. It carries a glass tube about 1 cm in diameter and 6 to 7 cm long, which is constricted at the bottom to pass through the stopper of the bottle. The tube is to hold glass wool, purified cotton, or similar material, moistened with a 10% solution of lead acetate. All tubes of a set of generators should be charged with equal amounts of this material. The solution serves to hold back any hydrogen sulfide generated in the bottle, and also helps to maintain a uniform content of moisture in the evolved gases.

Above the lead acetate tube is a narrow glass tube of 2.6- to 2.7-mm i.d. and 10 to 12 cm long that holds the strip of mercuric bromide paper. The diameter of this tube must not be large enough to permit curling of the paper strip. The paper strips are best obtained by the purchase of commercially cut strips which are of uniform width of 2.5 mm and are generally supplied in a manner that facilitates their preparation for use. The strips are soaked about an hour in a 5% solution of mercuric bromide in alcohol. They are drained and allowed to dry in clean air. It is essential that all strips used for a particular test be sensitized in the same manner and at the same time.

Procedure

Place the sample in the generator bottle with water and about 5 ml of acid. The acid may be either sulfuric or hydrochloric but in any set of determinations the kind and the amounts of acid in all the bottles must be the same. If acid is used up in dissolving the sample, the amount must be replaced and the bottles for preparation of standards must contain an amount of arsenic-free salt equal to that resulting from the action of the acid on the sample. Add 7.5 ml of potassium iodide solution (10 g in 100 ml) and 4 drops of stannous chloride solution (40 g of $SnCl_2 \cdot 2H_2O$ dissolved in hydrochloric acid to make 100 ml). Mix, allow to stand 30 min at not less than 25 °C, and dilute to 40 ml. Prepare the lead acetate tube, removing excess solution, and insert the paper strip into the small tube. Add to the bottle the required amount of zinc, which may be 10 to 15 g of stick zinc or 2 to 5 g of granulated zinc. Insert the stopper carrying the tubes and immerse the generator bottle in a water bath maintained at a constant temperature between 20 and 25 °C. At the end of 1.5 h, remove the paper strips and compare the stains.

The character of the stain is affected by variables that should be controlled as closely as possible. Moisture is a factor that is regulated in part by the lead acetate solution on the inert material in the tube next to the generator. A major factor is the rate of evolution of arsine and hydrogen. The kind and concentration of acid in the generator can be regulated fairly easily, but special pains must be taken to have the zinc the same in all generators of a set. Uniformity in a set is much more important than the form of the zinc. Good results may be obtained with pieces of stick zinc, mossy zinc, or granulated zinc. The best concentration of acid may depend on the form of the zinc and the amount used in each generator.

Attention to all the details is necessary to make certain that stains from equal amounts of arsenic in samples and standards shall be of equal length and appearance. This factor must be emphasized if the practice is followed of making a series of standard stains and using a graph based on the relation between amounts of arsenic and length of stain. A control of average arsenic content should give a stain whose length falls on the graph. A blank will show any significant amount of arsenic in the reagents used.

Chloride Determination

Dissolve x g of the compound under test in 30 to 40 ml of water and, if the solution is basic, neutralize to litmus with nitric acid before diluting to 50 ml. Filter the solution if it is not clear. Dilute 10 ml of the solution to 25 ml, add 1 ml each of concentrated HNO_3 and 0.1 M aqueous $AgNO_3$, and allow to stand for 10 min protected from direct sunlight. Any turbidity should not exceed that produced by 0.10 mg of chloride in an equal volume of solution treated exactly like the sample.

The quantity x depends on the chloride limit for the compound. For a limit of 50 ppm (0.005%), $x = 2$; for 200 ppm (0.02%), $x = 0.5$.

The standard solution of chloride should be prepared from reagent-grade sodium chloride. Dissolve 0.165 g of reagent-grade sodium chloride in water and dilute to 100.0 ml with water. Dilute a 10.0-ml aliquot of this solution to 1 liter with water. The resulting solution contains 0.01 mg of chloride per ml.

Sulfate Determination

Dissolve x g of the compound under test in 20 ml of water and, if the solution is not clear, filter. Add 2 ml of dilute hydrochloric acid (1 volume of concentrated HCl plus 19 volumes of H_2O) and 2 ml of 0.5 M barium chloride solution and allow to stand 10 min. Any turbidity should not exceed that produced by 0.10 mg of sulfate (SO_4) in an equal volume of solution, treated exactly like the sample.

The quantity x depends on the sulfate limit for the compound, but should not exceed about 2. For a limit of 50 ppm (0.005%), $x = 2$; for 200 ppm (0.02%), $x = 0.5$.

The standard solution of sulfate should be prepared from anhydrous sodium sulfate and contain 0.10 mg of sulfate per ml. Dissolve 0.148 g of ACS reagent-grade anhydrous sodium sulfate in water and dilute to 1 liter with water.

Calcium Determination

The method described in Section 36.310 and 3.011 of the *Official Methods of Analysis of the Association of Official Analytical Chemists,* 11th ed., AOAC, Washington, D.C. (1970), pp. 672, 35, is used.

Transfer a representative portion of well-mixed sample containing at least 50 mg of calcium to a 100-ml platinum or porcelain dish. Ash at a temperature not greater than 525 °C, until apparently carbon-free (gray to brown). Cool, moisten with 20 ml of H_2O, break up ash with stirring rod, and add 10 ml of concentrated HCl, cautiously, under watch glass. Rinse off watch glass into dish, and evaporate to dryness on steam bath. Add 50 ml of 4% HCl, heat on steam bath 15 min, and filter through quantitative paper into a 200-ml volumetric flask. Wash filter and dish thoroughly with hot H_2O, cool filtrate, dilute to mark, and mix. Transfer aliquot containing 20 to 40 mg of calcium to suitable beaker, dilute to 100 ml.

Transfer aliquot to a 200-ml beaker, add H_2O (if necessary) to make up to 50 ml, heat to boiling, and add 10 ml of saturated ammonium oxalate solution and a drop of methyl red (dissolve 1 g of methyl red in 200 ml of alcohol). Almost neutralize with 28% NH_4OH, and boil until precipitate is coarsely granular. Cool, add 6% NH_4OH until color is faint pink (pH 5.0), and let stand at least 4 h. Filter, and wash with H_2O at room temperature until filtrate is oxalate-free. Pierce the point of the filter with a platinum wire, and wash precipitate into the beaker in which the calcium was precipitated, using a stream of hot H_2O. Add about 10 ml of 20% H_2SO_4, heat to about 90 °C, add about 50 ml of hot H_2O, and titrate with 0.01 M $KMnO_4$. Finally, add the filter paper to the solution, and complete the titration. Correct for $KMnO_4$ consumed in the blank determination.

Potentiometric Determination of pH

The general directions provided in *The Pharmacopeia of the United States,* 18th revision (U.S.P. XVIII), Mack Publishing Co., Easton, Pa. (1970), p. 938, may be used as a guide.

Electronic instruments for measuring pH (pH meters) sense electrical potential differences between a reference electrode (usually of the calomel type) and some pH-dependent electrode (glass, hydrogen, or antimony, for example). The pH-dependent electrode is immersed in the solution of unknown pH, whereas the reference electrode is brought into contact with this solution through a special "salt bridge," usually a saturated solution of potassium chloride. By means of internal circuits involving potential standards, these instruments may be adjusted to read pH differences for any selected temperature. However, absolute pH values read from a pH meter are meaningful only shortly after the instrument has been calibrated against a standard buffer solution.

Readings obtained with a pH meter may not be assumed to represent the pH of the corresponding samples, unless (a) the readings can be reproduced after repeated rinsing of the electrode with portions of the sample and (b) two or more standard buffer solutions have been found to give a correct reading on the same instrument during the same period of operation. Readings obtained on unbuffered samples, such as solutions in distilled water, will have a considerably greater uncertainty than those for buffered solutions and will show a tendency for the reading to drift. Prolonged washing of the electrodes after standardization is necessary. Readings obtained from "flow-type" cells may be used if comparable evidence of validity is obtained.

Because of variations in the nature and operation of the available pH meters, it is not practicable to give detailed universally applicable directions for the potentiometric determination of pH. The instructions provided for each instrument by its manufacturer should be followed.

Special Tests

Certain special tests are required for individual compounds. Details of these are given in the appropriate specifications.

REFERENCES

1. E. Lederer and M. Lederer, *Chromatography*, 2nd ed., Elsevier Co., Amsterdam (1957).
2. G. N. Kowkabany, *Advan. Carbohydr. Chem.*, **9**, 303 (1954).
3. A. Jeanes, C. S. Wise, and R. J. Dimler, *Anal. Chem.*, **23**, 415 (1951).
4. S. M. Partridge, *Nature,* **164**, 443 (1949).
5. R. U. Lemieux and H. F. Bauer, *Anal. Chem.*, **26**, 920 (1954).
6. M. L. Wolfrom and J. B. Miller, *Anal. Chem.*, **28**, 1037 (1956).
7. R. J. Dimler, W. C. Schaefer, C. S. Wise, and C. E. Rist, *Anal. Chem.*, **24**, 1411 (1952).
8. E. Stahl, *Thin-Layer Chromatography*, Academic Press, New York (1962).
9. K. Randerath, *Thin-Layer Chromatography*, Academic Press, New York (1962).
10. J. M. Bobbitt, *Thin-Layer Chromatography*, Reinhold Publishing Co., New York (1963).
11. E. V. Truter, *Thin-Film Chromatography*, Cleaver Hume Press, Ltd., London (1963).
12. C. T. Bishop, *Methods Biochem. Anal.*, **10**, 1 (1962).
13. C. T. Bishop, *Advan. Carbohydr. Chem.*, **19**, 95 (1964).
14. C. C. Sweeley, R. Bentley, M. Makita, and W. W. Wells, *J. Am. Chem. Soc.*, **85**, 2497 (1963).
15. R. Bentley, C. C. Sweeley, M. Makita, and W. W. Wells, *Biochem. Biophys. Res. Commun.*, **11**, 14 (1963).

ACKNOWLEDGMENTS

The Subcommittee on Carbohydrates expresses its thanks to the American Chemical Society and the Association of Official Analytical Chemists. The use of portions of the text of The United States Pharmacopeia, Eighteenth Revision, official September 1, 1970, is by permission received from the Board of Trustees of The United States Pharmacopeial Convention, Inc. The said Convention is not responsible for any inaccuracy of quotation, or for any false or misleading implication that may arise by reason of the separation of excerpts from the original context.

Carbo-1

2-Acetamido-2-deoxy-D-galactopyranose (*N*-Acetyl-D-galactosamine)
Symbol: GalNAc

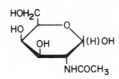

Formula: $C_8H_{15}NO_6$
Formula Wt: 221.21

Specific Rotation: $[\alpha]_D^{20}$ +90.0° ±1.0° ($\rho = 2$ g/100 ml) determined on a sample dried as described on p. 28.
Homogeneity: Determined by paper and gas–liquid chromatography.

Homogeneous by paper chromatography in solvent systems 1 and 2, detected by ammoniacal silver nitrate or by periodate–benzidine as described for Carbo-24.

The trimethylsilylated derivative is homogeneous by gas–liquid chromatography on a polyester column, isothermal at 170 °C.
Loss of Weight on Drying: Not more than 0.1%.
Water-Insoluble Material: Should give a clear colorless solution in water.
Ash: Not more than 0.1%.
Heavy Metals (as Cu): Not more than 10 ppm.
Iron (as Fe): Not more than 5 ppm.
Arsenic (as As): Not more than 0.5 ppm.
Chloride: Not more than 50 ppm.

Carbo-2

2-Acetamido-2-deoxy-D-glucopyranose (*N*-Acetyl-D-glucosamine)
Symbol: GlcNAc

Formula: $C_8H_{15}NO_6$
Formula Wt.: 221.21

Specific Rotation: $[\alpha]_D^{20}$ +41.2° ±0.3° ($\rho = 2$ g/100 ml) determined on a sample dried as described on p. 28.
Homogeneity: Determined by paper and gas–liquid chromatography.

Homogeneous by paper chromatography in solvent systems 1 and 2, detected by ammoniacal silver nitrate or periodate–benzidine as described for Carbo-24.

The trimethylsilylated derivative is homogeneous by gas–liquid chromatography on a polyester column, isothermal at 170 °C.
Loss of Weight on Drying: Not more than 0.1%.
Water-Insoluble Material: Should give a clear colorless solution in water.
Ash: Not more than 0.1%.
Heavy Metals (as Cu): Not more than 10 ppm.
Iron (as Fe): Not more than 5 ppm.
Arsenic (as As): Not more than 0.5 ppm.
Chloride: Not more than 50 ppm.

Carbo-3

2-Acetamido-2-deoxy-D-mannopyranose Monohydrate (*N*-Acetyl-D-mannosamine Monohydrate)
Symbol: ManNAc·H₂O

Formula: $C_8H_{15}NO_6 \cdot H_2O$
Formula Wt.: 239.23

Specific Rotation: $[\alpha]_D^{20}$ +10.2° ±0.3° ($\rho = 4$ g/100 ml) determined on a sample dried as described on p. 28.
Homogeneity: Determined by paper and gas–liquid chromatography.

Homogeneous by paper chromatography in solvent systems 1 and 2, detected by ammoniacal silver nitrate or by periodate–benzidine, as described for Carbo-24.

The trimethylsilylated derivative is homogeneous by gas–liquid chromatography on a polyester column, isothermal at 170 °C.
Loss of Weight on Drying: Not more than 0.1%.
Water-Insoluble Material: Should give a clear colorless solution in water.
Ash: Not more than 0.05%.
Heavy Metals (as Cu): Not more than 5 ppm.
Iron (as Fe): Not more than 5 ppm.
Arsenic (as As): Not more than 0.5 ppm.
Chloride: Not more than 50 ppm

Carbo-4

N-Acetylmuramic Acid [2-Acetamido-2-deoxy-3-*O*-(D-1-carboxyethyl)-D-glucopyranose]
Symbol: AcMur

Formula: $C_{11}H_{18}NO_8$
Formula Wt.: 292.19

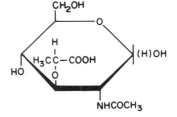

Specific Rotation: $[\alpha]_D^{20}$ +41.8° ±0.2° ($\rho = 1.58$ g/100 ml) at equilibrium (after 6 h).
Homogeneity: Determined by paper chromatography. Homogeneous by use of 5:5:1:3 pyridine–ethyl acetate–acetic acid–water, with detection by aniline hydrogen phthalate.
Loss of Weight on Drying: Not more than 0.1%.
Water-Insoluble Material: Should give a clear colorless solution in water.
Ash: Not more than 0.1%.
Infrared Spectrum: Consistent with structure.

Carbo-5

N-Acetylneuraminic Acid (5-Acetamido-3,5-dideoxy-D-*glycero*-α-D-*galacto*-nonulopyranosonic Acid)
Symbols: NeuAc or AcNeu

Formula: $C_{11}H_{19}NO_9$
Formula Wt.: 309.27

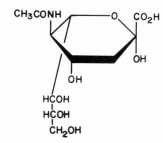

Specific Rotation: $[\alpha]_D^{25}$ −32.1° (ρ = 2 g/100 ml) found for a sample isolated from human milk.[1] Values for the compound prepared synthetically are $[\alpha]_D^{21}$ −33° (ρ = 0.8 g/100 ml)[2] and $[\alpha]_D^{24}$ −32.2° (ρ = 0.18 g/100 ml).[3]

Homogeneity: Determined by paper chromatography, thin-layer chromatography, and gas–liquid chromatography.

Homogeneous by paper chromatography in solvent system 1 on Whatman No. 1 paper, prewashed with 1% hydrochloric acid, and then distilled water, or on Schleicher and Schuell No. 589 paper when sprayed with orcinol reagent.*

Homogeneous by thin-layer chromatography on Kieselguhr G in 7:3 1-propanol–water when sprayed with orcinol reagent.

The per(trimethylsilyl) derivative, on a 2% OV-17 column at 250 °C, shows a single peak.[4]

Loss of Weight on Drying: Not more than 0.1%.
Ash: Not more than 0.05%.
Heavy Metals (as Cu): Not more than 10 ppm.
Iron (as Fe): Not more than 5 ppm.
Arsenic (as As): Not more than 0.5 ppm.

References

1. F. Zilliken and P. J. O'Brien, *Biochem. Prep.*, **7**, 1 (1960).
2. J. W. Cornforth, M. E. Firth, and A. Gottschalk, *Biochem. J.*, **68**, 57 (1958).
3. W. Wesemann and F. Zilliken, *Ann. Chem.*, **695**, 209 (1966).
4. D. A. Craven and C. W. Gehrke, *J. Chromatogr.*, **37**, 414 (1968).
5. R. Kleustrand and A. Nordal, *Acta Chem. Scand.*, **4**, 1320 (1950).

* The reagent is prepared from orcinol, 0.5 g, trichloroacetic acid, 15 g, and 1-butanol saturated with water, 100 ml.[5] The paper is dried and sprayed with the reagent, then heated at 105 °C for 15–20 min.

Carbo-6

2-Amino-2-deoxy-D-galactopyranose Hydrochloride
(D-Galactosamine Hydrochloride; Chondrosamine Hydrochloride)
Symbol: GalN·HCl

Formula: $C_6H_{13}NO_5$·HCl
Formula Wt.: 215.64

Specific Rotation: $[\alpha]_D^{20}$ +96.2° ±1° (ρ = 1 g/100 ml) determined on a sample dried as described on p. 28. Based on a measure-

ment reported by Gardell.[1] Ammonia should not be used to hasten the mutarotation of this compound as it affects the equilibrium value attained.

Homogeneity: Determined by paper chromatography.*

Homogeneous when chromatographed in system 2, or in phenol–water, and detected with ammoniacal silver nitrate. Two spots, one for the free base and one for the hydrochloride, are observed when the phenol–water system is used.

Loss of Weight on Drying: Not more than 0.05%.
Water-Insoluble Material: Should give a clear colorless solution in water.
Ash: Not more than 0.1%.
Heavy Metals (as Cu): Not more than 10 ppm.
Iron (as Fe): Not more than 5 ppm.
Arsenic (as As): Not more than 0.5 ppm.

References

1. S. Gardell, *Acta Chem. Scand.*, **7**, 207 (1953).
2. S. M. Partridge, *Biochem J.*, **42**, 238 (1948).

* Descending chromatography is used as described by Partridge.[2] The trough is filled with phenol saturated with water. In a dish at the bottom of the chromatography chamber is placed a humidifying solution consisting of water, saturated with phenol and containing 1% (w/v) of NH_3 and a few crystals of KCN. The phenol used in this work should be the best reagent-grade available.

Carbo-7

2-Amino-2-deoxy-D-glucopyranose Hydrochloride
(D-Glucosamine Hydrochloride)
Symbol: GlcN·HCl

Formula: $C_6H_{13}NO_5$·HCl
Formula Wt.: 215.64

Bibliography: J. C. Irvine and J. C. Earl, *J. Chem. Soc.*, **121**, 2370 (1922).

Specific Rotation: $[\alpha]_D^{20}$ +72.5° ±0.7° (ρ = 1 g/100 ml) determined on a sample dried as described on p. 28. Ammonia should not be used to hasten the mutarotation of this compound as it affects the equilibrium value attained.

Homogeneity: Determined by paper chromatography.

Homogeneous when chromatographed in system 2, or in phenol–water,* and detected by ammoniacal silver nitrate. Two spots, one for the free base and one for the hydrochloride, are observed when the phenol–water system is used.

Loss of Weight on Drying: Not more than 0.1%.
Water-Insoluble Material: Should give a clear colorless solution in water.
Ash: Not more than 0.1%.
Heavy Metals (as Cu): Not more than 10 ppm.
Iron (as Fe): Not more than 5 ppm.
Arsenic (as As): Not more than 0.5 ppm.

* The phenol–water system for chromatography is described under Carbo-6.

Carbo-8

D-Arabinitol (D-Arabitol)

CH_2OH
$HOCH$
$HCOH$
$HCOH$
CH_2OH

Formula: $C_5H_{12}O_5$
Formula Wt.: 152.15

Bibliography: N. K. Richtmyer and C. S. Hudson, *J. Am. Chem. Soc.*, **73**, 2249 (1951).

Specific Rotation: $[\alpha]_D^{20}$ $+130°$ $\pm1°$ ($\rho = 0.4$ g/100 ml in an excess of acidified molybdate) determined on a sample dried as described on p. 28. An accurately weighed sample (about 0.1 g) is dissolved in 20.0 ml of stock ammonium molybdate and made up to 25.0 ml with 0.5 M H_2SO_4.

Stock ammonium molybdate solution: 25.0 g of the commercial hydrated salt $(NH_4)_6Mo_7O_{24}\cdot4H_2O$ (Mallinckrodt analytical grade or the equivalent) is dissolved in distilled water to make 500 ml of solution, and filtered if necessary. Small amounts of crystalline material may separate from this solution on standing; the clear supernatant solution is used for rotatory measurements.

Melting Point:[1] 103.2–104.2 °C.
Homogeneity: Determined by gas–liquid chromatography.

The trimethylsilylated derivative gives only one peak when examined by gas–liquid chromatography on a polyester column, isothermal at 170 °C.
Reducing Material: A sample applied to filter paper, as in the standard procedure for paper chromatography, gives no coloration with aniline hydrogen phthalate spray.
Loss of Weight on Drying: Not more than 0.1%.
Water-Insoluble Material: Should give a clear colorless solution in water.
Ash: Not more than 0.05%.
Heavy Metals (as Cu): Not more than 10 ppm.
Iron (as Fe): Not more than 5 ppm.
Arsenic (as As): Not more than 0.5 ppm.

Reference

1. Communicated by Pfanstiehl Laboratories, Inc.

Carbo-9

L-Arabinitol (L-Arabitol)

CH_2OH
$HCOH$
$HOCH$
$HOCH$
CH_2OH

Formula: $C_5H_{12}O_5$
Formula Wt.: 152.15

Specific Rotation: $[\alpha]_D^{20}$ $-130°$ $\pm1°$ ($\rho = 0.4$ g/100 ml in an excess of acidified molybdate) determined on a sample dried as described on p. 28. The solution of the sample in excess acidified molybdate is prepared as described for D-arabinitol, Carbo-8.

Melting Point:[1] 102.6–103.6 °C.
Homogeneity: Determined by gas–liquid chromatography.

The trimethylsilylated derivative gives only one peak when examined by gas–liquid chromatography on a polyester column, isothermal at 170 °C.
Reducing Material: A sample applied to filter paper, as in the standard procedure for paper chromatography, gives no coloration with aniline hydrogen phthalate spray.

Loss of Weight on Drying: Not more than 0.1%.
Water-Insoluble Material: Should give a clear colorless solution in water.
Ash: Not more than 0.05%.
Heavy Metals (as Cu): Not more than 10 ppm.
Iron (as Fe): Not more than 5 ppm.
Arsenic (as As): Not more than 0.5 ppm.

Reference

1. Communicated by Pfanstiehl Laboratories, Inc.

Carbo-10

D-Arabinopyranose (D-Arabinose)
Symbol: Ara

Formula: $C_5H_{10}O_5$
Formula Wt.: 150.13

Specific Rotation: $[\alpha]_D^{20}$ $-104.5°$ $\pm0.5°$ ($\rho = 4$ g/100 ml) determined on a sample dried as described on p. 28.
Homogeneity: Determined by paper and gas–liquid chromatography.

Descending paper chromatography in system 2 for 48 h, or in system 1, shows no contaminants on treatment with aniline hydrogen phthalate.

Gas–liquid chromatography of the trimethylsilylated derivative shows no contaminants on a polyester column, isothermal at 170 °C.
Loss of Weight on Drying: Not more than 0.1%.
Water-Insoluble Material: Should give a clear colorless solution in water.
Ash: Not more than 0.05%.
Heavy Metals (as Cu): Not more than 10 ppm.
Iron (as Fe): Not more than 5 ppm.
Arsenic (as As): Not more than 0.5 ppm.

Carbo-11

L-Arabinopyranose (L-Arabinose)
Symbol: LAra

Formula: $C_5H_{10}O_5$
Formula Wt.: 150.13

Bibliography: H. S. Isbell and W. W. Pigman, *J. Res. Nat. Bur. Stand.*, **18**, 141 (1937); J. Rosin, *Reagent Chemicals and Standards*, 5th ed., D. Van Nostrand Co., Inc., Princeton, New Jersey (1967), p. 60.

Specific Rotation: $(\alpha)_D^{20}$ $+104.5°$ $\pm0.5°$ ($\rho = 4$ g/100 ml) determined on a sample dried as described on p. 28.
Homogeneity: Determined by paper and gas–liquid chromatography.

Descending paper chromatography in system 2 for 48 h, or in system 1, shows no contaminants when treated with aniline hydrogen phthalate.

Gas–liquid chromatography of the trimethylsilylated derivative shows no contaminants on a polyester column, isothermal at 170 °C.

Loss of Weight on Drying: Not more than 0.1%.
Ash: Not more than 0.05%.
Heavy Metals (as Cu): Not more than 10 ppm.
Iron (as Fe): Not more than 5 ppm.
Arsenic (as As): Not more than 0.5 ppm.

Carbo-12
L-Ascorbic Acid

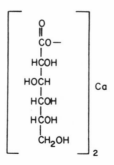

Formula: $C_6H_8O_6$
Formula Wt.: 176.13

Bibliography: *The Pharmacopeia of the United States of America*, 18th Revision (U.S.P. XVIII), Mack Publishing Co., Easton, Pennsylvania (1970), p. 51.

Specific Rotation: $[\alpha]_D^{20}$ +21.0° ±0.5° (ρ = 10 g/100 ml)* determined on a sample dried as described on p. 28.
Melting Point:[1] 190–192 °C.
Assay: Not less than 99.5%.
Loss of Weight on Drying: Not more than 0.1%.
Water-Insoluble Material: Should give a clear practically colorless solution in water.
Heavy Metals (as Cu): Not more than 10 ppm.
Method of Assay: Dissolve ∼400 mg of ascorbic acid, accurately weighed, in a mixture of 100 ml of water and 25 ml of dilute sulfuric acid. Titrate the solution at once with 0.1 N iodine, adding 3 ml of starch (T.S.) as the end-point is approached. Each ml of 0.1 N iodine is equivalent to 8.806 mg of $C_6H_8O_6$.

Reference

1. Data provided through the courtesy of A. J. Schmidtz of Charles Pfizer & Co., Inc., Brooklyn, New York.

* To be used for both criteria and specifications.

Carbo-13
Calcium D-Gluconate

Formula: $C_{12}H_{22}O_{14}Ca$
Formula Wt.: 430.38

Bibliography: *The Pharmacopeia of the United States of America*, 18th Revision (U.S.P. XVIII), Mack Publishing Co., Easton, Pennsylvania (1970), p. 91.

Assay: Minimum of 99.8%, calculated on the dried basis.
Loss of Weight on Drying: Not more than 0.2%.
Chloride: Not more than 10 ppm.
Sulfate: Not more than 100 ppm.
Heavy Metals (as Cu): Not more than 10 ppm.
Arsenic (as As): Not more than 1 ppm.
Sucrose: Negative to test given in U.S.P. XVIII, p. 12.

Method of Assay: Weigh accurately ∼800 mg of calcium D-gluconate, and dissolve in 150 ml of water containing 2 ml of dilute hydrochloric acid. While stirring, preferably with a magnetic stirrer, add about 30 ml of 0.05 M disodium (ethylenedinitrilo) tetraacetate from a 50-ml buret, then add 15 ml of sodium hydroxide (T.S.) and 300 mg of hydroxynaphthol blue indicator, and continue the titration to a blue end-point. Each ml of 0.05 M disodium (ethylenedinitrilo)tetraacetate is equivalent to 21.52 mg of $C_{12}H_{22}O_{14}Ca$.

Carbo-14
Calcium D-*glycero*-D-*gulo*-Heptonate Dihydrate

Formula: $C_{14}H_{26}O_{16}Ca \cdot 2H_2O$
Formula Wt.: 526.46

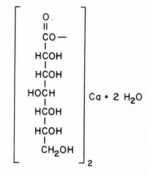

Homogeneity: Determined by gas–liquid chromatography.

The corresponding 1,4-lactone, derived from the salt by acidification and evaporation to dryness, contained two contaminants shown by gas–liquid chromatography of its trimethylsilylated derivative. On an SE-30 column, temperature-programmed from 150 to 250 °C at 4 °C per min, three components were detected in the approximate ratios of 2:42:7.
Loss of Weight on Drying: Not more than 5% when dried at 60 °C, 0.5 mmHg to constant weight. The theoretical value for the dihydrate is 6.85%.
Water-Insoluble Material: A 10% aqueous solution should be clear and colorless.
Calcium: 7.5 ±0.3%. The theoretical value for the dihydrate is 7.61%.
Heavy Metals (as Cu): Not more than 10 ppm.
Iron (as Fe): Not more than 100 ppm.
Arsenic (as As): Not more than 0.5 ppm.
Sulfate: Not more than 200 ppm.
Chloride: Not more than 200 ppm.
pH: A 10% solution should have a pH of 7.0–8.0.

Carbo-15
Cellobiose

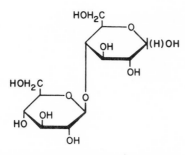

Formula: $C_{12}H_{22}O_{11}$
Formula Wt.: 342.30

Bibliography: F. J. Bates and Associates, *Polarimetry, Saccharimetry and the Sugars*, U.S. Govt. Printing Office, Washington, D.C. (1942), p. 710.

Specific Rotation: $[\alpha]_D^{20}$ +34.6° ±0.1° (ρ = 8 g/100 ml) determined on a sample dried as described on p. 28.
Homogeneity: Determined by paper and gas–liquid chromatography.
　　Descending paper chromatography for 48 h in system 1 or 2 shows no contaminants by permanganate–periodate spray.
　　Gas–liquid chromatography of the trimethylsilylated derivative shows no contaminants on an SE-30 column, temperature-programmed from 170° to 280 °C at 4 °C per min.
Loss of Weight on Drying: Not more than 0.1%.
Water-Insoluble Material: Should give a clear colorless solution in water.
Ash: Not more than 0.05%.
Heavy Metals (as Cu): Not more than 10 ppm.
Iron (as Fe): Not more than 5 ppm.
Arsenic (as As): Not more than 0.5 ppm.

Carbo-16
2-Deoxy-D-*arabino*-hexopyranose
(2-Deoxy-D-glucose)
Symbol: dGlc

Formula: $C_6H_{12}O_5$
Formula Wt.: 164.16

Specific Rotation: $[\alpha]_D^{20}$ +46.6° ±0.2° (ρ = 2 g/100 ml) determined on a sample dried as described on p. 28. Inspection of the rather extensive literature on 2-deoxy-D-glucose indicates that $[\alpha]_D^{20}$ +46.6° is the most probable equilibrium value for this sugar in water.
Homogeneity: Determined by paper and gas–liquid chromatography.
　　Descending paper chromatography for 24 h in system 1 or 2 shows no contaminants detectable by aniline hydrogen phthalate spray.
　　Gas–liquid chromatography of the trimethylsilylated derivative shows no contaminants on a polyester column, isothermal at 170 °C.
Loss of Weight on Drying: Not more than 0.1%.
Water-Insoluble Material: Should give a clear colorless solution in water.
Ash: Not more than 0.05%.
Heavy Metals (as Cu): Not more than 10 ppm.
Iron (as Fe): Not more than 5 ppm.
Arsenic (as As): Not more than 0.5 ppm.
Storage: Should be kept under an inert gas or in a vacuum desiccator after the package has been opened.

Carbo-17
2-Deoxy-D-*erythro*-pentopyranose
(2-Deoxy-D-ribose)
Symbol: dRib

Formula: $C_5H_{10}O_4$
Formula Wt.: 134.13

Specific Rotation: $[\alpha]_D^{20}$ −57.3° ±0.3° (ρ = 1 g/100 ml) determined on a sample dried as described in section below on "Loss of Weight on Drying."*
Homogeneity: Determined by paper and gas–liquid chromatography.
　　Descending paper chromatography for 24 h in system 1 or 2 shows no contaminants detectable by aniline hydrogen phthalate spray.
　　Gas–liquid chromatography of the trimethylsilylated derivative shows no contaminants on a polyester column, isothermal at 170 °C.
Loss of Weight on Drying: Not more than 0.1% when dried at 40 °C and not above 0.1 mmHg for 1.25 h.
Water-Insoluble Material: Should give a clear colorless solution in water.
Ash: Not more than 0.05%.
Heavy Metals (as Cu): Not more than 5 ppm.
Iron (as Fe): Not more than 5 ppm.
Arsenic (as As): Not more than 0.5 ppm.
Stability and Storage: When stored in glass bottles, samples have been found to show a change in specific rotation. Should be kept under an inert gas or in a vacuum desiccator after the package has been opened.

* Unpublished measurement by Harry W. Diehl, National Institutes of Health.

Carbo-18
Erythritol

Formula: $C_4H_{10}O_4$
Formula Wt.: 122.12

Specific Rotation: None.
Melting Point: Not below 118 or above 120 °C. The freezing point of erythritol has been found[1] to be 118.9 °C. Melting points from 120 to 126 °C have been reported for this compound.
Homogeneity: Determined by gas–liquid chromatography.
　　Gas–liquid chromatography of the trimethylsilylated derivative shows no contaminants on a polyester column, isothermal at 150 °C.
Reducing Material: A sample applied to paper, as in the standard procedure for paper chromatography, gives no coloration with aniline hydrogen phthalate spray.
Loss of Weight on Drying: Not more than 0.1%.
Water-Insoluble Material: Should give a clear colorless solution in water.
Ash: Not more than 0.05%.
Heavy Metals (as Cu): Not more than 10 ppm.
Iron (as Fe): Not more than 5 ppm.
Arsenic (as As): Not more than 0.5 ppm.

Reference
1. G. S. Parks and C. T. Anderson, *J. Am. Chem. Soc.*, **48**, 1506 (1926).

Carbo-19
D-Fructopyranose (D-Fructose)
Symbol: Fru

Formula: $C_6H_{12}O_6$
Formula Wt.: 180.16

Bibliography: H. S. Isbell and W. W. Pigman, *J. Res. Nat. Bur. Stand.*, **20**, 773 (1938); J. Rosin, *Reagent Chemicals and Standards*, 5th ed., D. Van Nostrand Co., Inc., Princeton, New Jersey (1967), p. 262.

Specific Rotation: $[\alpha]_D^{20}$ $-92°$ $\pm1°$ ($\rho = 10$ g/100 ml) determined on a sample dried as described on p. 28. Unlike most other sugars, D-fructose shows a marked change in specific rotation with changes in (a) temperature and (b) the wavelength of the light used.
Homogeneity: Determined by paper and gas–liquid chromatography.

Homogeneous by paper chromatography in system 1 or 2; sprayed with aniline hydrogen phthalate.

Homogeneous by gas–liquid chromatography of its trimethylsilylated derivative on a polyester column, isothermal at 170 °C.
Loss of Weight on Drying: Not more than 0.1%
Water-Insoluble Material: Should give a clear colorless solution in water.
Ash: Not more than 0.05%.
Heavy Metals (as Cu): Not more than 10 ppm.
Iron (as Fe): Not more than 5 ppm.
Arsenic (as As): Not more than 0.5 ppm.

Carbo-20
D-Fucopyranose (D-Fucose)
Symbol: Fuc

Formula: $C_6H_{12}O_5$
Formula Wt.: 164.16

Specific Rotation: $[\alpha]_D^{20}$ $+75.6°$ $\pm0.6°$ ($\rho = 4$ g/100 ml) determined on a sample dried as described on p. 28.
Homogeneity: Determined by paper and gas–liquid chromatography.

No contaminants detectable with either periodate–benzidine (see Carbo-24) or ammoniacal silver nitrate sprays, when chromatographed on paper in 1-butanol–acetic acid–water (4:1:1 v/v).

No contaminants are detected by gas–liquid chromatography of the trimethylsilylated derivative on a polyester column, isothermal at 170 °C.
Loss of Weight on Drying: Not more than 0.1%.
Water-Insoluble Material: Should give a clear colorless solution in water.
Ash: Not more than 0.05%.
Heavy Metals (as Cu): Not more than 5 ppm.
Iron (as Fe): Not more than 5 ppm.
Arsenic (as As): Not more than 0.5 ppm.

Carbo-21
L-Fucopyranose (L-Fucose)
Symbol: LFuc

Formula: $C_6H_{12}O_5$
Formula Wt.: 164.16

Bibliography: F. J. Bates and Associates, *Polarimetry, Saccharimetry and the Sugars*, U.S. Govt. Printing Office, Washington, D.C. (1942), p. 716.

Specific Rotation: $[\alpha]_D^{20}$ $-75.9°$ $\pm0.2°$ ($\rho = 4$ g/100 ml) determined on a sample dried as described on p. 28.
Homogeneity: Determined by paper and gas–liquid chromatography.*

No contaminants detectable with ammoniacal silver nitrate after paper chromatography in system 1 or 2.

Gas–liquid chromatography of the trimethylsilylated derivative shows no contaminants on a polyester column, isothermal at 170 °C.
Loss of Weight on Drying: Not more than 0.1%.
Water-Insoluble Material: Should give a clear colorless solution in water.
Ash: Not more than 0.05%.
Heavy Metals (as Cu): Not more than 10 ppm.
Iron (as Fe): Not more than 5 ppm.
Arsenic (as As): Not more than 0.5 ppm.

* If not purified through a crystalline derivative, L-fucose may be contaminated with D-mannitol. This impurity is, however, readily detectable on a paper chromatogram with ammoniacal silver nitrate.

Carbo-22
Galactaric Acid (Mucic Acid)

Formula: $C_6H_{10}O_8$
Formula Wt.: 210.14

Specific Rotation: None.
Melting Point: Not below 213 or above 215 °C.*
Homogeneity: Homogeneous by gas–liquid chromatography of the trimethylsilylated derivative on an SE-30 column, temperature-programmed from 170 to 258 °C at 4 °C per min.
Loss of Weight on Drying: Not more than 0.3%.
Water-Insoluble Material: A saturated aqueous solution (about 0.3%) should be clear and colorless.
Ash: Not more than 0.05%.
Reducing Material: A sample applied to filter paper as a saturated solution should give no color with aniline hydrogen phthalate.
Heavy Metals (as Cu): Not more than 10 ppm.
Iron (as Fe): Not more than 5 ppm.
Arsenic (as As): Not more than 0.5 ppm.
Nitrate: Nil.†

* Melting occurs with decomposition and is very dependent upon the rate of heating. The range quoted is obtained after heating to 205 °C in 10 min, followed by a heating rate of 1 °C per min.
† A saturated solution (1 ml) at 100 °C is cooled, and 1 M ferrous sulfate solution (3 ml) is added. After removal of any crystallized mucic acid, concentrated sulfuric acid is carefully added to the solution so that the two solutions do not mix. A brown ring will be seen at the junction of the two liquids if nitrate is present.

Carbo-23
Galactitol (Dulcitol)

    CH₂OH
    |
    HCOH
    |
   HOCH
    |
   HOCH
    |
    HCOH
    |
    CH₂OH

Formula: $C_6H_{14}O_6$
Formula Wt.: 182.17

Specific Rotation: None.
Melting Point: Not below 188 or above 189 °C. The melting range from 188 to 189 °C is most frequently quoted for this substance.
Homogeneity: Determined by gas–liquid chromatography.

No contaminants detectable by gas–liquid chromatography of the trimethylsilylated derivative on a polyester column, isothermal at 170 °C.
Reducing Material: A sample applied to filter paper, as in the standard procedure for paper chromatography, gives no coloration with aniline hydrogen phthalate.
Loss of Weight on Drying: Not more than 0.1%.
Water-Insoluble Material: Should give a clear colorless solution in water.

Ash: Not more than 0.05%.
Heavy Metals (as Cu): Not more than 10 ppm.
Iron (as Fe): Not more than 5 ppm.
Arsenic (as As): Not more than 0.5 ppm.

Carbo-24
D-Galactono-1,4-lactone (D-Galactono-γ-lactone)

Formula: $C_6H_{10}O_6$
Formula Wt.: 178.14

Specific Rotation: $[\alpha]_D^{20}$ −78.4° ±0.4° (ρ = 4 g/100 ml; initial rotation*) determined on a sample dried as described on p. 28.
Melting Point:[1] 133–135 °C.
Homogeneity: Determined by paper and gas–liquid chromatography.

Homogeneous when chromatographed for 20 h in 1-butanol–acetic acid–water (4:1:5 v/v, top layer), or in 2-butanone–acetic acid–boric acid (saturated, aqueous) (9:1:1 v/v) and sprayed with periodate–benzidine reagent.†

Homogeneous by gas–liquid chromatography of the trimethylsilylated derivative on an SE-30 column, temperature-programmed from 150 to 260 °C at 4 °C per min.
Loss of Weight on Drying: Not more than 0.1%.
Water-Insoluble Material: Should give a clear colorless solution at 10% concentration.
Ash: Not more than 0.05%.
Heavy Metals (as Cu): Not more than 10 ppm.
Iron (as Fe): Not more than 5 ppm.
Arsenic (as As): Not more than 0.5 ppm.

References

1. F. J. Bates and Associates, *Polarimetry, Saccharimetry and the Sugars*, U.S. Govt. Printing Office, Washington, D.C. (1942), p. 719.
2. N. K. Richtmyer, R. M. Hann, and C. S. Hudson, *J. Am. Chem. Soc.*, **61**, 340 (1939).

3. M. Viscontini, D. Hoch, and P. Karrer, *Helv. Chim. Acta*, **38**, 642 (1955).

* A value of $[\alpha]_D^{20}$ +78.4° (H₂O, ρ = 4 g/100 ml) for L-galactono-1,4-lactone has been reported.[2] The optical rotation changes so slowly that a rotation measured within 10 min after dissolution may be considered to be a valid initial rotation.
† The paper is first sprayed with a 0.5% (w/v) aqueous sodium metaperiodate solution and, after 8 min at room temperature, the paper is sprayed with a benzidine solution [0.5% w/v benzidine in ethanol–glacial acetic acid (4:1)]. A white spot appears at once, on a blue-to-gray background.[3]

Carbo-25
D-Galactopyranose (D-Galactose)
Symbol: Gal

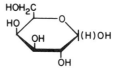

Formula: $C_6H_{12}O_6$
Formula Wt.: 180.16

Bibliography: H. S. Isbell and W. W. Pigman, *J. Res. Nat. Bur. Stand.*, **18**, 141 (1937); J. Rosin, *Reagent Chemicals and Standards*, 5th ed., D. Van Nostrand Co., Inc., Princeton, New Jersey (1967), p. 213.

Specific Rotation: $[\alpha]_D^{20}$ +80.2° ±0.4° (ρ = 5 g/100 ml) determined on a sample dried as described on p. 28.
Homogeneity: Determined by paper and gas–liquid chromatography.

Descending paper chromatography for 48 h in systems 1 or 2 shows no contamination detectable by aniline hydrogen phthalate spray.

Gas–liquid chromatography of the trimethylsilylated derivative should show no contaminants on a polyester column, isothermal at 170 °C.
Loss of Weight on Drying: Not more than 0.1%.
Water-Insoluble Material: Should give a clear colorless solution in water.
Ash: Not more than 0.05%.
Heavy Metals (as Cu): Not more than 10 ppm.
Iron (as Fe): Not more than 5 ppm.
Arsenic (as As): Not more than 0.5 ppm.

Carbo-26
D-Galactopyranuronic Acid Monohydrate (D-Galacturonic Acid Monohydrate)
Symbol: GalUA·H₂O

Formula: $C_6H_{10}O_7 \cdot H_2O$
Formula Wt.: 212.16

Specific Rotation: $[\alpha]_D^{20}$ +51.9° ±0.5° (ρ = 4 g/100 ml; at equilibrium, about 60 min) determined on a sample dried as described on p. 28.
Homogeneity: Determined by paper and gas–liquid chromatography.

When chromatographed on paper in 1-butanol–acetic acid–water (4:1:1 v/v), no contaminants detectable after spraying with periodate–benzidine (see Carbo-24).

When the compound is reduced by borohydride to L-galactonic acid, and the latter is lactonized, gas–liquid chromatography of the trimethylsilylated lactone should show no contaminants on a polyester column, isothermal at 200 °C.

Loss of Weight on Drying: Not more than 0.1%.

Water-Insoluble Material: Should give a clear colorless to pale yellow solution in water.

Ash: Not more than 0.2%.

Heavy Metals (as Cu): Not more than 5 ppm.

Iron (as Fe): Not more than 10 ppm.

Arsenic (as As): Not more than 0.5 ppm.

Carbo-27

D-Glucitol (Sorbitol)

Formula: $C_6H_{14}O_6$
Formula Wt.: 182.17

$$\begin{array}{c} CH_2OH \\ | \\ HCOH \\ | \\ HOCH \\ | \\ HCOH \\ | \\ HCOH \\ | \\ CH_2OH \end{array}$$

Bibliography: R. K. Ness, H. G. Fletcher, Jr., and C. S. Hudson, *J. Am. Chem. Soc.,* **73,** 4759 (1951); N. K. Richtmyer and C. S. Hudson, *J. Am. Chem. Soc.,* **73,** 2249 (1951).

Specific Rotation: $[\alpha]_D^{20}$ +103° ±1° (ρ = 0.4 g/100 ml in an excess of acidified molybdate) determined on a sample dried as described on p. 28. The solution of the sample in an excess of acidified molybdate is prepared as described for Carbo-8.

Melting Point:[1] 110–112 °C (anhydrous).

Homogeneity: Determined by gas–liquid chromatography.*

No contaminants are detected by gas–liquid chromatography of the trimethylsilylated derivative on a polyester column, isothermal at 170 °C.

Reducing Material: A sample applied to filter paper, as in the standard procedure for paper chromatography, gives no coloration with aniline hydrogen phthalate spray.

Loss of Weight on Drying: Not more than 0.1%.

Ash: Not more than 0.05%.

Heavy Metals (as Cu): Not more than 10 ppm.

Iron (as Fe): Not more than 5 ppm.

Arsenic (as As): Not more than 0.5 ppm.

Reference

1. P. G. Stecher (ed.), *The Merck Index,* 8th ed., Merck & Co., Inc., Rahway, New Jersey (1968), p. 971.

* Commercial samples may contain traces of D-mannitol, unless they have been purified through the pyridine addition compound.

Carbo-28

D-Glucono-1,5-lactone (D-Glucono-δ-lactone)

Formula: $C_6H_{10}O_6$
Formula Wt.: 178.14

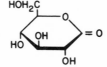

Specific Rotation: $[\alpha]_D^{20}$ +411° ±1° (ρ = 0.5 g/100 ml, in an excess of acidified molybdate) determined on a sample dried as described on p. 28.

An accurately weighed sample (about 0.5 g) is dissolved in 25 ml of water; 2.5 g of $(NH_4)_6Mo_7O_{24}\cdot4H_2O$ (Mallinckrodt analytical grade, or the equivalent) is dissolved in the solution. Glacial acetic acid (5.0 ml) is added, and the solution is made up with water to 100 ml. Change in rotation is complete after about 8 h, and the equilibrium rotation may then be determined.*

Melting Point:[1] 150–152 °C.

Homogeneity: Determined by paper and gas–liquid chromatography.

Homogeneous when chromatographed on paper for 20 h in 1-butanol–acetic acid–water (4:1:5 v/v, top layer) or 6 h in ethyl acetate–acetic acid–formic acid–water (18:3:1:4 v/v) and sprayed with periodate–benzidine as described for Carbo-24.

Homogeneous by gas–liquid chromatography of the trimethylsilylated derivative on an SE-30 column, temperature-programmed from 150 to 260 °C at 4 °C per min.

Loss of Weight on Drying: Not more than 0.1%.

Water-Insoluble Material: Should give a clear colorless solution in water at 10% concentration.

Ash: Not more than 0.05%.

Heavy Metals (as Cu): Not more than 10 ppm.

Iron (as Fe): Not more than 5 ppm.

Arsenic (as As): Not more than 0.5 ppm.

Reference

1. F. J. Bates and Associates, *Polarimetry, Saccharimetry and the Sugars,* U.S. Govt. Printing Office, Washington, D.C. (1942), p. 731.

* We are indebted to F. H. Hedger, Charles Pfizer & Co., Brooklyn, New York, for this method, which was devised for sodium D-gluconate. H. S. Isbell and H. L. Frush, *J. Res. Nat. Bur. Stand.,* **11,** 649 (1933) reported $[\alpha]_D^{20}$ +66.2° (initial) changing to +8.8° at 24 h (ρ = 5 g/100 ml in water). The initial change in rotation is too fast, and the equilibrium value too small, for practical use of optical rotation in water as a specification.

Carbo-29

D-Glucopyranose, Anhydrous (D-Glucose; Dextrose)
Symbol: Glu

Formula: $C_6H_{12}O_6$
Formula Wt.: 180.16

Bibliography: *Reagent Chemicals,* 4th ed., American Chemical Society, Washington, D.C. (1968), p. 267; F. J. Bates and Associates, *Polarimetry, Saccharimetry and the Sugars,* U.S. Govt. Printing Office, Washington, D.C. (1942), pp. 390, 551; J. Rosin, *Reagent Chemicals and Standards,* 5th ed., D. Van Nostrand Co., Inc., Princeton, New Jersey (1967), p. 167; *The Pharmacopeia of the United States of America,* 18th revision (U.S.P. XVIII), Mack Publishing Co., Easton, Pennsylvania (1970), p. 180.

CRITERIA: Standard Reference Material 917 of the National Bureau of Standards, Washington, D.C.

Purity	99.9%
α-D-Glucopyranose	>99.0%
β-D-Glucopyranose	<1.0%
Moisture	0.06%
Ash	0.002%
Insoluble matter	0.001 to 0.006%
Nitrogen	<0.001%

The value for the purity has an estimated inaccuracy of ±0.1%.

Specific Rotation:

$[\alpha]_D^{20}$ +53.2° (at equil., ρ = 20.1 g/100 ml in water)

$[\alpha]_{546}^{20}$ +62.8° (at equil., ρ = 20.1 g/100 ml in water)

$[\alpha]_D^{20}$ +112.6° (initial, ρ = 10.05 g/100 ml in methyl sulfoxide)

    The D-glucose used for this standard reference material was obtained from Pfanstiehl Laboratories, Inc., Waukegan, Illinois.

SPECIFICATIONS:

Specific Rotation: $[\alpha]_D^{20}$ +52.5° ±0.5° (ρ = 4 g/100 ml) determined on a sample dried as described on p. 28.

Homogeneity: Determined by paper and gas–liquid chromatography.

    Homogeneous by descending paper chromatography for 17 h in system 1 or 2, detected with aniline hydrogen phthalate spray. Should be homogeneous by gas–liquid chromatography of its trimethylsilylated derivative on a polyester column, isothermal at 170 °C.

Loss of Weight on Drying: Not more than 0.5%.

Water-Insoluble Material: Should give a clear colorless solution in water.

Ash: Not more than 0.05%.

Heavy Metals (as Cu): Not more than 5 ppm.

Iron (as Fe): Not more than 5 ppm.

Arsenic (as As): Not more than 0.5 ppm.

Carbo-30

β-D-Glucopyranose Pentaacetate

Symbol: GlcAc₅

Formula: $C_{16}H_{22}O_{11}$

Formula Wt.: 390.34

Specific Rotation: $[\alpha]_D^{25}$ +4.0° ±1.0° (ρ = 2.5 to 4.0 g/100 ml in chloroform).

Melting Point: Not below 130 and not above 133 °C. Melting points from 127 to 134 °C have been reported for this compound.

Homogeneity: Determined by paper, thin-layer, and gas–liquid chromatography.

    Homogeneous (and separable from the anomer) on chromatography on Whatman No. 1 paper containing *N,N*-dimethylformamide as the stationary phase, and developing with either isopropyl ether or light petroleum ether.[1]

    Homogeneous (and separable from the anomer) by thin-layer chromatography on silica gel containing methyl sulfoxide as the stationary phase, and developing with ether or 1:1 ether–isopropyl ether.[2]

    Homogeneous (and separable from the anomer) by gas–liquid chromatography on a fluoroalkylsilicone polymer, QF-1, at[3] 170 °C or on a 1,4-butanediol succinate, Dow-Corning grease on Chromosorb B column at[4] 213 °C.

Loss of Weight on Drying: Not more than 0.1%.

Ash: Not more than 0.05%.

Heavy Metals (as Cu): Not more than 10 ppm.

Iron (as Fe): Not more than 5 ppm.

Arsenic (as As): Not more than 0.5 ppm.

References

1. B. Wickberg, *Acta Chem. Scand.*, **12**, 615 (1958).
2. G. R. Inglis, *J. Chromatogr.*, **20**, 417 (1965).
3. W. J. A. Vanden Heuvel and E. C. Horning, *Biochem. Biophys. Res. Commun.*, **4**, 399 (1961).
4. S. W. Gunner, J. K. N. Jones, and M. B. Perry, *Can. J. Chem.*, **39**, 1892 (1961).

Carbo-31

D-Glucurono-6,3-lactone

(α-D-Glucofuranurono-6,3-lactone; D-Glucurone)

Formula: $C_6H_8O_6$

Formula Wt.: 176.13

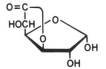

Specific Rotation: $[\alpha]_D^{20}$ +19.4° ±0.4° (ρ = 8 g/100 ml; at equilibrium, about 60 min)* determined on a sample dried as described on p. 28.

Homogeneity: Determined by paper and gas–liquid chromatography.

    Homogeneous when chromatographed on (a) paper for 16 h in 2-butanone–acetic acid–boric acid (saturated aqueous) (9:1:1 v/v) or (b) Whatman No. 4 paper for 6 h in 2-propanol–pyridine–acetic acid–water (8:8:1:4 v/v), and sprayed with periodate–benzidine as described for Carbo-24.

    Homogeneous by gas–liquid chromatography of the trimethylsilylated derivative on an SE-30 column, temperature-programmed from 150 to 250 °C at 4 °C per min.

Loss of Weight on Drying: Not more than 0.1%.

Water-Insoluble Material: An aqueous solution should be clear and colorless at 10% concentration.

Ash: Not more than 0.05%.

Heavy Metals (as Cu): Not more than 10 ppm.

Iron (as Fe): Not more than 5 ppm.

Arsenic (as As): Not more than 0.5 ppm.

* L. Zervas and P. Sessler, *Chem. Ber.*, **66**, 1326 (1933), reported $[\alpha]_D^{20}$ +19.4° (ρ = 8 g/100 ml, at equilibrium).

Carbo-32

Glycogen*

Formula: $(C_6H_{10}O_5)_n$

Bibliography: *The Merck Index*, 8th ed. (1968), p. 501.

Specific Rotation: $[\alpha]_D^{20}$ +198° ±3.0° (ρ = 2 g/100 ml).

Loss of Weight on Drying: Not more than 0.1%.

Water-Insoluble Material: A 2% solution in water should give a slightly opalescent solution, free from foreign matter.

Ash: Not more than 0.25%.

Reducing Sugars: Trace.

* Produced from mollusks.

Carbo-33
D-Gulono-1,4-lactone (D-Gulono-γ-lactone)

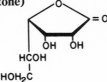

Formula: $C_6H_{10}O_6$
Formula Wt.: 178.14

Specific Rotation: $[\alpha]_D^{20}$ $-56.2°$ $\pm 1.0°$ ($\rho = 4$ g/100 ml; initial value) determined on a sample dried as described on p. 28.

Melting Point:[1] 187.3–189.3 °C.

Homogeneity: Determined by paper and gas–liquid chromatography.

Paper chromatography in 1-butanol–acetic acid–water (4:1:1), and spraying with periodate–benzidine as described in Carbo-24, revealed three spots. The weaker slower-moving spots are probably the 1,5-lactone and the free acid.

Gas–liquid chromatography of the trimethylsilylated derivative of the lactone reveals no contaminants on a polyester column, isothermal at 200 °C.

Loss of Weight on Drying: Not more than 0.1%.

Water-Insoluble Material: Should give a clear colorless solution in water.

Ash: Not more than 0.05%.

Heavy Metals (as Cu): Not more than 5 ppm.

Iron (as Fe): Not more than 5 ppm.

Arsenic (as As): Not more than 0.5 ppm.

Reference

1. Communicated by Pfanstiehl Laboratories, Inc.

Carbo-34
D-*glycero*-D-*gulo*-Heptono-1,4-lactone

Formula: $C_7H_{12}O_7$
Formula Wt.: 208.17

Specific Rotation: $[\alpha]_D^{20}$ $-51.8°$ $\pm 0.5°$ ($\rho = 4.3$ g/100 ml)* determined on a sample dried as described on p. 28.

Melting Point:[1] 152–154 °C.

Homogeneity: Determined by paper and gas–liquid chromatography.

When chromatographed on paper, using the same solvent systems and spray reagent as described for Carbo-24, all samples tested showed two spots. The weaker slower-moving spot is presumably the free acid.

No contaminants detectable by gas–liquid chromatography of the trimethylsilylated derivative on an SE-30 column, temperature-programmed from 150 to 260 °C at 4 °C per min.

Loss of Weight on Drying: Not more than 0.1%.

Water-Insoluble Material: At a concentration of 10%, a clear colorless solution in water should be obtained.

Ash: Not more than 0.05%.

Heavy Metals (as Cu): Not more than 10 ppm.

Iron (as Fe): Not more than 10 ppm.

Arsenic (as As): Not more than 0.5 ppm.

Chloride: Not more than 50 ppm.

Sulfate: Not more than 50 ppm.

Reference

1. Communicated by Pfanstiehl Laboratories, Inc.

* A. Thompson and M. L. Wolfrom, *J. Am. Chem. Soc.*, **68**, 1510 (1946), reported $[\alpha]_D^{20}$ $-51.8°$ ($\rho = 4.3$ g/100 ml in water). The change in rotation is slow, and a satisfactory initial rotation may be obtained within 5 min after dissolution is complete.

Carbo-35
D-*manno*-Heptulose

Formula: $C_7H_{14}O_7$
Formula Wt.: 210.19

Specific Rotation: $[\alpha]_D^{20}$ $+29.0°$ $\pm 0.5°$ ($\rho = 2$ g/100 ml) determined on a sample dried as described on p. 28.

Homogeneity: Determined by paper and gas–liquid chromatography.

No contaminants detectable by periodate–benzidine (see Carbo-24) or ammoniacal silver nitrate sprays, after paper chromatography in system 1 or 2.

No contaminants detectable by gas–liquid chromatography of the trimethylsilylated derivative on a polyester column, isothermal at 170 °C.

Loss of Weight on Drying: Not more than 0.1%.

Water-Insoluble Material: Should give a clear colorless solution in water.

Ash: Not more than 0.1%.

Heavy Metals (as Cu): Not more than 10 ppm.

Iron (as Fe): Not more than 5 ppm.

Arsenic (as As): Not more than 0.5 ppm.

Carbo-36
myo-Inositol*

Formula: $C_6H_{12}O_6$
Formula Wt.: 180.16

Bibliography: *The National Formulary*, 13th revision (NF XIII), J. B. Lippincott Co., Philadelphia, Pennsylvania (1970).

Specific Rotation: None.

Melting Point: Not below 225 or above 226 °C.*

Homogeneity: Determined by paper and gas–liquid chromatography.

No contaminants are detectable with periodate–permanganate spray after paper chromatography in system 1 or 2.

One commercial sample was homogeneous; another showed a small contaminant that moved faster than the trimethylsilylated derivative on gas–liquid chromatography on an SE-30 column, temperature-programmed from 150 to 250 °C at 4 °C per min.

Loss of Weight on Drying: Not more than 0.1%.

Water-Insoluble Material: Should give a clear colorless solution at a concentration of 10%.

Ash: Not more than 0.05%.

Heavy Metals (as Cu): Not more than 10 ppm.

Iron (as Fe): Not more than 5 ppm.

Arsenic (as As): Not more than 0.5 ppm.

* The usual commercial form is essentially anhydrous. *myo*-Inositol crystallizes from water below 50 °C as a dihydrate, from which the water of crystallization can be removed by heating *in vacuo* at 105 °C.

Carbo-37
Inulin

Formula: $(C_6H_{10}O_5)_n$

Bibliography: *The Merck Index*, 8th ed. (1968), p. 569; H. Kiliani, *Ann. Chem.* **205**, 145 (1880); J. R. Katz and J. C. Derksen, *Rec. Trav. Chim.*, **50**, 248 (1931); J. R. Katz and A. Weidinger, *ibid.*, 1133; E. L. Hirst, D. I. McGilvray, and E. G. V. Percival, *J. Chem. Soc.*, 1297 (1950).

Description: Polysaccharide; colorless, tasteless solid; amorphous, or spherocrystals that show an x-ray powder diffraction pattern. A fructan containing a small proportion (~6%) of D-glucose. Found in many tubers and roots; also in some cacti. The purest form so far investigated is found in dahlia (*Dahlia variabilis*) tubers in the autumn. Readily soluble in warm water; very slightly soluble in water at room temperature. The crystalline form is less soluble, and more stable, than the amorphous form. $[\alpha]_D^{20}$ varies from $-30°$ to $-40°$ ($\rho = 5$ g/100 ml, water), depending on the source and purity; the $-40°$ value was obtained on highly purified product from dahlia tubers.

CRITERIA:

Specific Rotation: $[\alpha]_D^{20} -40°$ ($\rho = 5$ g/100 ml).

SPECIFICATIONS:

Specific Rotation: $[\alpha]_D^{20} -35° \pm 5°$ ($\rho = 5$ g/100 ml).

Loss of Weight on Drying: Not more than 10%.

Ash: Not more than 0.2%.

Carbo-38
Lactose Monohydrate

Formula: $C_{12}H_{22}O_{11} \cdot H_2O$

Formula Wt.: 360.32

Bibliography: *The Pharmacopeia of the United States of America*, 18th revision (U.S.P. XVIII), Mack Publishing Co., Easton, Pennsylvania (1970), p. 358; H. S. Isbell and W. W. Pigman, *J. Res. Nat. Bur. Stand.*, **18**, 141 (1937); J. Rosin, *Reagent Chemicals and Standards*, 5th ed., D. Van Nostrand Co., Inc., Princeton, New Jersey (1967), p. 246.

Specific Rotation: $[\alpha]_D^{20} +52.6° \pm 0.5°$ ($\rho = 8$ g/100 ml) determined on a sample dried at 80 °C for 2 h.

Homogeneity: Determined by paper and gas–liquid chromatography.

No contaminants detectable with ammoniacal silver nitrate

after 48 h of descending paper chromatography in either system 1 or 2. Because of the relatively low solubility of lactose in water, only half the standard quantity of material is applied to the paper.

No contaminants detectable by gas–liquid chromatography of the trimethylsilylated derivative on an SE-30 column, temperature-programmed from 170 to 250 °C at 4 °C per min.

Loss of Weight on Drying: Not more than 0.1% after drying at 80 °C for 2 h.

Water-Insoluble Material: Should give a clear colorless solution in water.

Dextrins: A solution of 0.5 g in 10 ml of water should show no coloration when treated with a few drops of iodine solution.

Ash: Not more than 0.05%.

Heavy Metals (as Cu): Not more than 10 ppm.

Iron (as Fe): Not more than 5 ppm.

Arsenic (as As): Not more than 0.5 ppm.

Carbo-39
D-Lyxopyranose (D-Lyxose)
Symbol: Lyx

Formula: $C_5H_{10}O_5$

Formula Wt.: 150.13

Bibliography: H. S. Isbell and W. W. Pigman, *J. Res. Nat. Bur. Stand.*, **18**, 141 (1937).

Specific Rotation: $[\alpha]_D^{20} -13.8° \pm 0.4°$ ($\rho = 4$ g/100 ml) determined on a sample dried as described on p. 28.

Homogeneity: Determined by paper and gas–liquid chromatography.

No contaminants detectable by aniline hydrogen phthalate after paper chromatography in system 1 or 2.

No contaminants detectable by gas–liquid chromatography of the trimethylsilylated derivative on a polyester column, isothermal at 170 °C.

Loss of Weight on Drying: Not more than 0.1%.

Water-Insoluble Material: Should give a clear colorless solution in water.

Ash: Not more than 0.05%.

Heavy Metals (as Cu): Not more than 10 ppm.

Iron (as Fe): Not more than 5 ppm.

Arsenic (as As): Not more than 0.5 ppm.

Carbo-40
Maltose Monohydrate

Formula: $C_{12}H_{22}O_{11} \cdot H_2O$

Formula Wt.: 360.32

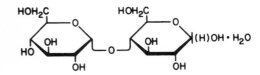

Bibliography: H. S. Isbell and W. W. Pigman, *J. Res. Nat. Bur. Stand.*, **18**, 141 (1937). For a study of the dehydration of maltose, see J. E. Cleland and W. R. Fetzer, *Ind. Eng. Chem., Anal. Ed.*, **14**, 27 (1942).

Specific Rotation: $[\alpha]_D^{20}$ +130.4° ±1.3° (ρ = 4 g/100 ml) determined on an undried sample and calculated on the basis of the monohydrate.

Homogeneity: Determined by paper and gas–liquid chromatography.

No commercial samples of maltose have yet been found to be homogeneous by paper chromatography in system 1 or 2.

No contaminants were detected by gas–liquid chromatography of the trimethylsilylated derivative on an SE-30 column, temperature-programmed from 170 to 250 °C at 4 °C per min.

Loss of Weight on Drying: Not more than 6% (to constant weight at 100 °C and not more than 0.5 mmHg).*

Water-Insoluble Material: Should give a clear colorless solution in water.

Dextrins: A solution of 0.5 g in 10 ml of water should give no coloration when treated with several drops of dilute iodine solution.

Ash: Not more than 0.05%.

Heavy Metals (as Cu): Not more than 10 ppm.

Iron (as Fe): Not more than 5 ppm.

Arsenic (as As): Not more than 0.5 ppm.

---

* Because of the drying methods used, commercial maltose may contain less water of crystallization than the 5.00% theoretically required for the monohydrate. This moisture is lost only slowly at 60 °C *in vacuo* but relatively rapidly at 100 °C and 0.5 mmHg. It should be noted that the anhydrous material is hygroscopic; weighings should therefore be made with the compound in a closed container.

Carbo-41
D-Mannitol

Formula: $C_6H_{14}O_6$
Formula Wt.: 182.17

$$
\begin{array}{c}
CH_2OH \\
HOCH \\
HOCH \\
HCOH \\
HCOH \\
CH_2OH
\end{array}
$$

Bibliography: N. K. Richtmyer and C. S. Hudson, *J. Am. Chem. Soc.*, **73**, 2249 (1951).

Specific Rotation: $[\alpha]_D^{20}$ +141° ±1° (ρ = 0.4 g/100 ml in an excess of acidified molybdate) determined on a sample dried as described on p. 28.

The solution in acidified molybdate is prepared as described for D-arabinitol, Carbo-8.

Reducing Material: A sample applied to filter paper, as in the standard procedure for paper chromatography, gives no coloration with aniline hydrogen phthalate spray.

Melting Point:[1] 166–168 °C.

Homogeneity: Determined by gas–liquid chromatography.

Homogeneous by gas–liquid chromatography of the trimethylsilylated derivative on a polyester column, isothermal at 170 °C.

Loss of Weight on Drying: Not more than 0.1%.

Ash: Not more than 0.05%.

Heavy Metals (as Cu): Not more than 10 ppm.

Iron (as Fe): Not more than 5 ppm.

Arsenic (as As): Not more than 0.5 ppm.

Reference

1. *The Merck Index*, 8th ed. (1968), p. 664.

Carbo-42
D-Mannopyranose (D-Mannose)
Symbol: Man

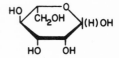

Formula: $C_6H_{12}O_6$
Formula Wt.: 180.16

Bibliography: H. S. Isbell and W. W. Pigman, *J. Res. Nat. Bur. Stand.*, **18**, 141 (1937).

Specific Rotation: $[\alpha]_D^{20}$ +14.2° ±0.4° (ρ = 4 g/100 ml) determined on a sample dried as described on p. 28.

Homogeneity: Determined by paper and gas–liquid chromatography.

No contaminants detectable by aniline hydrogen phthalate after paper chromatography in either system 1 or 2.

No contaminants detectable by gas–liquid chromatography of the trimethylsilylated derivative on a polyester column, isothermal at 170 °C.

Loss of Weight on Drying: Not more than 0.1%.

Water-Insoluble Material: Should give a clear colorless solution in water.

Ash: Not more than 0.05%.

Heavy Metals (as Cu): Not more than 10 ppm.

Iron (as Fe): Not more than 5 ppm.

Arsenic (as As): Not more than 0.5 ppm.

Carbo-43
L-Mannopyranose (L-Mannose)
Symbol: LMan

Formula: $C_6H_{12}O_6$
Formula Wt.: 180.16

Specific Rotation: $[\alpha]_D^{20}$ −14.2° ±0.4° (ρ = 4 g/100 ml) determined on a sample dried as described on p. 28.

Homogeneity: Determined by paper and gas–liquid chromatography.

No contaminants detectable by aniline hydrogen phthalate spray after paper chromatography in system 1 or 2.

No contaminants detectable by gas–liquid chromatography of the trimethylsilylated derivative on a polyester column, isothermal at 170 °C.

Loss of Weight on Drying: Not more than 0.1%.

Water-Insoluble Material: Should give a clear colorless solution in water.

Ash: Not more than 0.05%.

Heavy Metals (as Cu): Not more than 10 ppm.

Iron (as Fe): Not more than 5 ppm.

Arsenic (as As): Not more than 0.5 ppm.

Carbo-44
Melezitose Monohydrate

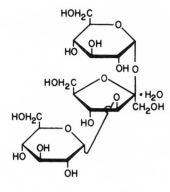

Formula: $C_{18}H_{32}O_{16} \cdot H_2O$
Formula Wt.: 522.46

Bibliography: N. K. Richtmyer and C. S. Hudson, *J. Org. Chem.*, **11**, 610 (1946).

Specific Rotation: $[\alpha]_D^{20}$ +91.7° ±0.5° (ρ = 2 g/100 ml) determined on a sample dried as in section below on "Loss of Weight on Drying" and calculated on the basis of the monohydrate.

Homogeneity: Determined by paper and gas–liquid chromatography.

No contaminants detectable by periodate–permanganate reagent after 48 h of paper chromatography in system 1 or 2.

No contaminants detectable by gas–liquid chromatography of the trimethylsilylated derivative on an SE-30 column, temperature-programmed from 170 to 250 °C at 4 °C per min.

Loss of Weight on Drying: Not over 3.6% at 110 °C and about 1 mmHg. These conditions were used by Richtmyer and Hudson, cited above. A value of 3.45% corresponds to the theoretical loss expected of a monohydrate.

Water-Insoluble Material: Should give a clear colorless solution in water.

Ash: Not more than 0.05%.
Heavy Metals (as Cu): Not more than 10 ppm.
Iron (as Fe): Not more than 5 ppm.
Arsenic (as As): Not more than 0.5 ppm.

Carbo-45
Melibiose Monohydrate*

Formula: $C_{12}H_{22}O_{11} \cdot H_2O$
Formula Wt.: 360.32

Bibliography: H. G. Fletcher, Jr., and H. W. Diehl, *J. Am. Chem. Soc.*, **74**, 5774 (1952).

Specific Rotation: $[\alpha]_D^{20}$ +135.2° ±0.7° (ρ = 4 g/100 ml) determined on a sample dried as in section below on "Loss of Weight on Dry-

ing" and calculated on the basis of the monohydrate. Fletcher and Diehl found $[\alpha]_D^{20}$ +142.3° (anhydrous basis) from which the value presented for the monohydrate has been calculated.

Homogeneity: Determined by paper and gas–liquid chromatography.

No contaminants detectable by ammoniacal silver nitrate after descending paper chromatography for 48 h in system 1 or 2.

No contaminants detectable by gas–liquid chromatography of the trimethylsilylated derivative on an SE-30 column, temperature-programmed from 170 to 250 °C at 4 °C per min.

Loss of Weight on Drying: Not more than 5.1% when dried to constant weight at 100 °C and not more than 0.5 mmHg.

Water-Insoluble Material: Should give a clear colorless solution in water.

Ash: Not more than 0.05%.
Heavy Metals (as Cu): Not more than 10 ppm.
Iron (as Fe): Not more than 5 ppm.
Arsenic (as As): Not more than 0.5 ppm.

* The α anomer may be crystallized from aqueous alcohol as the monohydrate or from dry methanol in essentially anhydrous form. On standing, the anhydrous modification absorbs moisture from the atmosphere and approaches the monohydrate in water content. The figure quoted is based on the 5% of water theoretically contained in the monohydrate plus the 0.1% tolerance for additional moisture normally acceptable in the case of unhydrated sugars.

Carbo-46
Methyl α-D-Glucopyranoside
Symbol: αMeGlc

Formula: $C_7H_{14}O_6$
Formula Wt.: 194.19

Specific Rotation: $[\alpha]_D^{20}$ +158.9° ±1.0° (ρ = 10 g/100 ml)[1] determined on a sample dried as described on p. 28.

Melting Point:[2] 168.8–169.3 °C.

Homogeneity: Determined by paper, thin-layer, and gas–liquid chromatography.

Homogeneous when chromatographed on paper for 48 h in 1-butanol–acetic acid–water (4:1:5 v/v, top layer) or in *tert*-pentyl alcohol–propyl alcohol–water (4:1:1.5 v/v),[3] and sprayed with periodate–permanganate.*

Homogeneous by thin-layer chromatography in ethyl acetate–acetic acid–water (9:2:2 v/v).

Homogeneous by gas–liquid chromatography of the trimethylsilylated derivative on a polyester column, isothermal at 200 °C.

Loss of Weight on Drying: Not more than 0.1%.

Water-Insoluble Material: Should give a clear colorless solution at 10% concentration.

Ash: Not more than 0.05%.
Heavy Metals (as Cu): Not more than 10 ppm.
Iron (as Fe): Not more than 5 ppm.
Arsenic (as As): Not more than 0.5 ppm.

References

1. F. J. Bates and Associates, *Polarimetry, Saccharimetry and the Sugars*, U.S. Govt. Printing Office, Washington, D.C. (1942), p. 730.
2. Communicated by Pfanstiehl Laboratories, Inc.
3. J. A. Cifonelli and F. Smith, *Anal. Chem.*, **23**, 1132 (1954).

* This glycoside is much slower than the reducing sugars in giving a visible spot with this reagent.

45

Carbo-47
Methyl β-D-Glucopyranoside
Symbol: βMeGlc

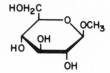

Formula: C₇H₁₄O₆
Formula Wt.: 194.19

Specific Rotation: $[\alpha]_D^{20}$ −34.2° ±0.5° ($\rho = 10$ g/100 ml)[1] determined on a sample dried as described on p. 28.
Melting Point:[2] 111.2–112.7 °C.
Homogeneity: Determined by paper, thin-layer, and gas–liquid chromatography.

Homogeneous when chromatographed on paper in the same solvent systems used for Carbo-46 and sprayed with periodate–permanganate.*

Homogeneous by thin-layer chromatography in ethyl acetate–acetic acid–water (9:2:2 v/v).

Homogeneous by gas–liquid chromatography of the trimethylsilylated derivative on a polyester column, isothermal at 200 °C.
Loss of Weight on Drying: Not more than 0.1%.
Water-Insoluble Material: A 10% solution in water should be clear and colorless.
Ash: Not more than 0.05%.
Heavy Metals (as Cu): Not more than 10 ppm.
Iron (as Fe): Not more than 5 ppm.
Arsenic (as As): Not more than 0.5 ppm.

References

1. F. J. Bates and Associates, *Polarimetry, Saccharimetry and the Sugars*, U.S. Govt. Printing Office, Washington, D.C. (1942), p. 730.
2. Communicated by Pfanstiehl Laboratories, Inc.

* This glycoside is much slower than the reducing sugars in giving a visible spot with this reagent.

Carbo-48
Methyl α-D-Mannopyranoside
Symbol: αMeMan

Formula: C₇H₁₄O₆
Formula Wt.: 194.19

Specific Rotation: $[\alpha]_D^{20}$ +79.2° ±1.0° ($\rho = 4$ g/100 ml) determined on a sample dried as described on p. 28.
Melting Point:[1] 193–194 °C.
Homogeneity: Determined by paper and gas–liquid chromatography.

No contaminants detectable by periodate–benzidine (see Carbo-24) or ammoniacal silver nitrate sprays after paper chromatography in system 1 or 2.

No contaminants detectable by gas–liquid chromatography of the trimethylsilylated derivative.
Loss of Weight on Drying: Not more than 0.1%.
Water-Insoluble Material: Should give a clear colorless solution in water.
Ash: Not more than 0.05%.
Heavy Metals (as Cu): Not more than 5 ppm.
Iron (as Fe): Not more than 5 ppm.
Arsenic (as As): Not more than 0.5 ppm.

Reference

1. F. J. Bates and Associates, *Polarimetry, Saccharimetry and the Sugars*, U.S. Govt. Printing Office, Washington, D.C. (1942), p. 746.

Carbo-49
Methyl β-D-Xylopyranoside
Symbol: βMeXyl

Formula: C₆H₁₂O₅
Formula Wt.: 164.16

Specific Rotation: $[\alpha]_D^{20}$ −65.5° ±1.0° ($\rho = 13$ g/100 ml)[1] determined on a sample dried as described on p. 28.
Melting Point:[2] 159.1–160.1 °C.
Homogeneity: Determined by paper, thin-layer, and gas–liquid chromatography.

Homogeneous when chromatographed on paper in (a) 1-butanol–acetic acid–water (4:1:5 v/v) for 20 h, (b) ethyl acetate–acetic acid–formic acid–water (18:3:1:4 v/v) for 20 h, or (c) ethyl acetate–formic acid–acetic acid–water (18:3:1:4 v/v) for 6 h, and sprayed with periodate–benzidine as described for Carbo-24.

Homogeneous by thin-layer chromatography in ethyl acetate–acetic acid–water (9:2:2 v/v).

Homogeneous by gas–liquid chromatography of the trimethylsilylated derivative on a polyester column, isothermal at 200 °C.
Loss of Weight on Drying: Not more than 0.1%.
Water-Insoluble Material: Should give a clear colorless solution in water.
Ash: Not more than 0.05%.
Heavy Metals (as Cu): Not more than 10 ppm.
Iron (as Fe): Not more than 5 ppm.
Arsenic (as As): Not more than 0.5 ppm.

References

1. C. S. Hudson, *J. Am. Chem. Soc.*, **47**, 265 (1925).
2. Communicated by Pfanstiehl Laboratories, Inc.

Carbo-50
Phenyl β-D-Glucopyranoside
Symbol: βPhGlc

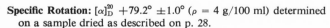

Formula: C₁₂H₁₆O₆
Formula Wt.: 256.26

Specific Rotation: $[\alpha]_D^{20}$ −72.5° ±1.0° ($\rho = 2$ g/100 ml) determined on a sample dried as described on p. 28.
Melting Point:[1] 175–176 °C.
Homogeneity: Determined by paper and gas–liquid chromatography.

The compound is not detectable on paper with periodate–benzidine or ammoniacal silver nitrate sprays. However, no contaminants can be detected by these reagents after chromatography in system 1 or 2.

The trimethylsilylated derivative gives a single peak on gas–liquid chromatography on a polyester column, isothermal at 170 °C.
Loss of Weight on Drying: Not more than 0.1%.
Water-Insoluble Material: Should give a clear colorless solution in water.
Ash: Not more than 0.05%.
Heavy Metals (as Cu): Not more than 5 ppm.
Iron (as Fe): Not more than 5 ppm.
Arsenic (as As): Not more than 0.5 ppm.

Reference

1. Communicated by Pfanstiehl Laboratories, Inc.

Carbo-51
Raffinose Pentahydrate

Formula: $C_{18}H_{32}O_{16} \cdot 5H_2O$
Formula Wt.: 594.52

Specific Rotation: $[\alpha]_D^{20}$ $+105.2°$ $\pm0.7°$ ($\rho = 4$ g/100 ml)[1] determined on a sample dried as in the section below on "Loss of Weight on Drying" and calculated on the basis of the pentahydrate.

Homogeneity: Determined by paper and gas–liquid chromatography.

No contaminants detectable by periodate–permanganate spray after 48 h of paper chromatography in system 1 or 2.

No contaminants detectable by gas–liquid chromatography of the trimethylsilylated derivative on an SE-30 column, temperature-programmed from 170 to 250 °C at 4 °C per min.

Loss of Weight on Drying: Not more than 15.3% (to constant weight at 78 °C and about 0.5 mmHg). Theoretical for the pentahydrate is 15.15%.

Water-Insoluble Material: Should give a clear colorless solution in water.

Ash: Not more than 0.05%.

Heavy Metals (as Cu): Not more than 10 ppm.

Iron (as Fe): Not more than 5 ppm.

Arsenic (as As): Not more than 0.5 ppm.

References

1. F. J. Bates and Associates, *Polarimetry, Saccharimetry and the Sugars*, U.S. Govt. Printing Office, Washington, D.C. (1942), p. 750.

Carbo-52
α-L-Rhamnopyranose Monohydrate

Formula: $C_6H_{12}O_5 \cdot H_2O$
Formula Wt.: 182.17

Specific Rotation: $[\alpha]_D^{20}$ $+8.2°$ $\pm0.4°$ ($\rho = 4$ g/100 ml) determined on a sample dried as described below in section on "Loss of Weight on Drying."

Homogeneity: Determined by paper and gas–liquid chromatography.

No contaminants detectable by aniline hydrogen phthalate spray after paper chromatography in system 1 or 2.

No contaminants detectable by gas–liquid chromatography of the trimethylsilylated derivative on a polyester column, isothermal at 170 °C.

Loss of Weight on Drying: Not more than 0.1% (to constant weight at 34 °C and 0.5 mmHg).*

Water-Insoluble Material: Should give a clear colorless solution in water.

Ash: Not more than 0.05%.

Heavy Metals (as Cu): Not more than 10 ppm.

Iron (as Fe): Not more than 5 ppm.

Arsenic (as As): Not more than 0.5 ppm.

* Attempts to remove water of crystallization from this hydrate at 64 °C *in vacuo* result in partial melting, the loss of water being less than the theoretical. Under the conditions specified here, water of crystallization does not appear to be lost.

Carbo-53
Ribitol (Adonitol)

Formula: $C_5H_{12}O_5$
Formula Wt.: 152.15

Specific Rotation: None.

Melting Point: Not below 101 or above 102 °C. The melting point of 102 °C, which E. Fischer reported[1] for ribitol, has been confirmed by various more recent authors.

Homogeneity: Determined by gas–liquid chromatography.

No contaminants detectable by gas–liquid chromatography of the trimethylsilylated derivative on a polyester column, isothermal at 170 °C.

Reducing Material: A sample applied to filter paper, as in the standard procedure for paper chromatography, gives no coloration with aniline hydrogen phthalate spray.

Loss of Weight on Drying: Not more than 0.1%.

Water-Insoluble Material: Should give a clear colorless solution in water.

Ash: Not more than 0.05%.

Heavy Metals (as Cu): Not more than 10 ppm.

Iron (as Fe): Not more than 5 ppm.

Arsenic (as As): Not more than 0.5 ppm.

Reference

1. E. Fischer, *Chem. Ber.*, **26**, 633 (1893).

Carbo-54
D-Ribopyranose (D-Ribose)
Symbol: Rib

Formula: $C_5H_{10}O_5$
Formula Wt.: 150.13

Specific Rotation: $[\alpha]_D^{20}$ $-20.4°$ $\pm0.4°$ ($\rho = 2$ g/100 ml)* determined on a sample dried as described on p. 28.

Homogeneity: Determined by paper and gas–liquid chromatography.

No contaminants detectable by aniline hydrogen phthalate after paper chromatography in system 1 or 2.

No contaminants detectable by gas–liquid chromatography of the trimethylsilylated derivative on a polyester column, isothermal at 170 °C.

Loss of Weight on Drying: Not more than 0.1%.
Water-Insoluble Material: Should give a clear colorless solution in water.
Ash: Not more than 0.05%.
Heavy Metals (as Cu): Not more than 10 ppm.
Iron (as Fe): Not more than 5 ppm.
Arsenic (as As): Not more than 0.5 ppm.

* Unpublished measurement by H. W. Diehl of the National Institutes of Health.

Water-Insoluble Material: Should give a clear colorless solution in water.
Ash: Not more than 0.1%.
Heavy Metals (as Cu): Not more than 10 ppm.
Iron (as Fe): Not more than 5 ppm.
Arsenic (as As): Not more than 0.5 ppm.

* Sedoheptulosan crystallizes also in an anhydrous form, into which the monohydrate can be converted by recrystallization from 15 parts of methanol. The anhydrous modification is metastable, whereas the hydrated form is stable in air.

Carbo-55
Salicin[1]

Formula: $C_{13}H_{18}O_7$
Formula Wt.: 286.28

Specific Rotation: $[\alpha]_D^{20}$ $-62.3°$ $\pm 0.6°$ (ρ = 5 g/100 ml) determined on a sample dried as described on p. 28.
Melting Point:[2] 199–202 °C (crystallized from water).
Homogeneity: Determined by paper and thin-layer chromatography.
   Homogeneous when chromatographed on paper in system 1 or 2, and detected by permanganate–periodate spray.*
   A faster-moving trace contaminant is detected by thin-layer chromatography in ethyl acetate–acetic acid–water (9:2:2 v/v).
Loss of Weight on Drying: Not more than 0.1%.
Water-Insoluble Material: A saturated aqueous solution (about 5%) should be clear and colorless.
Ash: Not more than 0.05%.
Heavy Metals (as Cu): Not more than 10 ppm.
Iron (as Fe): Not more than 5 ppm.
Arsenic (as As): Not more than 0.5 ppm.

References

1. Standards were set for salicin in *The National Forumlary*, 9th ed., J. B. Lippincott Co., Philadelphia, Pennsylvania (1950), p. 441.
2. *The Merck Index*, 8th ed. (1968), p. 929.

* In comparison with reducing sugars, the spot appears slowly.

Carbo-56
Sedoheptulosan Monohydrate

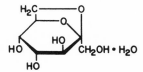

Formula: $C_7H_{12}O_6 \cdot H_2O$
Formula Wt.: 210.19

Specific Rotation: $[\alpha]_D^{20}$ $-134.0°$ $\pm 2.0°$ (ρ = 2 g/100 ml) determined on a sample dried as described on p. 28.
Homogeneity: Determined by paper and gas–liquid chromatography.
   No contaminants detectable by periodate–benzidine reagent (as described for Carbo-24) after paper chromatography in system 1 or 2.
   No contaminants detectable by gas–liquid chromatography of the trimethylsilylated derivative on a polyester column, isothermal at 170 °C.
Loss of Weight on Drying: Not more than 0.1%.*

Carbo-57
Sodium D-*glycero*-D-*gulo*-Heptonate Dihydrate

Formula: $C_7H_{13}O_8Na \cdot 2H_2O$
Formula Wt.: 284.20

Specific Rotation: $[\alpha]_D^{20}$ $+69.9°$ $\pm 0.5°$ (ρ = 0.5 g/100 ml); dissolve 0.5 g of sample in 25 ml of water, add 2.5 g of ammonium molybdate tetrahydrate $[(NH_4)_6Mo_7O_{24} \cdot 4H_2O]$, add 5 ml of glacial acetic acid, and dilute to 100 ml. Satisfactory readings may be obtained during the first 20 min.*
Homogeneity: Determined by paper and gas–liquid chromatography.
   Homogeneous when chromatographed on paper for 16 h in 2-butanone–acetic acid–boric acid (saturated aqueous) (9:1:1 v/v) and sprayed with peridoate–benzidine reagent as described for Carbo-24.
   Gas–liquid chromatography of the trimethylsilylated lactone derived from this compound showed two components in the approximate ratio of 1:30 on an SE-30 column, temperature-programmed from 150 to 250 °C at 4 °C per min.
Loss of Weight on Drying: Not more than 12.8% when dried at 100 °C for 3 h at about 0.5 mmHg. Theoretical value for the dihydrate is 12.68%.
Water-Insoluble Material: A 10% aqueous solution should be clear and colorless.
Ash: Not more than 8.5%. The theoretical value is 8.09% for the dihydrate.
Heavy Metals (as Cu): Not more than 10 ppm.
Iron (as Fe): Not more than 10 ppm.
Arsenic (as As): Not more than 0.5 ppm.
Chloride: Not more than 50 ppm.
Sulfate: Not more than 50 ppm.
pH: A 10% solution in water should have a pH of 7.0–8.0.

* We are indebted to F. H. Hedger, Charles Pfizer & Co., Inc., Brooklyn, New York, for this method, which was devised for sodium D-gluconate.

Carbo-58
L-Sorbopyranose (L-Sorbose)

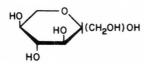

Formula: $C_6H_{12}O_6$
Formula Wt.: 180.16

Bibliography: H. S. Isbell and W. W. Pigman, *J. Res. Nat. Bur. Stand.*, 19, 443 (1937).

Specific Rotation: $[\alpha]_D^{20}$ −43.3° ±0.2° (ρ = 12 g/100 ml) determined on a sample dried as described on p. 28.
Homogeneity: Determined by paper and gas–liquid chromatography.

No contaminants detectable by ammoniacal silver nitrate after paper chromatography in system 1 or 2.

No contaminants detectable by gas–liquid chromatography of the trimethylsilylated derivative on a polyester column, isothermal at 170 °C.
Loss of Weight on Drying: Not more than 0.1%.
Water-Insoluble Material: Should give a clear colorless solution in water.
Ash: Not more than 0.05%.
Heavy Metals (as Cu): Not more than 10 ppm.
Iron (as Fe): Not more than 5 ppm.
Arsenic (as As): Not more than 0.5 ppm.

Carbo-59
Stachyose Tetrahydrate

Formula: $C_{24}H_{42}O_{21} \cdot 4H_2O$
Formula Wt.: 738.65

Bibliography: M. L. Wolfrom, R. C. Burrell, A. Thompson, and S. S. Furst, *J. Am. Chem. Soc.*, 74, 6299 (1952).

Specific Rotation: $[\alpha]_D^{20}$ +131.3° ±0.6° (ρ = 4 g/100 ml) determined on a sample dried as in the section below on "Loss of Weight on Drying" and calculated on the basis of the tetrahydrate.
Homogeneity: Determined by paper and gas–liquid chromatography.

No contaminants detectable by periodate–permanganate spray after 48 h of paper chromatography in system 1 or 2.

No contaminants detectable by gas–liquid chromatography of the trimethylsilylated derivative on a 3% SE-30 column,* isothermal at 275 °C.
Loss of Weight on Drying: Not more than 9.9%† (to constant weight at 78 °C and about 0.5 mmHg). The theoretical value for the tetrahydrate is 9.75%.

Water-Insoluble Material: Should give a clear colorless solution in water.
Ash: Not more than 0.05%.
Heavy Metals (as Cu): Not more than 10 ppm.
Iron (as Fe): Not more than 5 ppm.
Arsenic (as As): Not more than 0.5 ppm.

* For the higher oligosaccharides, columns that contain 3% (instead of 10%) of liquid phase are recommended. The higher content of liquid phase results in impractically large retention volumes for these compounds.
† Based on analytical evaluations by Dr. Edward S. Rorem, Western Utilization Research and Development Service, U.S. Department of Agriculture, Albany, California.

Carbo-60
Starch, Soluble

Formula: $(C_6H_{10}O_5)_n \cdot xH_2O$

Bibliography: J. C. Small, *J. Am. Chem. Soc.*, 41, 116 (1919); J. Rosin, *Reagent Chemicals and Standards*, 5th ed., Van Nostrand Co., Princeton, New Jersey (1967), p. 498; *The Pharmacopeia of the United States of America*, 18th revision (U.S.P. XVIII), Mack Publishing Co., Easton, Pennsylvania (1970), p. 682; *Reagent Chemicals*, 4th ed., American Chemical Society, Washington, D.C. (1968), p. 586.

Ash: 0.2%.
Reducing Sugars: Trace. Take 10 g in 100 ml and shake for 15 min at room temperature. Allow to settle and then filter, rejecting the first 10 ml of filtrate. Add 50 ml of Fehling's solution to 50 ml of the filtrate, bring to a boil in 4 min, and boil for 2 min. Filter at once through a tared Gooch crucible, wash successively with hot water, 10 ml of ethyl alcohol, and 15 ml of ether. Dry at 100 °C for 30 min, cool, and weigh. The weight of the cuprous oxide so obtained is not more than 0.047 g.
Freedom from Erythrodextrin: Erythrodextrin may be readily recognized by its iodine color test in solutions of the hydrolytic products of starch after the removal of unchanged starch, soluble starch, and amylodextrin. This is readily accomplished by the use of a reagent containing 2 g of iodine and 6 g of potassium iodide in 1 liter of saturated ammonium sulfate solution. The addition of an equal volume of this reagent to an approximately 1% solution of a soluble-starch sample immediately precipitates everything of molecular weight larger than that of the erythrodextrin, and a clear, red-brown filtrate results. To the filtrate in a test tube is added, dropwise, 0.05 M sodium thiosulfate solution until the color is just discharged. To this liquid is added a measured amount of an iodine solution sufficient to supply an excess of iodine. An equal volume of water, to which has been added the same measured volume of iodine solution, serves as a control. The dextrin red stands out on comparison. Very small amounts of erythrodextrin can be recognized in this way.

Carbo-61
Sucrose*

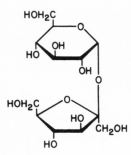

Formula: $C_{12}H_{22}O_{11}$
Formula Wt.: 342.30

Bibliography: Report of the Proceedings of the Tenth Session of the International Commission for Uniform Methods of Sugar Analysis, *Intern. Sugar J.* 52, 201 (1950); F. J. Bates and Associates, *Polarimetry, Saccharimetry and the Sugars,* U.S. Govt. Printing Office, Washington, D.C. (1942), pp. 392, 551; J. Rosin, *Reagent Chemicals and Standards,* 5th ed., D. Van Nostrand Co., Inc., Princeton, New Jersey (1967), p. 505; *The Pharmacopeia of the United States of America,* 18th revision (U.S.P. XVIII), Mack Publishing Co., Easton, Pennsylvania (1970), p. 692.

CRITERIA: Standard Reference Material 17A of the National Bureau of Standards, Washington, D.C.

Moisture   <0.01%
Ash   <0.001%
Reducing substances,
  estimated as invert
  sugar[1]   <0.02%

Each 100 ml of a "normal" sucrose solution contains 26.000 g of dried substance, weighed with brass weights in air (760 mmHg, 20 °C, 50% relative humidity). At 20 °C, this solution in a 2-dm polarimeter tube reads 100 °S (International sugar degrees). The illumination is white light filtered through a 15-mm layer of a 6% solution of potassium dichromate. The International Sugar Scale was defined and adopted by the International Commission for Uniform Methods of Sugar Analysis at the Eighth Session, Amsterdam, 1932.[2]

The rotation in circular degrees of the "normal" sucrose solution, observed in a 2-dm polarimeter tube, for wavelength 546.1 nm is 40.763° and for wavelength 589.25 nm is 34.617°.

The specific rotations of sucrose for the "normal" solution are

$[\alpha]_{546.1}^{20}$ + 78.342° (ρ = 26 g/100 ml of solution)
$[\alpha]_{589.25}^{20}$ + 66.529° (ρ = 26 g/100 ml of solution)

SPECIFICATIONS:
Homogeneity: Determined by paper and gas–liquid chromatography.

Homogeneous by paper chromatography for 17 h in system 1 or 2.

Homogeneous by gas–liquid chromatography of the trimethylsilylated derivative, SE-30 column, temperature-programmed from 170 to 250 °C at 4 °C per min.
Reducing Substances (as invert sugar): Not more than 0.004%.
Ash: Not more than 0.004%.
Moisture: Not more than 0.002%.

References

1. F. J. Bates and R. F. Jackson, *Bull. Bur. Stand.* 13, 67 (1916).
2. Proceedings of the Eighth Session, International Commission for Uniform Methods of Sugar Analysis, *Intern. Sugar J.* 35, 17 (1933); F. J. Bates and Associates, NBS Circular C440, pp. 79, 775 (1942).

* The sucrose was supplied by the California and Hawaiian Sugar Refining Corporation of Crockett, California.

Carbo-62
Tetra-*O*-acetyl-β-D-ribofuranose
Symbol: Rib*f*Ac₄

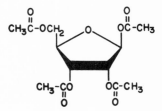

Formula: $C_{13}H_{18}O_9$
Formula Wt.: 318.28

Description: Crystalline solid (orthorhombic prisms).
Specific Rotation:
$[\alpha]_D^{24}$ −14.6° ±1.0° (ρ = 5 g/100 ml in methanol)
$[\alpha]_D^{21}$ −12.6° ±1.0° (ρ = 5 g/100 ml in chloroform)
Melting Point: 84–85 °C.
Loss of Weight on Drying: Not more than 0.1%.
Ash: Not more than 0.05%.
Heavy Metals (as Cu): Not more than 10 ppm.
Iron (as Fe): Not more than 5 ppm.
Arsenic (as As): Not more than 0.5 ppm.

References

1. H. Zinner, *Chem. Ber.,* **86**, 817 (1953).
2. G. B. Brown, J. Davoll, and B. A. Lowy, *Biochem. Prep.,* **4**, 70 (1955).

Carbo-63
Tetra-*O*-acetyl-β-D-ribopyranose
Symbol: Rib*p*Ac₄

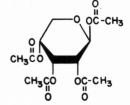

Formula: $C_{13}H_{18}O_9$
Formula Wt.: 318.28

Description: Crystalline solid (bipyramidal prisms).
Specific Rotation:
$[\alpha]_D^{24}$ −56.8° ±1° (ρ = 4.9 g/100 ml in methanol)
$[\alpha]_D^{21}$ −57.9° ±1° (ρ = 5.2 g/100 ml in chloroform)
Melting Point: 113–114 °C.
Loss of Weight on Drying: Not more than 0.1%.
Ash: Not more than 0.05%.
Heavy Metals (as Cu): Not more than 10 ppm.
Irons (as Fe): Not more than 5 ppm.
Arsenic (as As): Not more than 0.5 ppm.

Reference

1. G. B. Brown, J. Davoll, and B. A. Lowy, *Biochem. Prep.,* **4**, 74 (1955).

Carbo-64
α,α-Trehalose Dihydrate

Formula: $C_{12}H_{22}O_{11} \cdot 2H_2O$
Formula Wt.: 378.33

Specific Rotation: $[\alpha]_D^{20}$ +179.9° ±0.4° (ρ = 7 g/100 ml) determined on a sample dried as described in the section below on "Loss of Weight on Drying" and calculated on the basis of the dihydrate.*
Homogeneity: Determined by paper and gas–liquid chromatography.

No contaminants detectable by periodate–permanganate spray after 48 h of descending paper chromatography in system 1 or 2.

No contaminants detectable by gas–liquid chromatography of the trimethylsilylated derivative on an SE-30 column, temperature-programmed from 200 to 280 °C at 4 °C per min.
Loss of Weight on Drying: At 60 °C and not above 0.5 mmHg, not more than 9.6%. Theoretical moisture content for the dihydrate is 9.52%.
Water-Insoluble Material: Should give a clear colorless solution in water.
Ash: Not more than 0.05%.
Heavy Metals (as Cu): Not more than 10 ppm.
Iron (as Fe): Not more than 5 ppm.
Arsenic (as As): Not more than 0.5 ppm.

* This specific rotation is the average of the determinations made with a highly purified sample provided by Dr. N. K. Richtmyer, National Institutes of Health.

Carbo-65
Tri-*O*-acetyl-D-glucal (3,4,6-Tri-*O*-acetyl-1,5-anhydro-2-deoxy-D-arabino-hex-1-enitol)

Formula: $C_{12}H_{16}O_7$
Formula Wt.: 272.23

Bibliography: E. Fischer, *Chem. Ber.*, **47**, 196 (1914); P. A. Levene and R. S. Tipson, *J. Biol. Chem.*, **90**, 94 (1931); B. Helferich, E. M. Mulcahy, and H. Ziegler, *Ber.*, **87**, 233 (1954).

Description: Colorless crystals.
Specific Rotation: $[\alpha]_D^{25}$ −16.0° ±0.5° (ρ = 2 g/100 ml in anhydrous ethanol)
Melting Point: 57–58 °C.
Homogeneity: Homogeneous by thin-layer chromatography with 6:1 chloroform–butanone; detection by spraying with sulfuric acid and heating.
Loss of Weight on Drying: Not more than 0.1%.
Ash: Not more than 0.1%.

Carbo-66
Turanose

Formula: $C_{12}H_{22}O_{11}$
Formula Wt.: 342.30

Bibliography: H. S. Isbell and W. W. Pigman, *J. Res. Nat. Bur. Stand.*, **20**, 773 (1938).

Specific Rotation: $[\alpha]_D^{20}$ +75.8° ±0.4° (ρ = 4 g/100 ml) determined on a sample dried as described on p. 28.
Homogeneity: Determined by paper and gas–liquid chromatography.

No contaminants detectable by ammoniacal silver nitrate spray after paper chromatography in system 1 or 2.

No contaminants detectable by gas–liquid chromatography of the trimethylsilylated derivative on an SE-30 column, temperature-programmed from 170 to 260 °C at 4 °C per min.
Loss of Weight on Drying: Not more than 0.1%.
Water-Insoluble Material: Should give a clear colorless solution in water.
Ash: Not more than 0.05%.
Heavy Metals (as Cu): Not more than 10 ppm.
Iron (as Fe): Not more than 5 ppm.
Arsenic (as As): Not more than 0.5 ppm.

Carbo-67
Xylitol

Formula: $C_5H_{12}O_5$
Formula Wt.: 152.15

```
   CH2OH
   |
   HCOH
   |
   HOCH
   |
   HCOH
   |
   CH2OH
```

Specific Rotation: None.
Melting Point: Not below 93 and not above 95 °C.*
Homogeneity: Determined by gas–liquid chromatography.

No contaminants detectable by gas–liquid chromatography of the trimethylsilylated derivative on a polyester column, isothermal at 170 °C.
Reducing Material: A sample applied to filter paper, as in the standard procedure for paper chromatography, gives no coloration with aniline hydrogen phthalate spray.
Loss of Weight on Drying: Not more than 0.1%.
Water-Insoluble Material: Should give a clear colorless solution in water.
Ash: Not more than 0.05%.
Heavy Metals (as Cu): Not more than 10 ppm.
Iron (as Fe): Not more than 5 ppm.
Arsenic (as As): Not more than 0.5 ppm.

References

1. M. L. Wolfrom and E. J. Kohn, *J. Am. Chem. Soc.*, **64**, 1739 (1942).
2. J. F. Carson, S. W. Waisbrot, and F. T. Jones, *J. Am. Chem. Soc.*, **65**, 1777 (1943).

* Xylitol was first crystallized in a hygroscopic metastable form[1] melting at 61–61.5 °C. It was later obtained in a stable modification[2] melting at 93–94.5 °C.

Carbo-68
D-Xylopyranose (D-Xylose)
Symbol: Xyl

Formula: $C_5H_{10}O_5$
Formula Wt.: 150.13

Bibliography: H. S. Isbell and W. W. Pigman, *J. Res. Nat. Bur. Stand.*, **18**, 141 (1937); J. Rosin, *Reagent Chemicals and Standards*, 5th ed., D. Van Nostrand Co., Inc., Princeton, New Jersey (1967), p. 543.

Specific Rotation: $[\alpha]_D^{20}$ +18.8° ±0.6° (ρ = 4 g/100 ml) determined on a sample dried as described on p. 28.
Homogeneity: Determined by paper and gas–liquid chromatography.

No contaminants detectable by aniline hydrogen phthalate spray after paper chromatography in system 1 or 2.

No contaminants detectable by gas–liquid chromatography of the trimethylsilylated derivative on a polyester column, isothermal at 170 °C.
Loss of Weight on Drying: Not more than 0.1%.
Water-Insoluble Material: Should give a clear colorless solution in water.
Ash: Not more than 0.05%.
Heavy Metals (as Cu): Not more than 10 ppm.
Iron (as Fe): Not more than 5 ppm.
Arsenic (as As): Not more than 0.5 ppm.

Carbo-69
L-Xylopyranose (L-Xylose)

Formula: $C_5H_{10}O_5$
Formula Wt.: 150.13

Specific Rotation: $[\alpha]_D^{20}$ −18.8° ±0.6° (ρ = 4 g/100 ml) determined on a sample dried as described on p. 28.
Homogeneity: Determined by paper and gas–liquid chromatography.

No contaminants detectable by aniline hydrogen phthalate spray after paper chromatography in system 1 or 2.

No contaminants detectable by gas–liquid chromatography of the trimethylsilylated derivative on a polyester column, isothermal at 170 °C.
Loss of Weight on Drying: Not more than 0.1%.
Water-Insoluble Material: Should give a clear colorless solution in water.
Ash: Not more than 0.05%.
Heavy Metals (as Cu): Not more than 10 ppm.
Iron (as Fe): Not more than 5 ppm.
Arsenic (as As): Not more than 0.5 ppm.

Carotenoids and Related Compounds

The first step in the preparation of many pure carotenoids is extraction from a natural source. This extraction is made as gently as possible, to prevent the isomerization and destruction of the carotenoids. The preferred method uses extraction of the carotenoids with a chilled solvent. One quite satisfactory method involves the homogenization of plant material in a mixture of acetone and petroleum ether in a Waring Blendor. Frequently, solid carbon dioxide is added to keep the solution cold. Antioxidants and basic compounds (e.g., calcium carbonate) may also be added, to minimize oxidation of the carotenoids or their destruction by plant acids. Other methods involve grinding with solvent in a mortar or extracting with boiling solvent (methanol or ethanol, for example). The carotenoids are transferred to a hydrocarbon solvent (petroleum ether or benzene) after the addition of water and salt (sodium chloride or ammonium sulfate) to the alcohol or acetone phase. Xanthophylls are then separated from the carotenes by partition between immiscible solvents. Most commonly, this separation is effected with petroleum ether and 90% methanol. However, other solvent mixtures are also used. The xanthophylls are found in the methanol (hypophase), and the carotenes are present in the petroleum ether (epiphase). If xanthophyll esters are present, the carotene solution is saponified with alcoholic potassium hydroxide (preferably at room temperature or below), and the resultant xanthophylls are separated from the carotenes by partition between petroleum ether and aqueous alcohol. Following separation and thorough removal of water from the hydrocarbon solvent, the carotenes are separated from one another by chromatography. The xanthophylls are also transferred to petroleum ether, the solution is dried, and the pigments are separated by chromatography. The separated compounds (carotenes or xanthophylls) may then be crystallized or stored in a nonpolar solvent at −10 to −20 °C. Similar procedures are used for the purification of chemically synthesized carotenoids. To minimize *cis–trans* isomerization of the carotenoids, all operations should be carried out in dim light or semidarkness and as rapidly as possible. Pure solvents should be used, to minimize destruction of the carotenes, and, when possible, operations should be conducted in an inert atmosphere.

The most useful criteria of purity for carotenoids are chromatography and absorption spectra. Colored impurities in a carotenoid preparation are readily detected by chromatography. Thin-layer, paper, or column chromatography may be used, with preference to the first two because they are more rapid and require a smaller amount of material. Colored impurities are also detected by examination of the absorption spectrum. The presence of *cis*-isomers in a carotenoid preparation may be determined by chromatography or by examination of the absorption spectrum. Most *cis*-isomers have pronounced "*cis*-peak" absorption. Colorless impurities in a carotenoid preparation are detected by a comparison of the absorption coefficients of the carotenoid with those of the preparation.

Several other methods are quite useful for establishing the presence or absence of impurities in a carotenoid preparation. Infrared spectra will show the presence of colored or colorless impurities. However, this method may not reveal the presence of small proportions of impurities. Nuclear magnetic resonance and mass spectrometry are more useful in establishing the purity of a carotenoid preparation. Gas–liquid chromatography of hydrogenated carotenoids has also been used to establish purity; this method is particularly useful if the column is programmed from about 150 to 280 °C.

Carotenoids are decomposed by light, air, and acid solvents. Therefore, it is recommended that they be stored in the dark (preferably in brown vials), in an inert atmosphere (or a sealed ampoule), at −10 to −20 °C.

The colorless compounds related to the carotenes, or to precursors in their biosynthesis, are assayed for purity by a variety of methods. These methods are detailed in the specifications for each compound.

The presentation of physical characteristics of the carotenoids and related compounds is uniform throughout these specifications. Thus, all light-absorption wavelength maxima are given in nanometers (nm), and all melting points and boiling points are given in degrees Celsius (°C).

Melting points are reported as stated in the particular reference cited. Since the melting point of a carotenoid can vary appreciably, depending upon the method used for its determination, original articles should be consulted when a comparison is made between an experimentally determined melting point and a value reported in these specifications.

The references listed for a compound reported in these specifications should be consulted for the synthesis, isolation from natural materials, purification, and assay for purity of that compound. In addition, the following bibliography may be consulted for a wide range of information of value in the preparation of carotenoids and in the analysis of these compounds for purity.

Following the bibliography are the names of contributors and reviewers who particularly assisted the Subcommittee and whose work is hereby gratefully acknowledged.

ANALYSES OF COMMERCIAL PRODUCTS

Analyses were carried out, during 1969, to determine the purity of 16 commercially available carotenoids and related compounds. The compounds assayed were purchased on the open market, and they were ones for which criteria and specifications had been reported in Publication No. 1344 of the National Academy of Sciences, the previous edition of the present publication. These compounds varied considerably in purity. Seven carotenoids and related compounds had purities of 90–100% by the criteria reported in the earlier publication, and seven had purities of 80–90%. One compound had

a purity of 70–80%, and one had a purity of less than 10%. None of the commercial products contained colored impurities. Hence, the impurities were either colorless compounds that had not been removed in the preparation of the commercial product or colorless oxidation products that were formed during the time interval between their preparation by the commercial supplier and our receipt of the product. It seems probable that the product of very low purity was one in which much oxidation occurred after its preparation by the supplier. Such a product would, however, be almost worthless to the purchaser.

BIBLIOGRAPHY

M. S. Barber, J. B. Davis, L. M. Jackman, and B. C. L. Weedon, *J. Chem. Soc.*, 2870 (1960).

B. H. Davies, in *Chemistry and Biochemistry of Plant Pigments*, T. W. Goodwin, ed., Academic Press, London and New York (1965), p. 489.

S. P. Colowick and N. O. Kaplan, *Methods in Enzymology: Steroids and Terpenoids*, Vol. XV, R. B. Clayton, ed., Academic Press, New York (1969).

T. W. Goodwin, in *Carotine und Carotinoide*, K. Lang, ed., D. Steinkopff Verlag, Darmstadt (1963), p. 1.

T. W. Goodwin, in *Carotine und Carotinoide*, K. Lang, ed., *noids*, Chapman & Hall, London (1952).

O. Isler, *Carotenoids*, Birkhäuser, Basel (1971).

O. Isler, R. Rüegg, U. Schwieter, and J. Würsch, *Vitamins Hormones*, **18**, 295 (1960).

S. Liaaen-Jensen, *Pure Appl. Chem.*, **20**, 421 (1969).

S. Liaaen-Jensen and A. Jensen, *Prog. Chem. Fats Lipids*, **8**, 129 (1965).

P. Karrer and E. Jucker, *Carotenoids*, E. A. Braude (translator), Elsevier, New York (1950).

J. A. Olson, in *Newer Methods of Nutritional Biochemistry*, Vol. 2, A. A. Albanese, ed., Academic Press Inc., New York (1965), p. 345.

J. W. Porter and D. G. Anderson, in *Chromatography*, E. Heftmann, ed., Reinhold Publishing Corp., New York (1961), p. 465.

U. Schwieter, G. Englert, N. Rigassi, and W. Vetter, *Pure Appl. Chem.*, **20**, 365 (1969).

H. H. Strain, *Leaf Xanthophylls*, Carnegie Institution of Washington, Washington, D. C. (1938).

B. C. L. Weedon, in *Chemistry and Biochemistry of Plant Pigments*, T. W. Goodwin, ed., Academic Press Inc., New York (1965), p. 75.

B. C. L. Weedon, *Prog. Chem. Org. Nat. Prod.*, **27**, 81 (1969).

B. C. L. Weedon, *Pure Appl. Chem.*, **20**, 531 (1969).

L. Zechmeister, cis–trans-*Isomeric Carotenoids, Vitamins A and Arylpolyenes*, Springer-Verlag, Vienna (1962).

CONTRIBUTORS AND REVIEWERS

A. G. Andrewes, The Norwegian Institute of Technology, Trondheim, Norway

G. Britton, The University of Liverpool, Liverpool, England

Peter H. W. Butterworth, University College, London, England

C. J. Chesterton, Royal Postgraduate Medical School, London, England

B. H. Davies, University College, Aberystwyth, Wales

B. C. L. Weedon, Queen Mary College, London, England

Carot-1
Antheraxanthin
5,6-Epoxy-5,6-dihydro-β,β-carotene-3,3′-diol

Formula: $C_{40}H_{56}O_3$
Formula Wt.: 584.89
Calc. %: C, 82.14; H, 9.65; O, 8.21

Antheraxanthin has been characterized as 5,6-epoxy-3,3′-dihydroxy-β-carotene.[1,2] The stereochemistry of this compound is unknown.

Sources:

Natural Sources. Antheraxanthin is present as an ester in *Lilium tigrinum*,[3] from which it was originally isolated. It is also present in smaller proportions in some fruits.

Partial Synthesis. Antheraxanthin is obtained on oxidation of zeaxanthin with monoperphthalic acid.[1,2,4] Two isomers, having different optical rotatory dispersion properties, are obtained.[4]

Isolation Procedures:[1–3] Antheraxanthin is extracted from biological materials with acetone or ethanol and then transferred to benzene. This pigment is then separated from other carotenoids by chromatography on a calcium hydroxide–Celite column.

Methods of Purification:

Solvent Partition. A partial purification of antheraxanthin can be obtained by partition between several solvent combinations.[5]

Chromatography. Antheraxanthin has been purified by chromatography on columns of calcium hydroxide[3] and of zinc carbonate.[1]

Crystallization. Needles or thin plates are obtained on crystallization from benzene–methanol.[3]

Derivatives. No derivatives other than the furanoid rearrangement product (mutatoxanthin) have been reported.[2]

Methods of Assaying for Purity:

Chromatography. Assays for chromatographic homogeneity can be carried out on thin-layer plates[6] or on kieselguhr paper.[7]

Visible Spectrum. Carbon disulfide:[3] 478 and 510 nm. Chloroform:[3] 460.5 and 490.5 nm.

Mass Spectrum. The mass spectrum of antheraxanthin has been reported.[8]

Melting Point. Antheraxanthin melts[2] at 205 °C.

Derivatives. Mutatoxanthin is obtained on treatment of antheraxanthin with chloroform containing a trace of hydrochloric acid.[1]

Probable Impurities: Violaxanthin and mutatoxanthin.

Conditions of Storage: Darkness (brown vial), inert atmosphere (sealed ampoule), and low temperature (−20 °C).

References

1. P. Karrer and E. Jucker, *Helv. Chim. Acta*, **28**, 300 (1945).
2. P. Karrer and E. Jucker, *Carotenoids*, E. A. Braude (translator), Elsevier, New York (1950).
3. P. Karrer and A. Oswald, *Helv. Chim. Acta*, **18**, 1303 (1935).
4. L. Bartlett, W. Klyne, W. P. Mose, P. M. Scopes, G. Galasko, A. K. Mallams, B. C. L. Weedon, J. Szabolcs, and G. Tóth, *J. Chem. Soc. (C)*, 2527 (1969).
5. A. L. Curl and G. F. Bailey, *J. Agr. Food Chem.*, **2**, 685 (1954).
6. H. R. Bolliger, A. König, and U. Schwieter, *Chimia*, **18**, 136 (1964).
7. A. Jensen and S. Liaaen-Jensen, *Acta Chem. Scand.*, **13**, 1863 (1959).
8. H. Budzikiewicz, H. Brzezinka, and B. Johannes, *Monatsh. Chem.*, **101**, 579 (1970).

Carot-2
β-Apocarotenal
8′-Apo-β-caroten-8′-al
(β-Apo-8′-carotenal)

Formula: $C_{30}H_{40}O$
Formula Wt.: 416.65
Calc. %: C, 86.48; H, 9.68; O, 3.84

Sources:

Natural Sources. β-Apocarotenal has been reported to be present in citrus fruits,[1,2] various vegetables,[1] grass, [1,3] liver,[1] and duodenal mucosa.[4]

Chemical Synthesis. β-Apocarotenal has been synthesized from "β-C19-aldehyde."[5] It has also been prepared by oxidation of β-carotene with potassium permanganate.[6,7]

Methods of Purification: β-Apocarotenal is purified by recrystallization from organic solvents (petroleum ether or petroleum ether–ethyl acetate).

Methods of Assaying for Purity:

Chromatography. β-Apocarotenal may be assayed for purity by chromatography on a column of partially deactivated alumina. It may also be chromatographed on thin-layer plates of silica gel G (Merck) or secondary magnesium phosphate. Cyclohexane–ethyl ether (8:2), carbon tetrachloride, or petroleum ether–ethyl ether (19:1) may be used as the developing agent.

Visible Spectrum. Cyclohexane: $E_{1\,cm}^{1\%}$ 2640 at 461 nm; 2165 at 488 nm.

Nuclear Magnetic Resonance. The nuclear magnetic resonance spectrum has been reported.[8]

Melting Point. 139 °C, violet plates from methanol.[9]

Probable Impurities: *cis*-Isomers of β-apocarotenal.

Conditions of Storage: Darkness (brown vial), inert atmosphere (sealed ampoule), and low temperature (0 °C).

References

1. A. Winterstein, A. Studer, and R. Rüegg, *Chem. Ber.*, **93**, 2951 (1960).
2. H. Thommen, *Naturwissenschaften*, **49**, 517 (1962).
3. H. Thommen and O. Wiss, *Z. Ernaehrungswiss. Suppl.*, **3**, 18 (1963).
4. A. Winterstein, *Angew. Chem.*, **72**, 902 (1960).
5. R. Rüegg, M. Montavon, G. Ryser, G. Saucy, U. Schwieter, and O. Isler, *Helv. Chim. Acta*, **42**, 854 (1959).
6. P. Karrer and U. Solmssen, *Helv. Chim. Acta*, **20**, 682 (1937).
7. P. Karrer, U. Solmssen, and W. Gugelmann, *Helv. Chim. Acta*, **20**, 1020 (1937).
8. U. Schwieter, G. Englert, N. Rigassi, and W. Vetter, *Pure Appl. Chem.*, **20**, 365 (1969).
9. P. Karrer and E. Jucker, *Carotenoids*, E. A. Braude (translator), Elsevier, New York (1950).

Carot-3

β-Apocarotenoic Acid Ethyl Ester
Ethyl 8'-Apo-β-caroten-8'-oate
[β-Apo-8'-carotenoic Acid (C_{30}) Ethyl Ester]

Formula: $C_{32}H_{44}O_2$
Formula Wt.: 460.71
Calc. %: C, 83.42; H, 9.62; O, 6.96

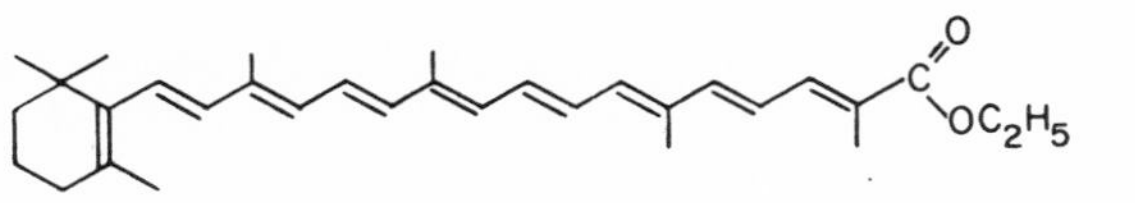

Sources:

Natural Sources. β-Apocarotenoic acid is a metabolic product of β-apocarotenal.[1,2] β-Apocarotenoic acid has been isolated from maize.[3]

Chemical Synthesis. Ethyl β-apocarotenoate is prepared in analogy to the methyl ester[4] from 15,15'-dehydro-10'-apo-β-carotenal (C_{27}) and (1-ethoxycarbonylethyl)triphenylphosphonium bromide.

Methods of Purification: Ethyl β-apocarotenoate is purified by crystallization from organic solvents (petroleum ether or petroleum ether–ethyl acetate).

Methods of Assaying for Purity:

Chromatography. Ethyl β-apocarotenoate is assayed for purity by chromatography on a column of partially deactivated alumina or on thin layers of secondary magnesium phosphate or alkaline silica gel G (Merck). Petroleum ether–ethyl ether (19:1) is used to develop the thin-layer chromatogram.

Visible Spectrum. Cyclohexane: $E_{1cm}^{1\%}$ values are 2550 at 449 nm and 2140 at 475 nm. The absorption spectra of β-apocarotenoic acid ethyl ester in hexane, ethanol, and petroleum ether have been published.

Melting Point. A range of 134–138 °C has been reported.

Probable Impurities: *cis*-Isomers of β-apocarotenoic acid ethyl ester.

Conditions of Storage: Darkness (brown vial), inert atmosphere (sealed ampoule), and low temperature (−20 °C).

References

1. O. Wiss and H. Thommen, *Carotine und Carotinoide*, K. Lang, ed., D. Steinkopff Verlag, Darmstadt (1963), p. 179.
2. H. Thommen, *Chimia*, 15, 433 (1961).
3. J. Baraud, F. Benitez, L. Genevois, and A. Maurice, *Compt. Rend.*, 260, 7045 (1965).
4. O. Isler, W. Guex, R. Rüegg, G. Ryser, G. Saucy, U. Schwieter, M. Walter, and A. Winterstein, *Helv. Chim. Acta*, 42, 864 (1959).

Carot-4

β-Apocarotenoic Acid Methyl Ester
Methyl 8'-Apo-β-caroten-8'-oate
[β-Apo-8'-carotenoic Acid (C_{30}) Methyl Ester]

Formula: $C_{31}H_{42}O_2$
Formula Wt.: 446.68
Calc. %: C, 83.36; H, 9.48; O, 7.16

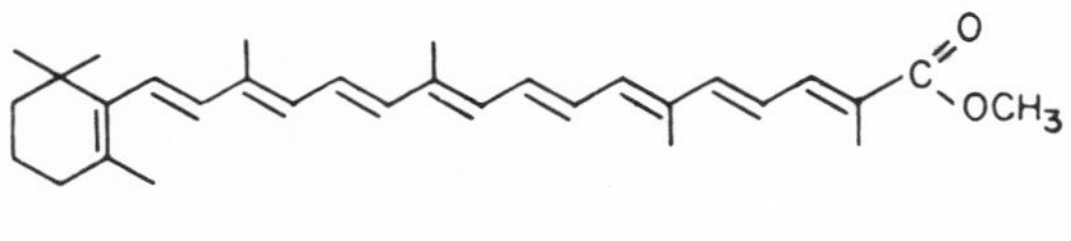

Sources:

Natural Sources. β-Apocarotenoic acid is a metabolic product of β-apocarotenal.[1,2]

Chemical Synthesis. The methyl ester of β-apo-8'-carotenoic acid (C_{30}) is prepared from 15,15'-dehydro-β-apo-10'-carotenal (C_{27}) and (1-methoxycarbonylethyl)triphenylphosphonium bromide.[3]

Methods of Purification: The methyl ester of β-apocarotenoic acid is purified by recrystallization from petroleum ether or petroleum ether–ethyl acetate.

Methods of Assaying for Purity:

Chromatography.[1–3] The methyl ester of β-apocarotenoic acid is assayed for purity by chromatography on a column of partially deactivated aluminum oxide or by chromatography on a thin layer of secondary magnesium phosphate or alkaline silica gel G (Merck). The thin-layer chromatograms are developed with a mixture of petroleum ether–ethyl ether (19:1).

Visible Spectrum. Petroleum ether: 446 and 471 nm. $E_{1cm}^{1\%}$ 2575 and 2160, respectively.[3]

Nuclear Magnetic Resonance. The nuclear magnetic resonance spectrum of the methyl ester has been reported.[4]

Melting Point.[3] The methyl ester of β-apocarotenoic acid melts at 136–137 °C.

Probable Impurities: Traces of *cis*-isomers of β-apocarotenoic acid methyl ester.

Conditions of Storage: Darkness (brown vial), inert atmosphere (sealed ampoule), and low temperature (−20 °C).

References

1. O. Wiss and H. Thommen, *Wiss. Veroeffentl. Deut. Ges. Ernaehrung*, 9, 188 (1963).
2. H. Thommen, *Chimia*, 15, 433 (1961).
3. O. Isler, W. Guex, R. Rüegg, G. Ryser, G. Saucy, U. Schwieter, M. Walter, and A. Winterstein, *Helv. Chim. Acta*, 42, 864 (1959).
4. U. Schwieter, G. Englert, N. Rigassi, and W. Vetter, *Pure Appl. Chem.*, 20, 365 (1969).

Carot-5

Astacin
3,3'-Dihydroxy-2,3,2',3'-tetrahydro-β,β-carotene-4,4'-dione ⇌ β,β-Carotene-3,4,3',4'-tetrone
(Astacene)

Formula: $C_{40}H_{48}O_4$
Formula Wt.: 592.83
Calc. %: C, 81.04; H, 8.16; O, 10.80

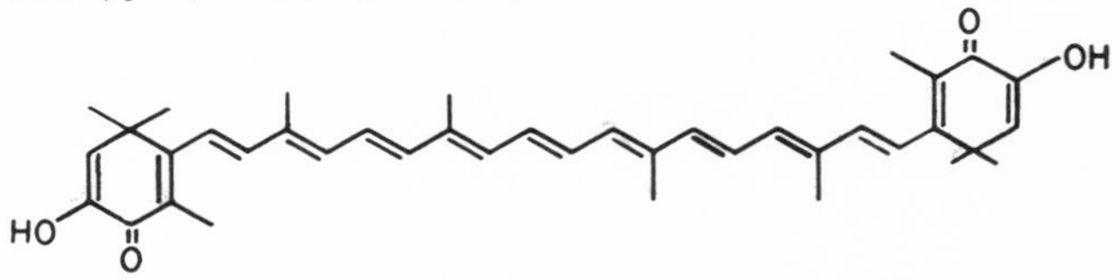

Astacin has been characterized as 3,3',4,4'-tetraketo-β-carotene. The enol form of this compound (shown above) preponderates.[1]

Sources:

Natural Sources. Astacin is not a commonly occurring natural pigment. However, it may be obtained on treatment of astaxanthin or astaxanthin esters with alkali.[2] Astaxanthin may be isolated from some algae and from a wide variety of animals.[3] One of the best sources of this compound is lobster shells.[4]

Chemical Synthesis. The synthesis of astacin from canthaxanthin has been reported.[5]

Isolation Procedures:[3,4] Shells of freshly killed lobsters are covered with 2 M HCl and left to stand therein until they turn red. They are then washed with water and the hypodermis is separated. Pigment is extracted with acetone at room temperature and then transferred to petroleum ether, with dilution of the extract with water. The petroleum ether solution is washed with water and 90% methanol, diluted with 2 M NaOH and sufficient ethanol to form a homogeneous solution, and kept in the dark at room temperature for 5 h. Sufficient water is then added to produce two layers. The ethanolic layer is separated, and covered with petroleum ether, and the astacene is precipitated by careful acidification with acetic acid. The pigment is washed with hot water, dissolved in a small volume of highly purified pyridine, and crystallized by the addition of a small proportion of water.

Methods of Purification:

Chromatography. Astacin may be chromatographed on such weak neutral adsorbents as alumina–fibrous clay[4] (1:4) or sucrose.[5] Other adsorbents, such as alumina, adsorb this compound too tightly, whereas such compounds as calcium carbonate do not adsorb it.

Solvent Partition. Astacin may be partially purified by partition between petroleum ether and alkaline methanol.[1,6]

Crystallization. Astacin crystallizes in needles from pyridine–water.[4]

Derivatives. Astacin forms a dioxime,[7] a bis-phenazine derivative,[7] and such esters as the diacetate[1,5,6] and dipalmitate.[2]

Methods of Assaying for Purity:

Chromatography. The purity of astacin may be determined by chromatography on kieselguhr paper.[1]

Visible Spectrum. Carbon disulfide,[2] broad maximum at 510 nm. Pyridine,[4] broad maximum at 500 nm. The $E_{1\,cm}^{1\%}$ (max), at 498 nm in pyridine is 1×10^{5}.[5]

Infrared Spectrum. The spectra for astacin and astacin diacetate in a potassium bromide pellet have been reported.[1]

Nuclear Magnetic Resonance Spectrum. The nuclear magnetic resonance spectrum of astacin has been reported.[1,5,8,9]

Mass Spectrum. The mass spectra of astacin and some of its derivatives have been determined.[10]

Melting Point. Several different melting points have been reported for this compound. These are 240–243 °C,[4] 241 °C,[11] 228 °C,[2] and 228–230 °C.[5]

Optical Rotation. Astacin is optically inactive.

Probable Impurity: Astaxanthin.

Conditions of Storage: Darkness (brown vial), inert atmosphere (sealed ampoule), and low temperature (−20 °C).

References

1. A. J. Aasen and S. Liaaen-Jensen, *Acta Chem. Scand.*, **20**, 1970 (1966).
2. R. Kuhn, J. Stene, and N. A. Sörensen, *Chem. Ber.*, **72**, 1688 (1939).
3. P. Karrer and E. Jucker, *Carotenoids*, E. A. Braude (translator), Elsevier, New York (1950).
4. R. Kuhn and E. Lederer, *Chem. Ber.*, **66**, 488 (1933).
5. J. B. Davis and B. C. L. Weedon, *Proc. Chem. Soc.*, 182 (1960).
6. R. Kuhn, E. Lederer, and A. Deutsch, *Z. Physiol. Chem.*, **220**, 229 (1933).
7. P. Karrer and L. Loewe, *Helv. Chim. Acta*, **17**, 745 (1934).
8. B. C. L. Weedon, in *Chemistry and Biochemistry of Plant Pigments*, T. W. Goodwin, ed., Academic Press, New York (1965).
9. B. C. L. Weedon, *Fortschr. Chem. Org. Naturstoffe*, **27**, 81 (1969).
10. J. Baldas, Q. N. Porter, A. P. Leftwick, R. Holzel, B. C. L. Weedon, and J. Szabolcs, *J. Chem. Soc.* (*D*), 415 (1969).
11. P. Karrer and F. Benz, *Helv. Chim. Acta*, **17**, 412 (1934).

Carot-6
Bixin
Methyl Hydrogen 9′-*cis*-6,6′-Diapocarotene-6,6′-dioate

Formula: $C_{25}H_{30}O_4$
Structure: See Kuhn and Winterstein,[1] Barber *et al.*[2,3]
Formula Wt.: 394.52
Calc. %: C, 76.11; H, 7.66; O, 16.23

Sources:

Natural Sources. Bixin has been isolated from the seeds of *Bixa orellana*[4] and all-*trans*-bixin has been found in the roots of *Aristolochia cymbifera*.[5] Bixin is extracted from commercial orlean, paté de rocou,[6] or bixa seeds[7] with organic solvents.

Chemical Synthesis. Methyl-*cis*-4-(natural) bixin[8,9] and all-*trans*-methyl bixin have been synthesized.[10–13]

Methods of Purification: Purification is achieved by recrystallization.

Methods of Assaying for Purity:

Chromatography. Methyl bixin and its geometrical isomers have been separated on chromatographic columns.[14] However, the column-chromatographic separation of the corresponding bixin isomers has not been reported. Methyl bixin has also been assayed for purity by thin-layer chromatography.[15]

Derivatives. Methyl bixin, norbixin, their dihydro and perhydro derivatives, and various other esters of bixin have been prepared.[4]

Visible Spectrum. Bixin:[4] carbon disulfide, 459, 489, and 523.5 nm; chloroform, 439, 469.5, and 503 nm; all-*trans*-bixin:[4] carbon disulfide, 457, 491, and 526.5 nm; chloroform, 443, 475, and 509.5 nm. Methyl bixin: benzene, 444, 471, and 502 nm. Absorption spectra for bixin in ethanol and methyl bixin in hexane have been reported.[4]

Infrared Spectrum. The infrared spectrum of methyl bixin has been reported.[16]

Mass Spectrum. The mass spectrum of bixindial has been reported.[17]

Nuclear Magnetic Resonance. The nuclear magnetic resonance spectra for methyl natural bixin,[3] all-*trans*-methyl bixin,[2,3] and *cis*-apo-1-norbixinal methyl ester[2,3] have been reported.

Melting Point. Bixin melts at 198 °C, all-*trans*-bixin[18] at 216–217 °C, and methyl bixin[4] at 163 °C.

Probable Impurity: all-*trans*-Bixin.

Conditions of Storage: The conditions of storage that are used for other carotenoids should be used for bixin. However, the compound is much more stable than many other carotenoids to air, heat, and light.[19]

References

1. R. Kuhn and A. Winterstein, *Chem. Ber.*, **65**, 646 (1932).
2. M. S. Barber, L. M. Jackman, and B. C. L. Weedon, *Proc. Chem. Soc.*, 23 (1960).
3. M. S. Barber, A. Hardisson, L. M. Jackman, and B. C. L. Weedon, *J. Chem. Soc.*, 1625 (1961).
4. P. Karrer and E. Jucker, *Carotenoids*, E. A. Braude (translator), Elsevier, New York (1950).
5. A. Green, C. H. Eugster, and P. Karrer, *Helv. Chim. Acta*, **37**, 1717 (1954).
6. L. Zechmeister, *Carotinoide*, Julius Springer, Berlin (1934).
7. R. Kuhn and L. Ehmann, *Helv. Chim. Acta*, **12**, 904 (1929).
8. G. Pattenden, J. E. Way, and B. C. L. Weedon, *J. Chem. Soc.* (*C*), 235 (1970).
9. B. C. L. Weedon, *Pure Appl. Chem.*, **20**, 531 (1969).
10. R. Ahmad and B. C. L. Weedon, *J. Chem. Soc.*, 3286 (1953).

11. H. H. Inhoffen and G. Raspé, *Ann. Chem.*, **592**, 214 (1955).
12. O. Isler, G. Gutman, M. Montavon, R. Rüegg, G. Ryser and P. Zeller, *Helv. Chim. Acta*, **40**, 1242 (1957).
13. E. Buchta and F. Andrée, *Chem. Ber.*, **92**, 3111 (1959).
14. L. Zechmeister and R. B. Escue, *J. Am. Chem. Soc.*, **66**, 322 (1944).
15. B. H. Davies, in *Chemistry and Biochemistry of Plant Pigments*, T. W. Goodwin, ed., Academic Press, New York (1965), p. 489.
16. B. C. L. Weedon, in *Chemistry and Biochemistry of Plant Pigments*, T. W. Goodwin, ed., Academic Press, New York (1965), p. 75.
17. C. R. Enzell, G. W. Francis, and S. Liaaen-Jensen, *Acta Chem. Scand.*, **23**, 727 (1969).
18. L. Zechmeister, *Fortschr. Chem. Org. Naturstoffe*, **18**, 223 (1960).
19. K. Lunde and L. Zechmeister, *J. Am. Chem. Soc.*, **77**, 1647 (1955).

References

1. H. Thommen and H. Wackernagel, *Naturwissenschaften*, **51**, 87 (1964).
2. H. Thommen and U. Gloor, *Naturwissenschaften*, **52**, 161 (1965).
3. H. Thommen and H. Wackernagel, *Biochim. Biophys. Acta*, **69**, 387 (1963).
4. D. L. Fox, *Comp. Biochem. Physiol.*, **6**, 1 (1962).
5. O. Voelker, *Naturwissenschaften*, **48**, 581 (1961).
6. D. L. Fox, *Comp. Biochem. Physiol.*, **5**, 31 (1962).
7. D. L. Fox, *Comp. Biochem. Physiol.*, **6**, 305 (1962).
8. F. Haxo, *Botan. Gaz.*, **112**, 228 (1950).
9. F. Leuenberger and H. Thommen, *J. Insect Physiol.*, **16**, 1855 (1970).
10. F. -C. Czygan, *Experientia*, **20**, 573 (1964).
11. F. J. Petracek and L. Zechmeister, *J. Am. Chem. Soc.*, **78**, 1427 (1956).
12. G. Gansser and L. Zechmeister, *Helv. Chim. Acta*, **40**, 1757 (1957).
13. R. Entschel and P. Karrer, *Helv. Chim. Acta*, **41**, 402 (1958).
14. O. Isler and P. Schudel, *Wiss. Veroeffentl. Deut. Ges. Ernaehrung*, **9**, 76 (1963).
15. C. K. Warren and B. C. L. Weedon, *J. Chem. Soc.*, 3972 (1958).
16. M. Akhtar and B. C. L. Weedon, *J. Chem. Soc.*, 4058 (1959).
17. H. R. Bolliger, in *Thin-Layer Chromatography*, E. Stahl, ed., Academic Press, New York (1964), p. 210.
18. B. C. L. Weedon, *Fortschr. Chem. Org. Naturstoffe*, **27**, 81 (1969).
19. U. Schwieter, G. Englert, N. Rigassi, and W. Vetter, *Pure Appl. Chem.*, **20**, 365 (1969).
20. C. R. Enzell, G. W. Francis, and S. Liaaen-Jensen, *Acta Chem. Scand.*, **23**, 727 (1969).
21. H. Budzikiewicz, H. Brzezinka, and B. Johannes, *Monatsh. Chem.*, **101**, 579 (1970).
22. J. C. J. Bart and C. H. MacGillavry, *Acta Crystallogr.*, Sect. B, **24**, 1587 (1968).

Carot-7

Canthaxanthin

β,β-Carotene-4,4'-dione

(β-Carotene-4,4'-dione)

Formula: $C_{40}H_{52}O_2$
Formula Wt.: 564.86
Calc. %: C, 85.06; H, 9.28; O, 5.66

Sources:

Natural Sources. Canthaxanthin has been reported to be present in crustaceae,[1] fishes (trout),[2] the organs of flamingos,[3,4] the feathers of several species of birds,[5-7] mushrooms,[8] insects,[9] and algae.[10]

Chemical Synthesis. Canthaxanthin has been synthesized from β-carotene,[11-13] 15,15'-didehydro-β-carotene,[14] crocetindialdehyde,[15] and dehydrocrocetindialdehyde.[16]

Methods of Purification: Canthaxanthin is separated from other pigments by chromatography on columns of magnesia or partially deactivated alumina. Further purification is achieved by recrystallization (dichloromethane or other solvents).

Methods of Assaying for Purity:

Chromatography. Purity of the pigment may be determined by chromatography on columns of deactivated alumina or magnesia, or by chromatography on a thin layer of silica gel G (Merck).[17] Dichloromethane–ethyl ether (9:1) is used to develop the chromatogram.

Visible Spectrum. Cyclohexane: The principal light-absorption maximum is found at 470 nm. An $E_{1\text{ cm}}^{1\%}$ value of 2200 is found at this wavelength.

Nuclear Magnetic Resonance. The nuclear magnetic resonance spectrum of canthaxanthin has been reported.[18,19]

Mass Spectrum. The mass spectrum of canthaxanthin has been reported.[20,21]

x-Ray Crystallography. The configuration of canthaxanthin has been determined.[22]

Melting Point. A melting point of 211–212 °C has been reported.[16]

Probable Impurities: Echinenone and *cis*-isomers.

Conditions of Storage: Darkness (brown vial), inert atmosphere (sealed ampoule), and low temperature (-20 °C).

Carot-8

Capsanthin

($3R,3'S,5'R$)3,3'-Dihydroxy-β,κ-caroten-6'-one

Formula: $C_{40}H_{56}O_3$
Formula Wt.: 584.89
Calc. %: C, 82.14; H, 9.65; O, 8.21

Sources:

Natural Sources. Fruits of red peppers (*Capsicum annum*) are a good source of capsanthin.[1-3]

Chemical Synthesis. The structure of capsanthin has been established through degradation,[4-6] and its synthesis from β-citraurin has been reported.[7] The absolute configuration of capsanthin has been established as 3R, 3'S, 5'R.[8,9,10]

Methods of Purification: The extraction, chromatography on calcium carbonate, and crystallization of capsanthin have been reported.[1,11]

Methods of Assaying for Purity:

Chromatography. Capsanthin may be assayed for purity by chromatography on columns of calcium carbonate or zinc carbonate. Carbon disulfide or benzene-ethyl ether is used to develop the column.[1,12] Purity may also be determined by chromatography on a thin layer of silica gel G (R_f 0.16 in a system of 20% ethyl acetate in dichloromethane),[13] or kielselguhr impregnated with vegetable oil (R_f 0.74 in a system of 20:4:3 methanol–acetone–water, saturated with vegetable oil).[14]

Derivatives. The following derivatives of capsanthin have been prepared: capsanthin diiodide; capsanthol; capsanthin diacetate, dipropionate, dibutyrate, divalerate, dicaproate, dicaprate, dimyristate, dipalmitate, distearate, and dibenzoate; capsanthone;

anhydrocapsanthone; capsanthylal; capsanthylal monoxime; and capsaldehyde.[15,16]

Solvent Partition. The solvent-partition ratio between petroleum ether and 90% methanol is 0:100.[15]

Visible Spectrum. Carbon disulfide: 503 and 542 nm. Petroleum ether: 475 and 505 nm. Benzene: 486 and 520 nm, $E_{1\,cm}^{1\%}$ 1790 at 486 nm.[16] Complete spectrum.[15–17] Iodine-isomerized capsanthin shows a "*cis* peak" at 355 nm.

Infrared Spectrum. The infared spectrum of capsanthin has been reported.[18]

Mass Spectrum. The mass spectrum of capsanthin has been reported.[19]

Nuclear Magnetic Resonance Spectrum. The nuclear magnetic resonance spectrum of capsanthin has been reported.[20]

Optical Rotatory Dispersion. The optical rotatory dispersion curves of capsanthin and related compounds have been reported.[10]

Melting Point. The melting point of capsanthin has been reported to be 176 °C (uncorrected)[21] and 175–176 °C (corrected).[11]

Optical Rotation. An $[\alpha]_{Cd}$ of +36° (chloroform) has been reported.[16]

Probable Impurities: *cis*-Isomers and traces of zeaxanthin and capsorubin.[16]

Conditions of Storage: Darkness (brown vial), inert atmosphere (sealed ampoule), and low temperature (−20° C).

References

1. L. Zechmeister and L. Cholnoky, *Ann. Chem.*, **454**, 54 (1927).
2. L. Zechmeister and L. Cholnoky, *Ann. Chem.*, **487**, 197 (1931).
3. L. Zechmeister and L. Cholnoky, *Ann. Chem.*, **489**, 1 (1931).
4. R. Entschel and P. Karrer, *Helv. Chim. Acta*, **43**, 89 (1960).
5. M. S. Barber, L. M. Jackman, C. K. Warren, and B. C. L. Weedon, *Proc. Chem. Soc.*, 19 (1960).
6. M. S. Barber, L. M. Jackman, C. K. Warren, and B. C. L. Weedon, *J. Chem. Soc.*, 4019 (1961).
7. B. C. L. Weedon, personal communication.
8. R. D. G. Cooper, L. M. Jackman and B. C. L. Weedon, *Proc. Chem. Soc.*, 215 (1962).
9. J. W. Faigle, H. Müller, W. von Philipsborn, and P. Karrer, *Helv. Chim. Acta*, **47**, 741 (1964).
10. L. Bartlett, W. Klyne, W. P. Mose, P. M. Scopes, G. Galasko, A. K. Mallams, B. C. L. Weedon, J. Szabolcs, and G. Tóth, *J. Chem. Soc. (C)*, 2527 (1969).
11. L. Zechmeister and L. Cholnoky, *Ann. Chem.*, **509**, 269 (1934).
12. P. Karrer and E. Jucker, *Helv. Chim. Acta*, **28**, 1143 (1945).
13. B. H. Davies, in *Chemistry and Biochemistry of Plant Pigments*, T. W. Goodwin, ed., Academic Press, New York (1965), p. 489.
14. K. Randerath, *Thin-Layer Chromatography*, Academic Press, New York (1963).
15. P. Karrer and E. Jucker, *Carotenoids*, E. A. Braude (translator), Elsevier, New York (1950).
16. L. Cholnoky, D. Szabo, and J. Szabolcs, *Ann. Chem.*, **606**, 194 (1957).
17. L. Zechmeister, *Fortschr. Chem. Org. Naturstoffe*, **18**, 223 (1960).
18. C. K. Warren and B. C. L. Weedon, *J. Chem. Soc.*, 3972 (1958).
19. H. Budzikiewicz, H. Brzezinka, and B. Johannes, *Monatsh. Chem.*, **101**, 579 (1970).
20. B. C. L. Weedon, *Fortschr. Chem. Org. Naturstoffe*, **27**, 81 (1969).
21. P. Karrer and A. Oswald, *Helv. Chim. Acta*, **18**, 1303 (1935).

Carot-9
Capsorubin
(3S,5R,3′S,5′R)3,3′-Dihydroxy-κ,κ-carotene-6,6′-dione

Formula: $C_{40}H_{56}O_4$
Formula Wt.: 600.89
Calc. %: C, 79.96; H, 9.39; O, 10.65

Sources:

Natural Sources. Capsorubin occurs as an ester in ripe paprika (*Capsicum annum*) fruits.[1,2] The absolute configuration of capsorubin has been established as 3S, 5R, 3′S, 5′R.[3,4,5,6]

Chemical Synthesis. The chemical synthesis of capsorubin has been reported.[3]

Isolation Procedures:[1,7] Paprika pods are pretreated with ethanol, and then extracted with petroleum ether. The combined extracts are concentrated in vacuum, and the residue is chromatographed on calcium carbonate. The capsorubin ester is saponified with methanolic potassium hydroxide. The unsaponifiable fraction, in carbon disulfide solution, is rechromatographed on calcium carbonate. The chromatographically purified pigment is then crystallized from benzene–petroleum ether.

Methods of Purification:

Chromatography. Capsorubin may be chromatographed on a column of calcium carbonate[7] or magnesia.[8]

Solvent Partition. A partial purification of capsorubin may be achieved by solvent partition.[7,8]

Crystallization. Capsorubin crystallizes as violet needles from benzene–petroleum ether and as plates from carbon disulfide.[7]

Derivatives. A number of esters of capsorubin have been prepared. These are the diacetate,[7,9] dipropionate, dibutyrate, divalerate, dicapronate, dicaprinate, dimyristate, dipalmitate, and distearate.[9] The reduction product, capsorubol,[8,10,11] and the oxidation product, capsorubone,[12–14] have also been prepared.

Methods of Assaying for Purity:

Chromatography. The homogeneity of capsorubin preparations can be ascertained by microchromatographic methods.[15,16] Chromatography on circular paper impregnated with a kieselguhr filler is recommended.[16]

Visible Spectrum. Hexane:[8] 443, 468, and 503 nm. Petroleum ether:[7] 444, 474, and 507 nm. Benzene: 455, 486, and 520 nm;[7,17] 460, 487, and 522 nm;[8] 463, 487, and 522 nm.[14] Carbon disulfide:[7] 468, 502, and 541.5 nm. Ethanol:[8] 482 nm. Spectral curves of capsorubin in hexane, benzene, and ethanol have been reported.[8] Molar absorption coefficients in benzene are the following:[14] 463 nm, 89.2×10^3; 487 nm, 129.8×10^3; and 522 nm, 119.1×10^3.

Infrared Spectrum. The infrared spectrum of capsorubin in chloroform has been reported.[11]

Nuclear Magnetic Resonance Spectrum. The nuclear magnetic resonance spectrum of capsorubin has been reported.[13,18]

Mass Spectrum. The mass spectrum of capsorubin has been reported.[19,20]

Melting Point.[9] Capsorubin melts at 218 °C.

Optical Rotation. The optical rotatory dispersion curve of capsorubin has been reported.[6]

Probable Impurities: Zeaxanthin and capsanthin.

Conditions of Storage: Darkness (brown vial), inert atmosphere (sealed ampoule), and low temperature (−20 °C).

References

1. L. Zechmeister and L. v. Cholnoky, *Ann. Chem.*, **509**, 269 (1934)
2. L. v. Cholnoky, K. Györgyfy, E. Nagy, and M. Pánczél, *Acta Chim. Acad. Sci. Hung.*, **6**, 143 (1955).
3. R. D. G. Cooper, L. M. Jackman, and B. C. L. Weedon, *Proc. Chem. Soc.*, 215 (1962).
4. H. Faigle and P. Karrer, *Helv. Chim. Acta*, **44**, 1257 (1961).
5. J. W. Faigle, H. Muller, W. von Philipsborn, and P. Karrer, *Helv. Chim. Acta*, **47**, 741 (1964).
6. L. Bartlett, W. Klyne, W. P. Mose, P. M. Scopes, G. Galasko, A. K. Mallams, B. C. L. Weedon, J. Szabolcs, and G. Tóth, *J. Chem. Soc. (C)*, 2527 (1969).
7. L. Zechmeister and L. v. Cholnoky, *Ann. Chem.*, **516**, 30 (1935).
8. A. L. Curl, *J. Agr. Food Chem.*, **10**, 504 (1962).
9. L. v. Cholnoky, D. Szabó, and J. Szabolcs, *Ann. Chem.*, **606**, 194 (1957).
10. L. v. Cholnoky and J. Szabolcs, *Naturwissenschaften*, **44**, 513 (1957).
11. C. K. Warren and B. C. L. Weedon, *J. Chem. Soc.*, 3972 (1958).
12. R. Entschel and P. Karrer, *Helv. Chim. Acta*, **43**, 89 (1960).
13. M. S. Barber, L. M. Jackman, C. K. Warren, and B. C. L. Weedon, *J. Chem. Soc.*, 4019 (1961).
14. L. v. Cholnoky and J. Szabolcs, *Acta Chim. Acad. Sci. Hung.*, **22**, 117 (1960).
15. H. R. Bolliger, A. König, and U. Schwieter, *Chimia*, **18**, 136 (1964).
16. A. Jensen and S. Liaaen-Jensen, *Acta Chem. Scand.*, **13**, 1863 (1959).
17. R. Ahmad and B. C. L. Weedon, *J. Chem. Soc.*, 3286 (1953).
18. B. C. L. Weedon, *Fortschr. Chem. Org. Naturstoffe*, **27**, 81 (1969).
19. J. Baldas, Q. N. Porter, A. P. Leftwick, R. Holzel, B. C. L. Weedon, and J. Szabolcs, *J. Chem. Soc, (D)*, 415 (1969).
20. H. Budzikiewicz, H. Brzezinka, and B. Johannes, *Monatsh. Chem.*, **101**, 579 (1970).

Carot-10
α-Carotene
(6′R)β,ε-Carotene

Formula: $C_{40}H_{56}$
Formula Wt.: 536.89
Calc. %: C, 89.49; H, 10.51

Sources:

Natural Sources. α-Carotene is present in much smaller proportions than β-carotene in most plant species. However, the best sources of this carotene are the same as the sources of β-carotene: carrots,[1] palm oil,[2,3] and green leaves of various species.[4] In some algae, α-carotene is the major carotene.[5]

Chemical Synthesis. The chemical synthesis of α-carotene has been reported.[6-9] The absolute configuration of α-carotene has been established as 6′R.[10]

Isolation Procedures: The extraction of α-carotene from plant sources, and its purification by distribution between immiscible solvents, chromatography, and crystallization have been reported.[1-4,11]

Methods of Purification:

Chromatography. α-Carotene has been purified by chromatography on columns of calcium hydroxide, alumina, or magnesia.[12]

Crystallization. α-Carotene may be crystallized from the same solvent pairs used to crystallize β-carotene and lycopene.[12]

Methods of Assaying for Purity:

Chromatography. The purity of α-carotene may be determined by chromatography on columns,[12] alumina paper,[13] or thin-layer plates of magnesia.[14]

Solvent Partition.[15] The partition ratio between hexane and 95% methanol is 100:0.

Visible Spectrum. Hexane (b.p. 65–67 °C): 422, 446, and 474 nm. $E_{1\,cm}^{1\%}$ 2725 (446 nm) and 2490 (474 nm). Spectral curve.[12] Light petroleum: 422, 444, and 473 nm. $E_{1\,cm}^{1\%}$ 2800 (444 nm).

Mass Spectrum. The mass spectrum of α-carotene has been reported.[16-19]

Nuclear Magnetic Resonance Spectrum. The nuclear magnetic resonance spectrum of α-carotene has been reported.[16,17,20]

Optical Rotatory Dispersion. The optical rotatory dispersion curve of α-carotene has been reported.[21]

Melting Point.[1,4] α-Carotene melts at 184–188 °C.

Optical Rotation. $[\alpha]_{643}^{18}$ +385° has been reported.[1]

Probable Impurities: Oxidation products, *cis*-isomers, and other carotenes (β-carotene and phytofluene).

Conditions of Storage: Darkness (brown vial), inert atmosphere (sealed ampoule), and low temperature (−20 °C).

References

1. P. Karrer and O. Walker, *Helv. Chim. Acta*, **16**, 641 (1933).
2. R. Kuhn and E. Lederer, *Z. Physiol. Chem.*, **200**, 246 (1931).
3. R. Kuhn and E. Lederer, *Chem. Ber.*, **64**, 1349 (1931).
4. H. H. Strain, *J. Biol. Chem.*, **105**, 523 (1934).
5. M. B. Allen, L. Fries, T. W. Goodwin, and D. M. Thomas, *J. Gen. Microbiol.* **34**, 259 (1964).
6. P. Karrer and C. H. Eugster, *Helv. Chim. Acta*, **33**, 1952 (1950).
7. C. H. Eugster and P. Karrer, *Helv. Chim. Acta*, **38**, 610 (1955).
8. H. H. Inhoffen, U. Schwieter, and G. Raspé, *Ann. Chem.*, **588**, 117 (1954).
9. C. Tscharner, C. H. Eugster, and P. Karrer, *Helv. Chim. Acta*, **40**, 1676 (1957).
10. C. H. Eugster, R. Buchecker, C. Tscharner, G. Uhde, and G. Ohloff, *Helv. Chim. Acta*, **52**, 1729 (1969).
11. R. Kuhn and H. Brockmann, *Z. Physiol. Chem.*, **200**, 255 (1931).
12. F. P. Zscheile, J. W. White, Jr., B. W. Beadle, and J. R. Roach, *Plant Physiol.*, **17**, 331 (1942).
13. A. Jensen, *Acta Chem. Scand.*, **14**, 2051 (1960).
14. H. R. Bolliger, A. König, and U. Schwieter, *Chimia*, **18**, 136 (1964).
15. F. J. Petracek and L. Zechmeister, *Anal. Chem.*, **28**, 1484 (1956).
16. U. Schwieter, H. R. Bolliger, L. H. Chopard-dit-Jean, G. Englert, M. Kofler, A. König, C. von Planta, R. Rüegg, W. Vetter, and O. Isler, *Chimia*, **19**, 294 (1965).
17. U. Schwieter, G. Englert, N. Rigassi, and W. Vetter, *Pure Appl. Chem.*, **20**, 365 (1969).
18. H. Budzikiewicz, H. Brzezinka, and B. Johannes, *Monatsh. Chem.*, **101**, 579, (1970).
19. C. R. Enzell, G. W. Francis, and S. Liaaen-Jensen, *Acta Chem. Scand.*, **23**, 727 (1969).
20. B. C. L. Weedon, *Fortschr. Chem. Org. Naturstoffe*, **27**, 81 (1969).
21. L. Bartlett, W. Klyne, W. P. Mose, P. M Scopes, G. Galasko, A. K. Mallams, B. C. L. Weedon, G. Szabolcs, and G. Tóth, *J. Chem. Soc. (C)*, 2527 (1969).

Carot-11
β-Carotene
β,β-Carotene

Formula: $C_{40}H_{56}$
Formula Wt.: 536.89
Calc. %: C, 89.49; H, 10.51

Sources:

Natural Sources. Excellent sources are carrots,[1] palm oil,[2] and green leaves of many plant species.[3]

Chemical Synthesis. Several chemical syntheses of β-carotene have been reported.[4-7]

Isolation and Methods of Purification: The extraction of β-carotene from plant material and the separation and purification of this compound by solvent partition, chromatography, and crystallization have been reported.[1-3,8,9] The standardization of some of the steps in these procedures has been effected through cooperative studies.[10]

Methods of Assaying for Purity:

Chromatography. Purity of the compound may be determined by column chromatography (usually on magnesia),[10] chromatography on kieselguhr paper,[11] or chromatography on a thin layer of magnesia.[12]

Solvent Partition.[13] The partition ratio between hexane and 95% methanol is 100:0.

Visible Spectrum.[6,9,14] Hexane (b.p. 65–67 °C): 450 and 478 nm. $E_{1\,cm}^{1\%}$ 2590 and 2280. Cyclohexane: 456 and 484 nm. $E_{1\,cm}^{1\%}$ 2500 and 2150.

Infrared Spectrum. The infrared absorption spectrum has been reported.[6]

Nuclear Magnetic Resonance Spectrum. The nuclear magnetic resonance spectrum has been reported.[15]

x-Ray Crystallography. The configuration of β-carotene has been determined.[16]

Mass Spectrum. The mass spectrum of β-carotene has been reported.[17-20]

Melting Point. β-Carotene melts at 178–180 °C.[3,5]

Probable Impurities: Oxidation products and *cis*-isomers; small proportions of related carotenes (α- and ζ-carotenes) may also be present if chromatographic purification has not been properly performed.

Conditions of Storage: Darkness (brown vial), inert atmosphere (sealed ampoule), low temperature (−20 °C).

References

1. P. Karrer and O. Walker, *Helv. Chim. Acta*, **16**, 641 (1933).
2. R. Kuhn and E. Lederer, *Z. Physiol. Chem.*, **200**, 246 (1931).
3. H. H. Strain, *J. Biol. Chem.*, **105**, 523 (1934).
4. H. H. Inhoffen, F. Bohlman, K. Bartram, G. Rummert and H. Pommer, *Ann. Chem.*, **570**, 54 (1950).
5. O. Isler, H. Lindlar, M. Montavon, R. Rüegg, and P. Zeller, *Helv. Chim. Acta*, **39**, 249 (1956).
6. O. Isler, L. H. Chopard-dit-Jean, M. Montavon, R. Rüegg, and P. Zeller, *Helv. Chim. Acta*, **40**, 1256 (1957).
7. O. Isler, *Carotenoids*, Birkhäuser, Basel (1971).
8. B. W. Beadle and F. P. Zscheile, *J. Biol. Chem.*, **144**, 21 (1942).
9. F. P. Zscheile, J. W. White, Jr., B. W. Beadle, and J. R. Roach, *Plant Physiol.*, **17**, 331 (1942).
10. F. T. Jones and E. M. Bickoff, *J. Assoc. Offic. Agr. Chem.*, **31**, 776 (1948).
11. A. Jensen and S. Liaaen-Jensen, *Acta Chem. Scand.*, **13**, 1863 (1959).
12. H. R. Bolliger, A. König, and U. Schwieter, *Chimia*, **18**, 136 (1964).
13. F. J. Petracek and L. Zechmeister, *Anal. Chem.*, **28**, 1484 (1956).
14. F. Stitt, E. M. Bickoff, G. F. Bailey, C. R. Thompson, and S. Friedlander, *J. Assoc. Offic. Agr. Chem.*, **34**, 460 (1951).
15. M. S. Barber, J. B. Davis, L. M. Jackman, and B. C. L. Weedon, *J. Chem. Soc.* 2870 (1960).
16. C. Sterling, *Acta Crystallogr.* **17**, 1224 (1964).
17. U. Schwieter, H. R. Bolliger, L. H. Chopard-dit-Jean, G. Englert, M. Kofler, A. König, C. von Planta, R. Rüegg, W. Vetter, and O. Isler, *Chimia*, **19**, 294 (1965).
18. U. Schwieter, G. Englert, N. Rigassi, and W. Vetter, *Pure Appl. Chem.*, **20**, 365 (1969).
19. H. Budzikiewicz, H. Brzezinka, and B. Johannes, *Monatsh. Chem.*, **101**, 579 (1970).
20. C. R. Enzell, G. W. Francis, and S. Liaaen-Jensen, *Acta Chem. Scand.*, **23**, 727 (1969).

Carot-12
γ-Carotene
β,ψ-Carotene

Formula: $C_{40}H_{56}$
Formula Wt.: 536.89
Calc. %: C, 89.49; H, 10.51

Sources:

Natural Sources. A strain of *Penicillium sclerotiorum* is one of the best sources of this pigment.[1] This compound is also present in small proportions in many plant materials, particularly fruits, that contain β-carotene.[2]

Chemical Synthesis. The chemical synthesis of γ-carotene has been reported by Garbers *et al.*[3] and Rüegg *et al.*[4]

Isolation Procedures: γ-Carotene is isolated from plant materials by the methods of solvent extraction, saponification, phasic separation, and chromatography commonly used for other carotenes.[2]

Methods of Purification:

Solvent Partition. γ-Carotene is epiphasic on partition between 90% aqueous methanol and petroleum ether.[2]

Chromatography. Purification of γ-carotene may be achieved by chromatography on a column of aluminum oxide[5,6] or calcium hydroxide.[5]

Crystallization. Benzene–methanol (2:1) has been used to crystallize γ-carotene.[5]

Methods of Assaying for Purity:

Chromatography. The purity of γ-carotene may be determined by column (aluminum oxide) or thin-layer (magnesium oxide) chromatography.[7]

Visible Spectrum.[2,4,7] Petroleum ether: 437, 462, and 494 nm. $E_{1\,cm}^{1\%}$ 2055, 3100, and 2720, respectively. Benzene: 448, 477, and 510 nm.

Infrared Spectrum. The infrared spectrum of γ-carotene has been reported.[7]

Nuclear Magnetic Resonance Spectrum. The nuclear magnetic resonance spectrum of γ-carotene has been reported.[7-9]

Mass Spectrum. The mass spectrum of γ-carotene has been reported.[7,8]

Melting Point. A melting point of 150 °C has been reported[10] for natural γ-carotene. Synthetic *trans*-γ-carotene melts[4,7] at 152–154 °C.

Probable Impurities: Oxidation products, *cis*-isomers, and lycopene (when isolated from natural sources).

Conditions of Storage: Darkness (brown vial), inert atmosphere (sealed ampoule), and low temperature (0 °C).

References

1. Y. Mase, W. J. Rabourn, and F. W. Quackenbush, *Arch. Biochem. Biophys.*, **68**, 150 (1957).
2. P. Karrer and E. Jucker, *Carotenoids*, E. A. Braude (translator), Elsevier, New York (1950).
3. G. F. Garbers, C. H. Eugster, and P. Karrer, *Helv. Chim. Acta*, **36**, 1783 (1953).
4. R. Rüegg, U. Schwieter, G. Ryser, P. Schudel, and O. Isler, *Helv. Chim. Acta*, **44**, 985 (1961).
5. R. Kuhn and H. Brockmann, *Chem. Ber.*, **66**, 407 (1933).
6. S. C. Kushwaha, C. Subbarayan, D. A. Beeler, and J. W. Porter, *J. Biol. Chem.*, **244**, 3635 (1969).
7. U. Schwieter, H. R. Bolliger, L. H. Chopard-dit-Jean, G. Englert, M. Kofler, A. König, C. von Planta, R. Rüegg, W. Vetter, and O. Isler, *Chimia*, **19**, 294 (1965).

8. B. C. L. Weedon, *Fortschr. Chem. Org. Naturstoffe*, **27**, 81 (1969).
9. U. Schwieter, G. Englert, N. Rigassi, and W. Vetter, *Pure Appl. Chem.*, **20**, 365 (1969).
10. L. Zechmeister and W. A. Schroeder, *Arch. Biochem.*, **1**, 231 (1942).

Carot-13

ζ-Carotene

7,8,7′,8′-Tetrahydro-ψ,ψ-carotene
(7,8,7′,8′-Tetrahydrolycopene)

Formula: $C_{40}H_{60}$
Formula Wt.: 540.90
Calc. %: C, 88.82; H, 11.18

Sources:

Natural Sources. Good sources of ζ-carotene are carrot roots,[1] fruit of certain varieties of tomato,[2] fruit of *Lonicera japonica*, the petals of *Calendula officinalis*, and certain fungi.[3] An unsymmetrical ζ-carotene (7,8,11,12-tetrahydrolycopene) is found in several photosynthetic bacteria.[4]

Chemical Synthesis. The chemical synthesis of the symmetrical and unsymmetrical ζ-carotenes has been reported.[5,6]

Isolation Procedures: ζ-Carotene is isolated from plant materials by the methods of solvent extraction, saponification, phasic separation, and chromatography commonly used for other carotenes.[7]

Methods of Purification:

Solvent Partition. ζ-Carotene is epiphasic on partition between 90% aqueous methanol and petroleum ether.

Chromatography. ζ-Carotene may be purified from other carotenes by chromatography on 50% magnesia–Hyflo Supercel.[7] Columns are developed with hexane, and ζ-carotene is eluted with 10% ethanol in hexane. Purification of ζ-carotene may also be effected on a column of alumina.[7] The chromatogram is developed with hexane containing 2–5% of diethyl ether.

Crystallization. The crystallization of ζ-carotene has been reported.[5]

Methods of Assaying for Purity:

Chromatography. The purity of ζ-carotene may be determined by column[7] or thin-layer[8] chromatography.

Visible Spectrum. Petroleum ether: 378, 400, and 425 nm.[5,7,8] $E_{1\,cm}^{1\%}$ 2270 (400 nm).[5,7,8] Carbon disulfide:[9] 402, 424, and 452 nm. The unsymmetrical ζ-carotene (7,8,11,12-tetrahydrolycopene) in petroleum ether has wavelength maxima[4] at 374, 394.5, and 418.5 nm.

Infrared Spectrum. The infrared spectra for the two forms of ζ-carotene have been reported.[5,6,10]

Nuclear Magnetic Resonance. The nuclear magnetic resonance spectra of ζ-carotene and its unsymmetrical isomer have been reported.[5,6]

Mass Spectrum. The mass spectra of symmetrical and unsymmetrical ζ-carotene have been reported.[4,11,12]

Melting Point. A melting-point range of 38–42 °C has been reported for all-*trans*-ζ-carotene.[5,6]

Probable Impurities: Oxidation products.

Conditions of Storage: Darkness (brown vial), inert atmosphere (sealed vial), and low temperature (−20 °C).[9]

References

1. H. H. Strain and W. M. Manning, *J. Am. Chem. Soc.*, **65**, 2258 (1943).
2. J. W. Porter and R. E. Lincoln, *Arch. Biochem.*, **27**, 390 (1950).
3. T. W. Goodwin, *Ann. Rev. Biochem.*, **24**, 497 (1955).
4. B. H. Davies, E. A. Holmes, D. E. Loeber, T. P. Toube, and B. C. L. Weedon, *J. Chem. Soc.* (C), 1266 (1969).
5. J. B. Davis, L. M. Jackman, P. T. Siddons, and B. C. L. Weedon, *Proc. Chem. Soc.*, 261 (1961).
6. J. B. Davis, L. M. Jackman, P. T. Siddons, and B. C. L. Weedon, *J. Chem. Soc.* (C), 2154 (1966).
7. H. A. Nash and F. P. Zscheile, *Arch. Biochem.*, **7**, 305 (1945).
8. B. H. Davies, in *Chemistry and Biochemistry of Plant Pigments*, T. W. Goodwin, ed., Academic Press, New York (1965).
9. H. A. Nash, F. W. Quackenbush, and J. W. Porter, *J. Am. Chem. Soc.*, **70**, 3613 (1948).
10. F. B. Jungalwala and J. W. Porter, *Arch. Biochem. Biophys.*, **110**, 291 (1965).
11. O. B. Weeks, A. G. Andrewes, B. O. Brown, and B. C. L. Weedon, *Nature*, **224** 879 (1969).
12. B. C. L. Weedon, *Fortschr. Chem. Org. Naturstoffe*, **27**, 81 (1969).

Carot-14

Citranaxanthin

5′,6′-Dihydro-5′-apo-18′-nor-β-caroten-6′-one

Formula: $C_{33}H_{44}O$
Formula Wt.: 456.72
Calc. %: C, 86.79; H, 9.71; O. 3.50

Sources:

Natural Sources. Citranaxanthin is found in the peel of the trigeneric hybrid *Sinton citrangequat*.[1]

Chemical Synthesis. The chemical synthesis of citranaxanthin from β-apo-8′-carotenal (C_{30}) and acetone has been reported.[1]

Isolation Procedures: The extraction of citranaxanthin from the peel of *Sinton citrangequat* fruit, the partial purification of this compound by distribution between immiscible solvents and its complete purification by chromatography and crystallization have been reported.[1]

Methods of Purification:

Chromatography. Citranaxanthin is purified by chromatography on a column of 1:1 (w/w) magnesium oxide–Hyflo Supercel.[1]

Crystallization. Citranaxanthin has been crystallized from petroleum ether.[1]

Methods of Assaying for Purity:

Chromatography. The purity of citranaxanthin may be determined by column chromatography[1] or by chromatography on thin-layer plates of silica gel G. The chromatoplates are developed with 2:3 petroleum ether–benzene.

Visible Spectrum. Hexane: 349 and 466 nm; $E_{1\,cm}^{1\%}$ 410 and 2575, respectively. Cyclohexane: 352 and 471 nm; $E_{1\,cm}^{1\%}$ 400 and 2475, respectively. Benzene: 360 and 482 nm; $E_{1\,cm}^{1\%}$ 365 and 2275, respectively.[1]

Infrared Spectrum. The infrared absorption spectrum of citranaxanthin has been determined.[1]

Nuclear Magnetic Resonance Spectrum. The nuclear magnetic resonance spectrum of citranaxanthin has been determined.[1]

Melting Point.[1] Citranaxanthin melts at 155–156 °C.

Derivatives.[1] Citranaxanthin forms an oxime that melts at 196–197 °C.

Probable Impurities: Oxidation products and *cis*-isomers.

Conditions of Storage: Darkness (brown vial), inert atmosphere (sealed ampoule), and low temperature (0 °C).

Reference

1. H. Yokoyama and M. J. White, *J. Org. Chem.*, 30, 2481 (1965).

Carot-15
Crocetin
8,8′-Diapocarotene-8,8′-dioic Acid

Formula: $C_{20}H_{24}O_4$
Formula Wt.: 328.41
Calc. %: C, 73.13; H, 7.37; O, 19.51

Sources:

Natural Sources. Crocetin (unesterified) is present in trace amounts in some plant species. Most of this compound is found as digentiobiose ester (crocin).[1-3] Crocetin has been isolated as the dimethyl ester; this compound is formed through the action of dilute sodium hydroxide in methanol on an extract of saffron, *Crocus sativus*.[4] Crocin is also found in *Cedrela toona, Nyctanthes arbor-tristis*, and *Verbascum phlomoides* petals and in the fruit of *Gardenia grandiflora*.[5]

Chemical Synthesis. The total synthesis of all-*trans*-crocetin dimethyl ester has been reported.[6,7]

Methods of Purification: Crocetin from saffron[6] and from the petals of *Crocus luteus*,[8] crocin from saffron,[9] and crocetin dimethyl ester from saffron[10] are purified largely by chromatography followed by crystallization, usually as the dimethyl ester.

Methods of Assaying for Purity:

Chromatography. Crocetin dimethyl ester may be assayed for purity by column chromatography.[11]

Derivatives. Mono- and dimethyl esters of crocetin, perhydrocrocetin and its dimethyl ester and diamide, dihydrocrocetin and its dimethyl ester, and crocetin tetrabromide have been prepared.[5]

Visible Spectrum.[5] Crocetin, in carbon disulfide: 426, 453, and 482 nm; pyridine: 411, 436, and 464 nm; chloroform: 434.5 and 463 nm; petroleum ether (b.p. 40–60 °C): 424.5 and 450.5 nm; hexane: 400, 420, and 444.5 nm. Crocetin dimethyl ester–petroleum ether (b.p. 40–60 °C): 422 and 448 nm;[6] 422 and 450 nm;[7] 420 and 444.5 nm.[10] $E_{1\,cm}^{1\%}$ values are given as follows: Crocetin dimethyl ester in petroleum ether, 4750 at 448 nm; in ethanol,[5] 5000 at 425 nm.

Infrared Spectrum. The infrared spectrum of crocetin dimethyl ester has been reported.[12,13]

Mass Spectrum. The mass spectrum of crocetindial has been reported.[14]

Nuclear Magnetic Resonance. The nuclear magnetic resonance spectra of crocetin dimethyl ester and crocetindial have been reported.[15]

Melting Point. The following melting points have been reported: crocetin, 285 °C; and crocetin dimethyl ester,[16] 223 °C.

Probable Impurity: Picrocrocin, a colorless glycoside closely related to crocin.

Conditions of Storage: Darkness (brown vial), inert atmosphere (sealed ampoule), and low temperature (−20 °C).

References

1. P. Karrer, F. Benz, R. Morf, H. Raudnitz, M. Stoll, and T. Takahashi, *Helv. Chim. Acta*, 15, 1399 (1932).
2. P. Karrer and H. Salomon, *Helv. Chim. Acta*, 11, 513 (1928).
3. P. Karrer and K. Miki, *Helv. Chim. Acta*, 12, 985 (1929).
4. R. Kuhn and A. Winterstein, *Chem. Ber.*, 67, 344 (1934).
5. P. Karrer and E. Jucker, *Carotenoids*, E. A. Braude, (translator) Elsevier, New York (1950).
6. H. H. Inhoffen, O. Isler, G. von der Bey, G. Raspé, P. Zeller, and R. Ahrens, *Ann. Chem.*, 580, 7 (1953).
7. O. Isler, H. Gutman, H. Lindlar, M. Montavon, R. Rüegg, G. Ryser, and P. Zeller, *Helv. Chim. Acta*, 39, 463 (1956).
8. P. Karrer and H. Salomon, *Helv. Chem. Acta*, 10, 397 (1927).
9. R. Kuhn, A. Winterstein, and W. Wiegand, *Helv. Chim. Acta*, 11, 716 (1928).
10. R. Kuhn and A. Winterstein, *Chem. Ber.*, 66, 209 (1933).
11. R. Kuhn and H. Brockmann, *Z. Physiol. Chem.*, 206, 41 (1932).
12. R. Kuhn, H. H. Inhoffen, H. A. Staab, and W. Otting, *Chem. Ber.*, 86, 965 (1953).
13. K. Lunde and L. Zechmeister, *J. Am. Chem. Soc.*, 77, 1647 (1955).
14. C. R. Enzell, G. W. Francis, and S. Liaaen-Jensen, *Acta Chem. Scand.*, 23, 727 (1969).
15. B. C. L. Weedon, *Fortschr. Chem. Org. Naturstoffe*, 27, 81 (1969).
16. R. Kuhn and F. L'Orsa, *Chem. Ber.*, 64, 1732 (1931).

Carot-16
Crocetin Diethyl Ester
Diethyl 8,8′-Diapocarotene-8,8′-dioate

Formula: $C_{24}H_{32}O_4$
Formula Wt.: 384.52
Calc. %: C, 74.97; H, 8.39; O, 16.64

Sources:

Natural Sources. Crocetin is present as the digentiobiose ester (crocin) in saffron (*Crocus sativus*) flowers.[1-4]

Chemical Synthesis. The chemical synthesis of crocetin diethyl ester has been reported.[5,6]

Isolation Procedures:[7] Crocin and crocetin are normally extracted from dried saffron flowers with ethanol after a pre-extraction of the flowers with ether. Crocin is obtained by crystallization from the alcoholic extract by the addition of ether. Crystalline crocetin may be obtained by addition of ether to an alcoholic extract of saffron flowers after prior saponification and acidification.

Methods of Purification:

Chromatography. Crocetin diethyl ester may be purified by chromatography on a column of silica gel G.[7]

Crystallization. Crocetin diethyl ester has been crystallized from benzene.[5]

Methods of Assaying for Purity:

Chromatography. The purity of crocetin diethyl ester may be determined by chromatography on a column or thin-layer plate of silica gel G. Dichloromethane is the solvent used in these separations.[8]

Visible Spectrum. Petroleum ether: 400, 422, and 450 nm. $E_{1\,cm}^{1\%}$ 2340, 3820, and 3850, respectively.[5] Cyclohexane: 402, 425, and 452 nm. $E_{1\,cm}^{1\%}$ 2190, 3590, and 3640, respectively. Benzene: 411, 435, and 462 nm. $E_{1\,cm}^{1\%}$ 1970, 3110, and 3000, respectively. Chloroform: 412, 434, and 462 nm. $E_{1\,cm}^{1\%}$ 2190, 3365, and 3200, respectively.[6]

Melting Point. Melting points of 218–219 °C and 216–218 °C have been reported[5] for crocetin diethyl ester.

Probable Impurities: Oxidation products and *cis*-isomers.

Conditions of Storage: Darkness (brown vial), inert atmosphere (sealed ampoule), and low temperature (0 °C).

References

1. P. Karrer and H. Salomon, *Helv. Chim. Acta*, **10**, 397 (1927).
2. P. Karrer and K. Miki, *Helv. Chim. Acta*, **12**, 985 (1929).
3. P. Karrer, F. Benz, R. Morf, H. Raudnitz, M. Stoll, and T. Takahashi, *Helv. Chim. Acta*, **15**, 1218 (1932).
4. P. Karrer, F. Benz, R. Morf, H. Raudnitz, M. Stoll, and T. Takahashi, *Helv. Chim. Acta*, **15**, 1399 (1932).
5. O. Isler, H. Gutmann, M. Montavon, R. Rüegg, G. Ryser, and P. Zeller, *Helv. Chim. Acta*, **40**, 1242 (1957).
6. U. Schwieter, H. Gutmann, H. Lindlar, R. Marbet, N. Rigassi, R. Rüegg, S. F. Schaeren, and O. Isler, *Helv. Chim. Acta*, **49**, 369 (1966).
7. P. Karrer and E. Jucker, *Carotenoids*, E. A. Braude (translator), Elsevier, New York, 1950.
8. R. Kuhn and H. Brockmann, *Z. Physiol. Chem.*, **206**, 41 (1932).

Carot-17

Cryptoxanthin

$(3R)\beta,\beta$-Caroten-3-ol

(β-Caroten-3-ol)

Formula: $C_{40}H_{56}O$
Formula Wt.: 552.89
Calc. %: C, 86.90; H, 10.21; O, 2.89

Sources:

Natural Sources. Cryptoxanthin may be isolated from maize seeds[1] or calyces of *Physalis alkekengi*.[2] It also occurs in some fruits[3] and in milk and butter.[4] The absolute configuration of cryptoxanthin has been established[5,6] as 3*R*.

Chemical Synthesis. The synthesis of cryptoxanthin has been reported.[7,8]

Isolation Procedures: The extraction, chromatography, and crystallization of cryptoxanthin have been reported.[2,9,10]

Methods of Purification:

Chromatography. Cryptoxanthin may be purified by chromatography on magnesia, calcium carbonate, or deactivated alumina. Ethanol in ethyl ether is used to develop the column.[2]

Crystallization. Cryptoxanthin may be crystallized from a mixture of chloroform and ethanol.[2,7]

Methods of Assaying for Purity:

Chromatography. Purity of the compound may be determined by column chromatography on magnesia[2] or by chromatography on kieselguhr paper.[11]

Solvent Partition. The partition ratio between hexane and 90% methanol is 87:3.[12]

Visible Spectrum. Petroleum ether[7] (b.p. 40–60 °C): 452 and 480 nm. $E_{1\,cm}^{1\%}$ 2370 and 2080. Ethanol: 451.5 and 478 nm. $E_{1\,cm}^{1\%}$ 2460 and 2165. Hexane: 450 and 478 nm. $E_{1\,cm}^{1\%}$ 2460 at 450 nm.

Infrared Spectrum. The infrared spectrum of cryptoxanthin has been reported.[7]

Optical Rotatory Dispersion. The optical rotatory dispersion curve of cryptoxanthin has been reported.[6]

Melting Point. Melting points of 165–169 °C[2] and 158–160 °C[7] have been reported.

Probable Impurities: Oxidation products and *cis*-isomers.

Conditions of Storage: Darkness (brown vial), inert atmosphere (sealed ampoule), low temperature (−20 °C).

References

1. R. Kuhn and C. Grundmann, *Chem. Ber.*, **66**, 1746 (1933).
2. F. P. Zscheile, J. W. White, Jr., B. W. Beadle, and J. R. Roach, *Plant Physiol.*, **17**, 331 (1942).
3. W. Diemair and W. Postel, *Wiss. Veroeffentl. Deut. Ges Ernaehrung*, **9**, 356 (1963).
4. J. C. Drummond, E. Singer, and R. J. MacWalter, *Biochem. J.*, **29**, 456 (1935).
5. T. E. DeVille, M. B. Hursthouse, S. W. Russell, and B. C. L. Weedon, *J. Chem. Soc. (D)*, 1311 (1969).
6. L. Bartlett, W. Klyne, W. P. Mose, P. M. Scopes, G. Galasko, A. K. Mallams, B. C. L. Weedon, J. Szabolcs, and G. Tóth, *J. Chem. Soc. (C)*, 2527 (1969).
7. O. Isler, H. Lindlar, M. Montavon, R. Rüegg, G. Saucy, and P. Zeller, *Helv. Chim. Acta*, **40**, 456 (1957).
8. D. E. Loeber, S. W. Russell, T. P. Toube, B. C. L. Weedon, and J. Diment, *J. Chem. Soc. (C)*, 404 (1971).
9. H. H. Strain, *Leaf Xanthophylls*, Carnegie Institution of Washington, Washington, D.C. (1938).
10. L. Zechmeister, *Carotinoide*, Julius Springer, Berlin (1934).
11. A. Jensen and S. Liaaen-Jensen, *Acta Chem. Scand.*, **13**, 1863 (1959).
12. J. W. White, Jr., and F. P. Zscheile, *J. Am. Chem. Soc.*, **64**, 1440 (1942).

Carot-18

2,2'-Diketospirilloxanthin

1,1'-Dimethoxy-3,4,3',4'-tetrahydro-1,2,1',2'-tetrahydro-ψ,ψ-carotene-2,2'-dione

Formula: $C_{42}H_{56}O_4$
Formula Wt.: 624.91
Calc. %: C, 80.73; H, 9.03; O, 10.24

Sources:

Natural Sources. 2,2'-Diketospirilloxanthin has been isolated from *Rhodopseudomonas spheroides*[1] and *Rhodopseudomonas gelatinosa*.[1,2]

Chemical Synthesis. The chemical synthesis of 2,2'-diketospirilloxanthin has been reported.[3,4]

Isolation Procedures: 2,2'-Diketospirilloxanthin is extracted from *Rhodopseudomonas* species with acetone, the material is saponified (after removal of acetone), and the nonsaponifiable compounds are transferred to a benzene–petroleum ether mixture.[1,2,5]

Methods of Purification:

Chromatography. 2,2'-Diketospirilloxanthin is purified by chromatography on a column of partially deactivated, neutral aluminum oxide.[5]

Crystallization. 2,2'-Diketospirilloxanthin has been crystallized from acetone–petroleum ether.[5]

Methods of Assaying for Purity:

Chromatography. The purity of 2,2'-diketospirilloxanthin may be determined by chromatography on filter paper containing a kieselguhr filler, by thin-layer chromatography on silica gel G plates, or by chromatography on a column of partially deactivated, neutral aluminum oxide.[3]

Solvent Partition.[5] The partition ratio between petroleum ether and 95% methanol is 41:9.

Visible Spectrum. Petroleum ether:[5] 487.5, 518, and 555 nm. Hexane: 349, 422, 488, 516, and 551 nm. $E_{1\,cm}^{1\%}$ 550, 820, 2125, 2725, and 2150, respectively. Cyclohexane: 352, 432, 495, 523, and 560 nm. $E_{1\,cm}^{1\%}$ 570, 800, 2055, 2580, and 2025, respectively. Benzene: 361, 539, and 576 nm. $E_{1\,cm}^{1\%}$ 550, 2365, and 1822, respectively. Carbon disulfide:[5] 530, 561, and 603 nm.

Infrared Spectrum. The infrared spectrum of 2,2'-diketospirilloxanthin has been determined.[5]

Nuclear Magnetic Resonance. The nuclear magnetic resonance spectrum of 2,2'-diketospirilloxanthin has been determined.[1]

Mass Spectrum. The mass spectrum of 2,2'-diketospirilloxanthin has been published.[6]

Melting Point. A melting point of 222 °C has been reported for naturally occurring 2,2'-diketospirilloxanthin, and of 225–227 °C for the synthetic compound.[3]

Probable Impurities: Oxidation products and *cis*-isomers.

Conditions of Storage: Darkness (brown vial), inert atmosphere (sealed ampoule), and low temperature (0 °C).

References

1. L. M. Jackman and S. Liaaen-Jensen, *Acta Chem. Scand.*, **18**, 1403 (1964).
2. T. W. Goodwin, *Arch. Mikrobiol.*, **24**, 313 (1956).
3. U. Schwieter, R. Rüegg, and O. Isler, *Helv. Chim. Acta*, **49**, 992 (1966).
4. P. S. Manchand and B. C. L. Weedon, *Tetrahedron Lett.*, 989 (1966).
5. S. Liaaen-Jensen, *Acta Chem. Scand.*, **17**, 303 (1963).
6. C. R. Enzell, G. W. Francis, and S. Liaaen-Jensen, *Acta Chem. Scand.*, **23**, 727 (1969).

Carot-19

Echinenone

β,β-Caroten-4-one

(β-Caroten-4-one)

Formula: $C_{40}H_{54}O$
Formula Wt.: 550.88
Calc. %: C, 87.21; H, 9.88; O, 2.90

Sources:

Natural Sources. Echinenone is found in echinoideae,[1,2] in crustacea,[3] and in blue-green algae.[4]

Chemical Synthesis, Echinenone is prepared synthetically from β-apo-8'-carotenal (Carot-2).[5,6]

Methods of Purification: Echinenone is purified by column chromatography on partially deactivated alumina or magnesia.[1-3] Further purification is achieved through crystallization.

Methods of Assaying for Purity:

Chromatography. The purity of echinenone may be determined by chromatography on partially deactivated alumina or magnesia,[1,2] or by chromatography on a thin layer of silica gel G with 4:1 cyclohexane–ethyl ether as the developing solvent.

Visible Spectrum. The principal maximum is found at 472 nm (benzene), 461 nm (cyclohexane), and 458 nm (petroleum ether,

b.p. 80–105 °C). The $E_{1\,cm}^{1\%}$ values are 2040, 2110, and 2160, respectively. Treatment of echinenone with sodium borohydride causes a shift in the absorption maximum (in ethanol) from 470 nm to 423, 451, and 478 nm.

Mass Spectrum. The mass spectrum of echinenone has been reported.[7,8]

Nuclear Magnetic Resonance. The nuclear magnetic resonance spectrum of echinenone has been reported.[9]

Melting Point. A melting point of 178–179 °C has been reported.[5]

Probable Impurities: Oxidation products and traces of *cis*-isomers.

Conditions of Storage: Darkness (brown vial), inert atmosphere (sealed ampoule), and low temperature (−20 °C).

References

1. E. Lederer, *Compt. Rend. Soc. Biol.*, **117**, 411 (1934).
2. T. W. Goodwin and M. M. Taha, *Biochem. J.*, **47**, 244 (1950).
3. H. Thommen and H. Wackernagel, *Naturwissenschaften*, **51**, 87 (1964).
4. S. Hertzberg and S. Liaaen-Jensen, *Phytochemistry*, **5**, 565 (1966).
5. M. Akhtar and B. C. L. Weedon, *J. Chem. Soc.*, 4058 (1959).
6. C. K. Warren and B. C. L. Weedon, *J. Chem. Soc.*, 3986 (1958).
7. C. R. Enzell, G. W. Francis, and S. Liaaen-Jensen, *Acta Chem. Scand.*, **23**, 727 (1969).
8. H. Budzikiewicz, H. Brzezinka, and B. Johannes, *Monatsh. Chem.*, **101**, 579 (1970).
9. B. C. L. Weedon, *Fortschr. Chem. Org. Naturstoffe*, **27**, 81 (1969).

Carot-20

Farnesyl Pyrophosphate

Formula: $C_{15}H_{28}O_7P_2$
Formula Wt: 382.34
Calc %: C, 47.13; H, 7.38; O, 29.29; P, 16.20

Sources:

Natural Sources. Farnesyl pyrophosphate does not normally accumulate in significant quantity in biological materials. However, this compound has been synthesized enzymically from mevalonic acid[1] and from isopentenyl and dimethylallyl pyrophosphates.[2,3]

Chemical Synthesis. The chemical synthesis of farnesyl pyrophosphate is effected through the pyrophosphorylation of farnesol.[4-6]

Methods of Purification:

Derivative Formation. Chemical synthesis yields a mixture of farnesyl phosphate and farnesyl pyrophosphate. These compounds can be selectively crystallized by treatment of an aqueous solution of the mixture with cyclohexylamine followed by lithium chloride.[4,5] The products are the dicyclohexylammonium salt of farnesyl phosphate and the lithium salt of farnesyl pyrophosphate. The water-insoluble lithium salt can be converted into the ammonium salt by passage[6] through a column of Dowex-50 ion-exchange resin.

Chromatography. Farnesyl pyrophosphate may be purified[7] by chromatography on Whatman No. 3 MM paper in a system of (40:20:1:39) (v/v) isopropyl alcohol–isobutyl alcohol–ammonia–water. An R_f value of 0.87 is obtained. The presence of phosphate at this R_f value is demonstrated by spraying with the Rosenberg reagent.[8]

Methods of Assaying for Purity:

Phosphate Determination Standard assays for the determination of phosphate are used to assay for the purity of farnesyl pyrophosphate.[5,6]

Chromatography. The purity of farnesyl pyrophosphate may be determined by paper[2,6] or ion-exchange[9] chromatography. Farnesyl pyrophosphate may be cleaved by treatment with 1 M HCl. The product (nerolidol) is extracted with petroleum ether, and assayed by gas–liquid chromatography.[1,10] Farnesyl pyrophosphate may also be cleaved by bacterial alkaline phosphatase or snake-venom diesterase. The liberated farnesol is extracted into petroleum ether and assayed by gas–liquid chromatography.[1,5,11]

Enzymic Assay. Radioactive farnesyl pyrophosphate may be converted into radioactive squalene by a liver enzyme system,[1,5] a yeast extract,[6] or a plant extract.[11] The radioactive squalene is extracted from the incubation mixture with petroleum ether, and then assayed by gas–liquid chromatography.

Probable Impurities: Geranyl pyrophosphate and farnesyl phosphate.

Conditions of Storage: Farnesyl pyrophosphate is stored as the lithium or ammonium salt at low temperature (0 °C).

References

1. G. Krishna, H. W. Whitlock, Jr., D. H. Feldbruegge, and J. W. Porter, *Arch. Biochem. Biophys.*, **114**, 200 (1966).
2. C. R. Benedict, J. Kett, and J. W. Porter, *Arch. Biochem. Biophys.*, **110**, 611 (1965).
3. J. K. Dorsey, J. A. Dorsey, and J. W. Porter, *J. Biol. Chem.*, **241**, 5353 (1966).
4. F. Cramer and W. Böhm, *Angew. Chem.*, **71**, 775 (1959).
5. G. Popják, J. W. Cornforth, R. H. Cornforth, R. Ryhage, and D. S. Goodman, *J. Biol. Chem.*, **237**, 56 (1962).
6. C. R. Childs, Jr., and K. Bloch, *J. Biol. Chem.*, **237**, 62 (1962).
7. D. G. Anderson, M. S. Rice, and J. W. Porter, *Biochem. Biophys. Res. Commun.*, **3**, 591 (1960).
8. H. Rosenberg, *J. Chromatogr.*, **2**, 487 (1959).
9. R. E. Dugan, E. Rasson, and J. W. Porter, *Anal. Biochem.*, **22**, 249 (1968).
10. L. A. Witting and J. W. Porter, *J. Biol. Chem.*, **234**, 2841 (1959).
11. D. A. Beeler, D. G. Anderson, and J. W. Porter, *Arch. Biochem. Biophys.*, **102**, 26 (1963).

Carot-21
Geraniol
(3,7-Dimethyl-2,6-octadien-1-ol)

Formula: $C_{10}H_{18}O$
Formula Wt.: 154.24
Calc. %: C, 77.87; H, 11.76; O, 10.37

Sources:

Natural Sources. Geraniol is a constituent of many essential oils, such as palmarosa,[1] citronella,[2] ginger grass,[3] geranium,[2] and attar of roses.[3] It also occurs in oils of *Andropogon schoenanthus*,[1] *Pelargonium odoratissum*,[1] *Anthocephalus cadamba*,[4] *Artemisia campestris*,[5] citrus leaves,[6] carrots,[7] coriander,[8] lavender,[9] and *Juniperus sabina*[10] It also is present in some algae[11] and seaweeds.[12] Geraniol may be obtained from citronella and other essential oils by fractional distillation.

Chemical Synthesis. Geraniol has been synthesized by reduction

of methyl geranate with lithium aluminum hydride.[13] It has also been synthesized in ether under pressure from 6-methyl-5-hepten-2-one, potassium hydroxide, and acetylene.[14] This reaction yields dehydrolinalool, which is then hydrogenated in the presence of palladium-on-calcium carbonate to linalool (96% yield), and this is converted into geraniol.[14]

Methods of Purification:

Solvent Extraction. The techniques of countercurrent distribution and liquid–liquid extraction have been used for the isolation of geraniol from geranium oil.[2]

Chromatography. Geraniol may be purified by ascending paper chromatography[15] or by thin-layer chromatography on plates of kieselguhr G, with 130:70:1 acetone–water–liquid paraffin as the solvent system;[16] hexane–ethyl acetate (1:4) may also be used as a solvent system.[17] Geraniol may be purified by gas–liquid chromatography on a silicone-treated column of Carbowax 20 M (10%) on Chromosorb W (60–80 mesh).[18] Other gas–liquid chromatographic systems have been used.[10,18,20]

Methods of Assaying for Purity:

Chromatography. The aforementioned techniques of paper, thin-layer, and gas–liquid chromatography may be used to assay for the purity of geraniol.

Derivatives. Geraniol may be identified, and assayed for purity by preparation of the 3,5-dinitrobenzoate[21] (m.p. 63 °C), the phenylurethan[22] (m.p. 82 °C), or the allophanate[22] (m.p. 124–124.5 °C).

Ultraviolet Spectrum. The ultraviolet absorption maximum, in cyclohexane, is at 190–195 nm.[23]

Infrared Spectrum. The infrared spectrum of geraniol has been reported.[24]

Nuclear Magnetic Resonance. The nuclear magnetic resonance spectrum of geraniol has been reported.[25]

Mass Spectrum. The fragmentation pattern for geraniol has been reported.[26]

Refractive Index.[22] The n_D^{20} is 1.4766.

Density.[22] The density of geraniol is 0.8894 g/ml at 20 °C.

Solubility. Soluble in alcohol and insoluble in water.[27]

Boiling Point.[22] Geraniol has been reported to boil at 230 °C.

Probable Impurity. The *cis*-isomer (nerol) is the most common impurity.

Conditions of Storage. Geraniol should be stored in full, tightly sealed containers, in a cool place, protected from light.

References

1. P. G. Stecher, ed., *Merck Index*, 8th edition (1968), p. 487.
2. A. M. Burger, *Parfuem Kosmetik*, **40**, 610 (1959); *Chem. Abstr.*, **56**, 3578c (1962).
3. *Scientific Section, Essential Oil Association U.S.A.*, No. 16 (1956).
4. R. Bahadur, G. N. Gupta, and M. C. Nigam, *Parfuem Kosmetik*, **47**, 198 (1966); *Chem. Abstr.*, **65**, 13451d (1966).
5. K. C. Guven, *Folia Pharm.*, **5**, 586 (1963); *Chem. Abstr.*, **59**, 7851e (1963).
6. J. A. Attaway, A. P. Pieringer, and L. J. Barabas, *Phytochemistry*, **5**, 141 (1966).
7. G. V. Pigulevskiĭ, D. T. Motskus, and L. L. Rodina, *Zh. Prikl. Khim.*, **35**, 1143 (1962); *Chem. Abstr.*, **57**, 7396i (1962).
8. G. M. Makarova and Yu. G. Borisyuk, *Farmatsevt. Zh.* (Kiev), **14**, 43 (1956); *Chem. Abstr.*, **58**, 2320f (1963).
9. R. Jaspersen Schib and H. Fluek, *Congr. Sci. Farm. Conf. Commun.*, **21**, *Pisa, 1961*, 608 (1962); *Chem. Abstr.*, **60**, 1533g (1964).
10. E. V. Rudloff, *Can. J. Chem.*, **41**, 2876 (1963).
11. T. Katayama, *Kagoshima Daigaku, Suisan Gakubu Kiyo*, **13**, 58 (1964); *Chem. Abstr.*, **62**, 10822h (1965).
12. T. Katayama, *Nippon Suisan Gakkaishi*, **27**, 75 (1961); *Chem. Abstr.*, **56**, 7710b (1962).
13. J. W. K. Burrell, R. F. Garwood, L. M. Jackman, E. Oskay, and B. C. L. Weedon, *J. Chem. Soc.*, 2144 (1966).
14. I. N. Nazarov, B. P. Gussev, and V. I. Gunar, *Zh. Obshch. Khim.*, **28**, 1444 (1958); *Chem. Abstr.*, **53**, 1102i (1959).
15. L. Syper, *Dissertationes Pharm.*, **17**, 33 (1965); *Chem. Abstr.*, **63**, 9035e (1965).
16. G. P. McSweeney, *J. Chromatogr.*, **17**, 183 (1965).

17. T. Okinoga, *Hiroshima Nogyo Tanki Daigaku Kenkyu Hokoku*, **2**, 237 (1965); *Chem. Abstr.*, **65**, 4240g (1966).
18. J. W. Porter, *Pure Appl. Chem.*, **20**, 449 (1969).
19. S. Geyer, W. Zieger, S. Helm, and R. Mayer, *Z. Chem.*, **5**, 309 (1965).
20. K. Laats and A. Erm, *Eesti NSV Teaduste Akad. Toimetised Fuusikalis Mat ja. Tehn. Teaduste Seer.*, **13**, 57 (1964); *Chem. Abstr.*, **61**, 8133g (1964).
21. L. A. Witting and J. W. Porter, *Biochem. Biophys. Research Commun.*, **1**, 341 (1959).
22. I. Heilbron, *Dictionary of Organic Compounds*, 4th Ed., Oxford University Press, New York, Vol. 3, (1965), p. 1504.
23. C. v. Planta, *Helv. Chim. Acta*, **45**, 84 (1962).
24. R. Boch and D. A. Shearer, *Nature*, **194**, 705 (1962).
25. M. S. Barber, J. B. Davis, L. M. Jackman, and B. C. L. Weedon, *J. Chem. Soc.*, 2870 (1960).
26. E. Stenhagen, S. Abrahamsson, and F. W. McLafferty, *Atlas of Mass Spectral Data*, Interscience Publishers, New York, Vol. 2 (1969), p. 931.
27. *Handbook of Physics and Chemistry*, 52nd Ed., Chemical Rubber Co., Cleveland, Ohio (1971), p. C-309.

Carot-22
Geranyl Pyrophosphate

Formula: $C_{10}H_{20}O_7P_2$
Formula Wt.: 314.22
Calc. %: C, 38.23; H, 6.42; O, 35.64; P, 19.71

Sources:

Natural Sources. Geranyl pyrophosphate does not normally accumulate in significant quantity in biological material. Neither has an enzymic synthesis of geranyl pyrophosphate been developed that would result in the synthesis of an appreciable quantity of this compound.

Chemical Synthesis. The chemical synthesis of geranyl pyrophosphate is effected through the pyrophosphorylation of geraniol.[1]

Methods of Purification:

Derivative Formation. Chemical synthesis yields a mixture of geranyl phosphate and geranyl pyrophosphate. These compounds can be selectively crystallized by treatment of an aqueous mixture with cyclohexylamine followed by lithium chloride.[1] The products are geranyl phosphate dicyclohexylammonium salt and the lithium salt of geranyl pyrophosphate. The water-insoluble lithium salt can be converted into the ammonium salt by passage[2] through Dowex-50 ion-exchange resin.

Chromatography. Geranyl pyrophosphate may be purified by paper chromatography on Whatman No. 3 MM paper in a system of 40:20:1:39 (v/v) isopropyl alcohol–isobutyl alcohol–ammonia–water.[2,3] An R_f value of 0.77–0.82 is obtained. The presence of phosphate at this R_f may be shown by spraying with the Rosenberg reagent.[4]

Methods of Assaying for Purity:

Phosphate Determination. Standard assays for the determination of phosphate are used to assay for the purity of geranyl pyrophosphate.[5]

Chromatography. The purity of geranyl pyrophosphate may be determined by paper chromatography.[2,3] Geranyl pyrophosphate may be cleaved with acid, or with bacterial alkaline phosphatase, or snake-venom diesterase.[2,3] The resultant linalool or geraniol is extracted with petroleum ether, and assayed by gas–liquid chromatography.[3]

Enzymic Assay. A mixture of geranyl pyrophosphate plus isopentenyl pyrophosphate is converted into farnesyl pyrophosphate by farnesyl pyrophosphate synthetase.[2,3] The product of the reaction is treated with 1 M HCl or alkaline phosphatase, as mentioned, and the liberated terpenols are extracted with petroleum ether and assayed by gas–liquid chromatography.[3]

Probable Impurity: Geranyl phosphate.

Conditions of Storage: Geranyl pyrophospate is stored as the lithium or ammonium salt at low temperature (0 °C).

References

1. F. Cramer and W. Böhm, *Angew. Chem.*, **71**, 775 (1959).
2. C. R. Benedict, J. Kett, and J. W. Porter, *Arch. Biochem. Biophys.*, **110**, 611 (1965).
3. J. K. Dorsey, J. A. Dorsey, and J. W. Porter, *J. Biol. Chem.*, **241**, 5353 (1966).
4. H. Rosenberg, *J. Chromatogr.*, **2**, 487 (1959).
5. G. Popják, J. W. Cornforth, R. H. Cornforth, R. Ryhage, and D. S. Goodman, *J. Biol. Chem.*, **237**, 56 (1962).

Carot-23
Geranylgeranyl Pyrophosphate

Formula: $C_{20}H_{36}O_7P_2$
Formula Wt.: 450.40
Calc. %: C, 53.34; H, 8.04; O, 24.87; P, 13.75

Sources:

Natural Sources. Geranylgeranyl pyrophosphate does not normally accumulate in significant quantity in biological material. However, this compound is synthesized from mevalonate by a homogenate of the endosperm of immature, wild cucumber seeds.[1]

Chemical Synthesis. The synthesis of geranylgeranyl pyrophosphate by pyrophosphorylation of geranylgeraniol has been reported.[2]

Methods of Purification:

Chromatography. Geranylgeranyl pyrophosphate may be purified by countercurrent distribution between the two phases of 15:5:1:19 (v/v) butyl alcohol–isopropyl ether–ammonia–water.[2] Geranylgeranyl pyrophosphate may be further purified by chromatography on DEAE-cellulose. A linear gradient of 0.02 M potassium chloride in 1 mM tris buffer (pH 8.9) is used.[2]

Methods of Assaying for Purity:

Phosphate Determination. Geranylgeranyl pyrophosphate is treated[2] with 1 M HCl for 15 min at 100 °C. The liberated phosphate (2 mol/mol of geranylgeranyl pyrophosphate) is then determined by assay.[3]

Chromatography. Geranylgeranyl pyrophosphate may be assayed for purity by chromatography on DEAE-cellulose.[1,4] It may also be cleaved by treatment with 1 M HCl. The geranyllinalool and geranylgeraniol released are extracted into benzene, and then assayed by thin-layer chromatography on silica gel G plates in a solvent system of 9:1 (v/v) benzene–ethyl acetate.[2] Geranylgeranyl pyrophosphate may also be cleaved by treatment with bacterial alkaline phosphatase. The liberated geranylgeraniol is extracted into petroleum ether and assayed by gas–liquid chromatography.[5]

Enzymic Assay. Geranylgeranyl pyrophosphate is enzymically converted into kaurene by a cell-free extract of *Echinocystis macrocarpa*.[2] Kaurene is extracted into acetone, and then assayed by thin-layer chromatography. Geranylgeranyl pyrophosphate is enzymically converted into phytoene by an enzyme system obtained from tomato fruit plastids.[6]

Probable Impurities: Geranylgeranyl phosphate and *cis*-isomers of geranylgeranyl pyrophosphate.

Conditions of Storage: Geranylgeranyl pyrophosphate is stored as a dry powder at low temperature (0 °C).

References

1. M. O. Oster and C. A. West, *Arch. Biochem. Biophys.*, **127**, 112 (1968).
2. C. D. Upper and C. A. West, *J. Biol. Chem.*, **242**, 3285 (1967).
3. B. B. Marsh, *Biochim. Biophys. Acta*, **32**, 357 (1959).
4. R. E. Dugan, E. Rasson, and J. W. Porter, *Anal. Biochem.*, **22**, 249 (1968).
5. D. L. Nandi and J. W. Porter, *Arch Biochem. Biophys.*, **105**, 7 (1964).
6. D. V. Shah, D. H. Feldbruegge, A. R. Houser, and J. W. Porter, *Arch. Biochem. Biophys.*, **127**, 124 (1968).

Carot-24
Isopentenyl Pyrophosphate
(3-Methyl-3-buten-1-yl Pyrophosphate)

Formula: $C_5H_{12}O_7P_2$
Formula Wt.: 246.09
Calc. %: C, 24.41; H, 4.91; O, 45.51; P, 25.17

$$H_2C=\overset{\overset{\displaystyle CH_3}{|}}{C}-CH_2-CH_2-O-\overset{\overset{\displaystyle O}{\|}}{\underset{\underset{\displaystyle OH}{|}}{P}}-O-\overset{\overset{\displaystyle O}{\|}}{\underset{\underset{\displaystyle OH}{|}}{P}}-OH$$

Sources: Synthesis of isopentenyl pyrophosphate is effected[1-3] through pyrophosphorylation of 3-methyl-3-buten-1-ol. Also, isopentenyl pyrophosphate may be synthesized enzymically from mevalonic acid.[4,5]

Methods of Purification:

Derivative Formation. Isopentenyl pyrophosphate is converted into the monocyclohexylammonium salt by passage through a column of Dowex-50 (cyclohexylammonium form) ion-exchange resin.[2] Isopentenyl pyrophosphate may also be converted into the lithium salt.[3]

Ion-Exchange Chromatography. Dowex-1 (formate form) is used to purify isopentenyl pyrophosphate.[4,5] Formic acid and ammonium formate are used as eluants. Isopentenyl pyrophosphate may also be purified by chromatography on a column of DEAE-cellulose.[6]

Paper Chromatography. Isopentenyl pyrophosphate has an R_f value of 0.60 when chromatographed on paper (Whatman No. 1) in a system of 20:5:8 (v/v) *tert*-butyl alcohol–formic acid–water.[2] An R_f value of 0.48 is found in 6:3:1 (v/v) 1-propanol–ammonia–water.[4]

Methods of Assaying for Purity:

Phosphate Determination. A standard assay may be used for the determination of phosphate.[3]

Chromatography. Isopentenyl pyrophosphate may be assayed for purity by paper[4] or ion-exchange chromatography.[4,5] The Rosenberg color reagent (for phosphate) is used to detect isopentenyl pyrophosphate on paper.[7]

Enzymic Assay. Radioactive isopentenyl pyrophosphate is converted into dimethylallyl pyrophosphate by isopentenyl pyro-

phosphate isomerase.[4] After acidification of the incubation mixture, this product is extracted into petroleum ether, and the extract is assayed for radioactivity. Isopentenyl pyrophosphate may also be cleaved by bacterial alkaline phosphatase or by snake-venom diesterase. The liberated alcohol may then be assayed by gas–liquid chromatography.[4]

Melting Point.[8] The tricyclohexylammonium salt melts at 145–147 °C.

Infrared Spectrum. The infrared spectrum of the cyclohexylammonium salt of isopentenyl pyrophosphate has been reported.[2]

Conditions of Storage: Store as the tricyclohexylammonium or the lithium salt.

References

1. F. Lynen, H. Eggerer, U. Henning, and I. Kessel, *Angew. Chem.* **70**, 738 (1958).
2. C. Yuan and K. Bloch, *J. Biol. Chem.*, **234**, 2605 (1959).
3. C. D. Foote and F. Wold, *Biochemistry*, **2**, 1254 (1963).
4. D. H. Shah, W. W. Cleland, and J. W. Porter, *J. Biol. Chem.*, **240**, 1946 (1965).
5. K. Bloch, S. Chaykin, A. H. Philips, and A. de Waard, *J. Biol. Chem.*, **234**, 2595 (1959).
6. R. E. Dugan, E. Rasson, and J. W. Porter, *Anal. Biochem.*, **22**, 249 (1968).
7. H. Rosenberg, *J. Chromatogr.*, **2**, 487 (1959).
8. Mann Research Laboratories, *Publication No. 202*, New York (1966), p. 66.

Carot-25
Lutein
(3R,3'S,6'R)-β,ε-Carotene-3,3'-diol
(Xanthophyll; α-Carotene-3,3'-diol)

Formula: $C_{40}H_{56}O_2$
Formula Wt.: 568.89
Calc. %: C, 84.45; H, 9.92; O, 5.63

Sources:

Natural Sources. Lutein is a major constituent of the xanthophyll fraction of many plants. It is present in appreciable proportions in green leaves,[1,2] red and yellow flowers (partly as the dipalmitate, helenien),[3] and in egg yolk.[4]

Chemical Synthesis. The synthesis of this compound has not yet been reported.

Isolation Procedures: The extraction, chromatography, and crystallization of lutein have been reported.[1-6]

Methods of Purification:

Chromatography. Lutein may be purified by chromatography on a column of magnesia or calcium hydroxide.[2,5,6]

Crystallization. A solvent pair frequently used for the crystallization of lutein is carbon disulfide–ethanol.[2,5]

Derivative Formation. The dipalmitate (helenien), other esters, ethers, and the ketone and perhydro derivatives, have been reported.[1,7]

Methods of Assaying for Purity:

Chromatography. The purity of lutein may be determined by chromatography on magnesia or calcium hydroxide[2,6] or on kieselguhr paper.[8]

Visible Spectrum. Ethanol: 423, 446.5, and 477.5 nm. $E_{1cm}^{1\%}$ values 1750, 2560, and 2340, respectively. Spectral curve.[2]

Mass Spectrum. The mass spectrum of lutein has been reported.[9,10]

Nuclear Magnetic Resonance. The nuclear magnetic resonance spectrum of lutein has been reported.[11,12]

Optical Rotatory Dispersion. The optical rotatory dispersion curve of lutein has been reported.[12]

Melting Point.[2] Lutein melts at 151 °C.

Optical Rotation. $[\alpha]_{cd}^{18}$ +160° (chloroform) has been reported.[3]

Probable Impurities: Oxidation products, *cis*-isomers, and possibly zeaxanthin.

Conditions of Storage: Darkness (brown vial), inert atmosphere (sealed ampoule), and low temperature (-20 °C).

References

1. R. Willstätter and W. Mieg, *Ann. Chem.,* **355,** 1 (1907).
2. F. P. Zscheile, J. W. White, Jr., B. W. Beadle, and J. R. Roach, *Plant Physiol.,* **17,** 331 (1942).
3. R. Kuhn, A. Winterstein, and E. Lederer, *Z. Physiol. Chem.,* **197,** 141 (1931).
4. A. E. Gillam and I. M. Heilbron, *Biochem. J.,* **29,** 1064 (1935).
5. H. H. Strain, *Leaf Xanthophylls,* Carnegie Institution of Washington, Washington, D.C. (1938).
6. J. B. Moster, F. W. Quackenbush, and J. W. Porter, *Arch. Biochem. Biophys.,* **38,** 287 (1952).
7. S. Liaaen-Jensen and S. Hertzberg, *Acta Chem. Scand.,* **20,** 1703 (1966).
8. A. Jensen and S. Liaaen-Jensen, *Acta Chem. Scand.,* **13,** 1863 (1959).
9. H. Budzikiewicz, H. Brzezinka, and B. Johannes, *Monatsh. Chem.,* **101,** 579 (1970).
10. C. R. Enzell, G. W. Francis, and S. Liaaen-Jensen, *Acta Chem. Scand.,* **23,** 727 (1969).
11. B. C. L. Weedon, *Fortschr. Chem. Org. Naturstoffe,* **27,** 81 (1969).
12. L. Bartlett, W. Klyne, W. P. Mose, P. M. Scopes, G. Galasko, A. K. Mallams, B. C. L. Weedon, J. Szabolcs, and G. Tóth, *J. Chem. Soc. (C),* 2527 (1969).

Carot-26
Lycopene
ψ,ψ-Carotene

Formula: $C_{40}H_{56}$
Formula Wt.: 536.89
Calc. %: C, 89.48; H, 10.52

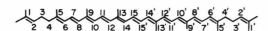

Sources:

Natural Sources. Tomato[1,2] and various other fruits and vegetables, animals,[3,4] and photosynthetic bacteria.[5]

Chemical Synthesis. The total synthesis of lycopene has been reported.[6–10]

Isolation Procedures: The isolation of lycopene has been described.[1–4,11,12]

Methods of Purification:

Chromatography. Column chromatography of the nonsaponifiable fraction is normally employed. The adsorbents most commonly used are deactivated alumina, calcium carbonate, calcium hydroxide, or magnesium oxide.[3,11]

Crystallization. Lycopene may be crystallized from the following solvent pairs:[3] carbon disulfide–methanol; ethyl ether–petroleum ether; acetone–petroleum ether. Lycopene is almost insoluble in methanol, moderately soluble in petroleum ether, benzene, or chloroform, and very soluble in carbon disulfide.

Methods of Assaying for Purity:

Chromatography. The purity of the compound may be determined by chromatography on circular filter paper having a suitable filler,[13,14] thin-layer chromatography,[15] or column chromatography.[1]

Solvent Partition.[16] The partition ratio between petroleum ether and 95% methanol is 100:0.

Visible Spectrum. Petroleum ether (b.p. 40–60 °C): 446, 472, and 505 nm. $E_{1\,cm}^{1\%}$ 2250, 3450, and 3150, respectively. Spectral curve.[1,7,12] Benzene:[3] 455, 487, and 522 nm. Isomerization with iodine results in "*cis*-peak" absorption at 345 nm (minor peak) and 362 nm in petroleum ether.[17]

Infrared Spectrum. Infrared absorption spectra in chloroform and carbon disulfide have been reported.[7] The spectrum in a potassium bromide pellet has also been reported.[18]

Nuclear Magnetic Resonance Spectrum. The nuclear magnetic resonance spectrum has been reported.[19]

Mass Spectrum. The mass spectrum of lycopene has been reported.[20,21]

Melting Point. Lycopene melts at 172–173 °C in an evacuated tube.[7]

Probable Impurities: Oxidation products and *cis*-isomers; small proportions of related carotenes (neurosporene and hydroxylated carotenes) may also be present when lycopene is isolated from natural sources.

Conditions of Storage: Darkness (brown vial), inert atmosphere (sealed ampoule), and a low temperature (-20 °C).

References

1. L. Zechmeister, A. L. LeRosen, W. A. Schroeder, A. Polgár, and L. Pauling, *J. Am. Chem. Soc.,* **65,** 1940 (1943).
2. J. W. Porter and R. E. Lincoln, *Arch. Biochem. Biophys.,* **27,** 390 (1950).
3. P. Karrer and E. Jucker, *Carotenoids,* E. A. Braude (translator), Elsevier, New York (1950).
4. T. W. Goodwin, *The Comparative Biochemistry of the Carotenoids,* Chapman & Hall, London (1952).
5. S. Liaaen-Jensen, in *Bacterial Photosynthesis,* H. Gest, A. San Pietro, and L. P. Vernon, eds., Antioch Press, Yellow Springs, Ohio (1963), p. 19.
6. P. Karrer, C. H. Eugster, and E. Tobler, *Helv. Chim. Acta,* **33,** 1349 (1950).
7. O. Isler, H. Gutmann, H. Lindlar, M. Montavon, R. Rüegg, G. Ryser, and P. Zeller, *Helv. Chim. Acta,* **39,** 463 (1956).
8. H. Pommer, *Angew. Chem.,* **72,** 911 (1960).
9. C. D. Robeson, U. S. Patent 2,932,674 (1960); *Chem. Abstr.,* **54,** 24852g (1960).
10. A. J. Chechak, M. H. Stern, and C. D. Robeson, *J. Org. Chem.,* **29,** 187 (1964).
11. S. Liaaen-Jensen and A. Jensen, *Prog. Chem. Fats Lipids,* **8,** 129 (1965).
12. F. P. Zscheile and J. W. Porter, *Anal. Chem.,* **19,** 47 (1947).
13. A. Jensen and S. Liaaen-Jensen, *Acta Chem. Scand.,* **13,** 1863 (1959).
14. A. Jensen, *Acta Chem. Scand.,* **14,** 2051 (1960).
15. H. R. Bolliger, A. König, and U. Schwieter, *Chimia,* **18,** 136 (1964).
16. F. J. Petracek and L. Zechmeister, *Anal. Chem.,* **28,** 1484 (1956).
17. L. Zechmeister, cis-trans *Isomeric Carotenoids, Vitamins A and Arylpolyenes,* Springer-Verlag, Vienna (1962).
18. S. Liaaen-Jensen, *Kgl. Norske Videnskab. Selskabs Skrifter,* No. 8, (1962).
19. M. S. Barber, J. B. Davis, L. M. Jackman, and B. C. L. Weedon, *J. Chem. Soc.,* 2870 (1960).
20. U. Schwieter, H. R. Bolliger, L. H. Chopard-dit-Jean, G. Englert, M. Kofler, A. König, C. von Planta, R. Rüegg, W. Vetter, and O. Isler, *Chimia,* **19,** 294 (1965).
21. C. R. Enzell, G. W. Francis, and S. Liaaen-Jensen, *Acta Chem. Scand.,* **23,** 727 (1969).

Carot-27
Lycoxanthin
ψ,ψ-Caroten-16-ol
(Lycopen-16-ol)

Formula: $C_{40}H_{56}O$
Formula Wt.: 552.90
Calc. %: C, 86.90; H, 10.21; O, 2.89

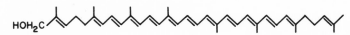

Lycoxanthin, previously thought to be lycopen-3-ol,[1] has recently been characterized as lycopen-16-ol.[2,3]

Sources:

Natural Sources. Lycoxanthin is found in fruits of *Solanum dulcamara*,[2–4] *Solanum esculentum*,[2,4,5] and *Tamus communis*.[4]

Chemical Synthesis. The chemical synthesis of lycoxanthin has been reported.[6]

Isolation Procedures: Lycoxanthin may be extracted from berries or fruit with ether, after dehydration of the tissues with ethanol.[4] The pigments of the extract are then transferred to benzene, and the lycoxanthin is purified by column chromatography.

Methods of Purification:

Chromatography. Lycoxanthin is purified by chromatography on a column of calcium carbonate, calcium hydroxide, deactivated alumina,[5,6] or alumina. The chromatogram is developed with benzene, and lycoxanthin is eluted with 3:1 benzene–methanol.[4]

Crystallization. Lycoxanthin can be crystallized from carbon disulfide or from a mixture of benzene and petroleum ether.[4]

Methods of Assaying for Purity:

Visible Spectrum.[4,7,8] Carbon disulfide: 473, 507, and 547 nm. Benzene:[4] 456, 487, and 521 nm. Petroleum ether:[7] 444, 472.5, and 503 nm. $E_{1\ cm}^{1\%}$ 3360 at 472.5 nm.

Infrared Spectrum. The infrared spectra of lycoxanthin and some of its derivatives have been reported.[2,5]

Nuclear Magnetic Resonance. The nuclear magnetic resonance spectra of lycoxanthin and lycoxanthin acetate have been reported.[2,5]

Mass Spectrum. The mass spectrum of lycoxanthin has been reported.[9]

Melting Point. Lycoxanthin has been reported to melt at 168 °C,[8] and, after crystallization from ethyl ether–light petroleum, at 173–174 °C (uncorr.).[3,6]

Derivatives. The acetate of lycoxanthin has been prepared.[4] This compound crystallizes from benzene–methanol and it has a melting point of 137 °C.[4] The aldehyde derivative of lycoxanthin has also been prepared.[3]

Probable Impurities: Lycophyll and oxidation products.

Conditions of Storage: Darkness (brown vial), an inert atmosphere (sealed ampoule), and cold temperature (-20 °C).

References

1. J. C. Sadana and B. S. Ahmad, *J. Sci. Ind. Res.* (India), **7B**, 172 (1948).
2. M. C. Markham and S. Liaaen-Jensen, *Phytochemistry*, **7**, 839 (1968).
3. M. Kelly, S. Authén-Andresen, and S. Liaaen-Jensen, *Acta Chem. Scand.* **25**, 1607 (1971).
4. L. Zechmeister and L. v. Cholnoky, *Chem. Ber.* **69**, 422 (1936).
5. L. v. Cholnoky and J. Szabolcs, *Tetrahedron Lett.*, 1931 (1968).
6. H. Kjøsen and S. Liaaen-Jensen, *Acta Chem. Scand.* **25**, 1500 (1971).
7. B. H. Davies in *Chemistry and Biochemistry of Plant Pigments*, T. W. Goodwin, ed., Academic Press, New York (1965).
8. P. Karrer and E. Jucker, *Carotenoids*, E. A. Braude (translator), Elsevier, New York (1950).

9. C. R. Enzell, G. W. Francis and S. Liaaen-Jensen, *Acta Chem. Scand.*, **23**, 727 (1969).

Carot-28
Mevalonic Acid
(3,5-Dihydroxy-3-methylpentanoic Acid)

Formula: $C_6H_{12}O_4$
Formula Wt.: 148.16
Calc. %: C, 48.63; H, 8.17; O, 43.20

$$HO-CH_2-CH_2-\overset{\overset{\displaystyle CH_3}{|}}{\underset{\underset{\displaystyle OH}{|}}{C}}-CH_2-CO_2H$$

Sources: Prepared synthetically from 4-acetoxy-2-butanone and ethyl bromoacetate.[1,2]

Methods of Purification:

Derivative Formation. Synthesis yields a racemic mixture of ethyl 5-acetoxy-3-hydroxy-3-methylpentanoate.[2] This product is hydrolyzed to a mixture of ($\pm$)-mevalonic acid and ($\pm$)-mevalono-1,5-lactone. The lactone may then be isolated by short-path distillation, and crystallized from acetone–ether.[2] It may also be converted into a benzhydrylamide derivative.[3] The *N,N'*-dibenzylethylenediammonium (DBED) salt of mevalonic acid may also be prepared and crystallized.[1,2] Only the natural ($+$) optical isomer of mevalonic acid is biologically active.[4] The biologically active isomer of mevalonolactone has the (R)-($-$) configuration.[5]

Methods of Assaying for Purity:

Bioassay. Mevalonic acid and mevalonolactone may be determined quantitatively by microbiological assay with *Lactobacillus acidophilus* ATCC 4963.[6]

Enzymic Assay. Mevalonic acid may be determined quantitatively by spectrophotometric assay in a system in which mevalonic acid is converted into phosphomevalonic acid with mevalonic acid kinase.[7]

Hydroxamate Assay. Mevalonolactone may be converted into the hydroxamate, and the quantity of this compound may be determined spectrophotometrically.[8]

Chromatography. Mevalonic acid may be chromatographed on paper or on a Dowex-1 (formate) column.[9] These methods are particularly useful if the mevalonic acid is radioactive. Mevalonic acid may also be assayed (as the lactone) by gas–liquid chromatography.[10]

Melting Point. Mevalonolactone[2] 27–28 °C; DBED salt[2] 124–125 °C; benzhydrylamide derivative[3] 93–95 °C.

Infrared Spectra. Strong bands are observed at 2.90–2.95 μm (hydroxyl function) and at 5.78 μm (ester function), for mevalonolactone in chloroform.[3]

Conditions of Storage: As the DBED salt, or as the lactone, in a sealed container in a refrigerator.

References

1. C. H. Hoffman, A. F. Wagner, A. N. Wilson, E. Walton, C. H. Shunk, D. E. Wolf, F. W. Holly, and K. Folkers, *J. Am. Chem. Soc.*, **79**, 2316 (1957).
2. K. Folkers, C. H. Shunk, B. O. Linn, F. M. Robinson, P. E. Wittreich, J. W. Huff, J. L. Gilfillan, and H. R. Skeggs, in *Biosynthesis of Terpenes and Sterols*, G. W. Wolstenholme and M. O'Connor, eds., Little, Brown & Co., Boston (1959), p. 20.
3. D. E. Wolf, C. H. Hoffman, P. E. Aldrich, H. R. Skeggs, L. D. Wright, and K. Folkers, *J. Am. Chem. Soc.*, **79**, 1486 (1957).

4. R. H. Cornforth, K. Fletcher, H. Hellig, and G. Popják, *Nature*, **185**, 923 (1960).
5. M. Eberle and D. Arigoni, *Helv. Chim. Acta*, **43**, 1508 (1960).
6. H. R. Skeggs, L. D. Wright, E. L. Cresson, G. D. E. McRae, C. H. Hoffman, D. E. Wolf, and K. Folkers, *J. Bacteriol.*, **72**, 519 (1956).
7. T. T. Tchen, *J. Biol. Chem.*, **233**, 1100 (1958).
8. H. J. Knauss, J. D. Brodie, and J. W. Porter, *J. Lipid Res.*, **3**, 197 (1962).
9. K. Bloch, S. Chaykin, A. H. Philips, and A. de Waard, *J. Biol. Chem.*, **234**, 2595 (1959).
10. R. B. Guchhait and J. W. Porter, *Anal. Biochem.*, **15**, 509 (1966).

Carot-29

Mevalonic Acid 5-Phosphate

(3,5-Dihydroxy-3-methylpentanoic Acid 5-Phosphate)

Formula: $C_6H_{13}O_7P$
Formula Wt.: 228.14
Calc. %: C, 31.59; H, 5.72; O, 49.09; P, 13.58

Sources: Prepared synthetically by phosphorylation of mevalonic benzhydrylamide[1] or reduced mevalonolactone, followed by oxidation.[2] Mevalonic acid 5-phosphate may also be synthesized enzymically from mevalonic acid.[3,4] Only the natural (R) optical isomer of mevalonic acid is converted into the phosphate.

Methods of Purification:

Derivative Formation. Mevalonic acid phosphate is converted into the tricyclohexylammonium salt by treatment with cyclohexylamine. The salt is then crystallized from water–acetone at −15 °C.[2]

Chromatography. Mevalonic acid phosphate may be purified by ion-exchange chromatography.[1,5] Purification may also be effected by paper chromatography (Whatman No. 1) in a system of isobutyric acid–ammonia–water (66:3:30 v/v).[2] An R_f of 0.42 is obtained. The presence of phosphate at this R_f may be shown by spraying with Rosenberg's reagent.[6]

Methods of Assaying for Purity:

Phosphate Determination. A standard assay may be used for the determination of phosphate.[2,5]

Enzymic Assay. Mevalonic acid phosphate may be assayed enzymically through conversion into mevalonic acid pyrophosphate by phosphomevalonic kinase. A spectrophotometric method has been developed for this assay.[3,4]

Chromatography. The purity of mevalonic acid phosphate may be determined by paper chromatography.[2]

Melting Point. Melting points of 145–147 °C for the cyclohexylammonium salt[7] and 154–156 °C for the tricyclohexylammonium salt[8] have been reported.

Conditions of Storage: Store as the cyclohexylammonium salt.

References

1. K. Folkers, C. H. Shunk, B. O. Linn, F. M. Robinson, P. E. Wittreich, J. W, Huff, J. L. Gilfillan, and H. R. Skeggs, in *Biosynthesis of Terpenes and Sterols*, G. W. Wolstenholme and M. O'Connor, eds., Little, Brown & Co., Boston (1959), p. 20.
2. C. D. Foote and F. Wold, *Biochemistry*, **2**, 1254 (1963).
3. T. T. Tchen, *J. Biol. Chem.*, **233**, 1100 (1958).
4. H. R. Levy and G. Popják, *Biochem. J.*, **75**, 417 (1960).
5. K. Bloch, S. Chaykin, A. H. Phillips, and A. de Waard, *J. Biol. Chem.*, **234**, 2595 (1959).

6. H. Rosenberg, *J. Chromatogr.*, **2**, 487 (1959).
7. F. Lynen, in *Biosynthesis of Terpenes and Sterols*, G. W. Wolstenholme and M. O'Connor, eds., Little, Brown & Co., Boston (1959), p. 95.
8. Mann Research Laboratories, *Publication No. 202*, New York (1966), p. 66.

Carot-30

Mevalonic Acid 5-Pyrophosphate

(3,5-Dihydroxy-3-methylpentanoic Acid 5-Pyrophosphate)

Formula: $C_6H_{12}O_9P_2$
Formula Wt.: 308.13
Calc. %: C, 23.38; H, 4.58; O, 51.93; P, 20.11

Sources:

Natural Sources. Mevalonic acid 5-pyrophosphate does not normally accumulate in plant or animal tissue. However, this compound can be synthesized in small amounts by phosphomevalonic kinase. The synthesis of this compound from mevalonic acid or mevalonic acid 5-phosphate by partially purified enzymes from yeast,[1,2] pig liver,[3] and rat liver[4] has been reported.

Chemical Synthesis. The chemical synthesis of mevalonic acid 5-pyrophosphate has not yet been reported.

Isolation Procedures:[4,5] The incubation mixture is deproteinized with acid, and the precipitated protein is washed thoroughly with water. The supernatant solution is then subjected to chromatography.

Methods of Purification:

Chromatography. Mevalonic acid 5-pyrophosphate may be purified by ion-exchange chromatography on Dowex-1 formate[2] or DEAE-cellulose[5–7] or by paper chromatography.[1–4,8]

Methods of Assaying for Purity:

Chromatography. The foregoing methods may be used to assay for the purity of a preparation of mevalonic acid 5-pyrophosphate.

Chemical Methods. Mevalonic acid 5-pyrophosphate may be cleaved with alkaline phosphatase.[8] Assays may then be made for mevalonic acid by gas–liquid chromatography[9] or for phosphate.[10] Quantitative assays for these compounds may be performed.[9,10]

Probable Impurities: Adenosine triphosphate and mevalonic phosphate.

Conditions of Storage: As a dry powder, or in a slightly alkaline, aqueous solution (pH 7–9) at a low temperature (−20 °C).

References

1. W. Henning, E. M. Möslein, and F. Lynen, *Arch. Biochem. Biophys.*, **83**, 259 (1959).
2. K. Bloch, S. Chaykin, A. H. Phillips, and A. de Waard, *J. Biol. Chem.*, **234**, 2595 (1959).
3. H. Hellig and G. Popják, *J. Lipid Res.*, **2**, 235 (1961).
4. L. A. Witting and J. W. Porter, *J. Biol. Chem.*, **243**, 2841 (1959).
5. R. E. Dugan, E. Rasson, and J. W. Porter, *Anal. Biochem.*, **22**, 249 (1968).
6. D. N. Skilleter and R. G. O. Kekwick, *Biochem. J.*, **108**, 11P (1968).
7. D. N. Skilleter and R. G. O. Kekwick, *Anal. Biochem.*, **20**, 171 (1967).
8. L. J. Rogers, S. P. J. Shah, and T. W. Goodwin, *Biochem. J.*, **99**, 381 (1966).
9. R. B. Guchhait and J. W. Porter, *Anal. Biochem.*, **15**, 509 (1966).
10. G. R. Bartlett, *J. Biol. Chem.*, **234**, 466 (1959).

Carot-31

Nerolidol

(3,7,11-Trimethyl-1,6,10-dodecatrien-3-ol)

Formula: $C_{15}H_{26}O$
Formula Wt.: 222.37
Calc. %: C, 81.02; H, 11.78; O, 7.20

Sources:

Natural Sources. Nerolidol is obtained from cajaput,[1] camphor,[2] grapefruit,[3] lime,[4] neroli,[5] and coriander fruit oils,[6] orange blossoms,[7] and Peru balsam.[7] Nerolidol is obtained from these oils by fractional distillation.

Chemical Synthesis. Nerolidol has been synthesized through the condensation of linalool with diketene, and subsequent hydrolysis and reduction of the intermediate. A 96% yield of pure product has been reported.[8] It has also been synthesized by condensation of cyclopropyl methyl ketone with the magnesium derivative of 1-bromo-4,8-dimethyl-3,7-nonadiene at 75 °C. A 20% yield of product was reported.[9]

Methods of Purification:

Chromatography. Nerolidol may be purified[10] by thin-layer chromatography on plates of kieselguhr G. It may also be purified by thin-layer chromatography on a plate of silica gel that has been impregnated with increasing concentrations of silver nitrate. 1,2-Dichloroethane–chloroform–ethyl acetate–propyl alcohol (10:10:1:1) is used as the solvent system.[11] Thin-layer plates (26 × 76 mm) of silica gel, 250 μm thick, may also be used, with 4:1 ethyl acetate–hexane.[12] Separation and purification of nerolidol has been effected[13,14] by gas–liquid chromatography on butanediol succinate (20%) on Chromosorb W.

Methods of Assaying for Purity:

Chromatography. The foregoing techniques of thin-layer and gas–liquid chromatography may be used to assay for the purity of nerolidol.

Ultraviolet Spectrum. The ultraviolet absorption spectrum of nerolidol in cyclohexane shows a maximum at 187–192 nm.[15,16]

Infrared Spectrum. The infrared spectrum of nerolidol has been reported.[17]

Refractive Index. The n_D^{20} for the isomers of nerolidol are[18] *dextro*, 1.4898; *levo*, 1.4799; DL, 1.4801.

Specific Rotation. The $[\alpha]_D^{20}$ for the isomers of nerolidol are *dextro*,[18] +142°; *levo*,[18] −6.5°; DL,[7] +15.5°.

Density. The density of nerolidol at 20 °C is 0.8778 g/ml.[7]

Solubility. Nerolidol is soluble in alcohol, ether, and other organic solvents.[18]

Derivatives. The phenylurethan (m.p. 37–38 °C, b.p. 145–146 °C), semicarbazone (m.p. 134–135 °C) and acetate (b.p. 128–129 °C at 1.6 mmHg) derivatives of nerolidol have been reported.[7]

Boiling Point.[7] Nerolidol has been reported to boil at 276 °C.

Probable Impurities. cis-Isomers and farnesol.

Conditions of Storage. Nerolidol should be stored in tightly sealed containers, protected from light, in a cool place.

References

1. V. K. Sood, *Perfum. Essent. Oil Rec.*, **57**, 362 (1966); *Chem. Abstr.*, **65**, 8661g (1966).
2. N. Hirota and M. Hiroi, *Koryo*, **70**, 23 (1963); *Chem. Abstr.*, **60**, 9096e (1964).
3. G. L. K. Hunter and M. G. Moshonas, *J. Food Sci.*, **31**, 167 (1966).
4. L. Peyron, *Soap Perfum. Cosmet.*, **39**, 633 (1966); *Chem. Abstr.*, **65**, 1842f (1966).
5. M. Calvarano, *Essenze Deriv. Agrumari*, **33** (1), 5 (1963); *Chem. Abstr.*, **59**, 11184F (1963).
6. E. Schratz and S. M. J. S. Quadry, *Planta Med.*, **14**, 310 (1966); *Chem. Abstr.*, **65**, 13530g (1966).
7. I. Heilbron, *Dictionary of Organic Compounds*, 4th ed. Vol. 4, Oxford University Press, New York (1965), p. 2418.
8. I. N. Nazarov, B. P. Gussev, and V. I. Gunar, *Zh. Obshch. Khim.*, **28**, 1444 (1958); *Chem. Abstr.*, **53**, 1102i (1959).
9. M. Julia, S. Julia, and R. Guegan, *Bull. Soc. Chim.*, 1072 (1960).
10. G. P. McSweeney, *J. Chromatogr.*, **17**, 183 (1965).
11. E. Stahl and H. Vollmann, *Talanta*, **12**, 525 (1965).
12. T. Okinaga, *Hiroshima Nogyo Tanki Daigaku Kenkyu Hokoku*, **2**, (4), 237 (1965); *Chem. Abstr.*, **65**, 4240f (1966).
13. J. W. Porter, *Pure Appl. Chem.*, **20**, 449 (1969).
14. C. R. Benedict, J. Kett, and J. W. Porter, *Arch. Biochem. Biophys.*, **110**, 611 (1965).
15. Y. R. Naves and C. Frei, *Helv. Chim. Acta*, **46**, 2551 (1963).
16. C. v. Planta, *Helv. Chim. Acta*, **45**, 84 (1962).
17. A. Ofner, W. Kimel, A. Holmgren, and F. Forrester, *Helv. Chim. Acta*, **42**, 2581 (1959).
18. *Handbook of Physics and Chemistry*, 47th ed., Chemical Rubber Co., Cleveland, Ohio (1966), p. C-430.

Carot-32

Neurosporene

7,8-Dihydro-ψ,ψ-carotene

Formula: $C_{40}H_{58}$
Formula Wt.: 538.91
Calc. %: C, 89.15; H, 10.85

Sources:

Natural Sources. Neurosporene is widely distributed in small amounts in fungi,[1,2] fruits and vegetables,[3,4] and photosynthetic bacteria.[5]

Chemical Synthesis. The total synthesis of neurosporene has been reported.[6,7]

Isolation Procedures: Neurosporene is extracted from natural sources[2,3] with a solvent such as acetone, ethanol, or methanol. The extract is then transferred to petroleum ether, with or without saponification, and subjected to column chromatography.

Methods of Purification:

Chromatography. Neurosporene is purified by chromatography on 1:1 magnesium oxide–Celite,[3] calcium hydroxide–Celite,[2] or alumina.[8]

Methods of Assaying for Purity:

Chromatography. Assays for the purity of neurosporene may be conducted by chromatography on calcium hydroxide–Celite,[2] magnesium oxide–Celite,[3] or alumina.[8] Thin-layer chromatography may also be performed.[9]

Visible Spectrum. Petroleum ether:[9] 416, 440, and 470 nm. $E_{1\,cm}^{1\%}$ 2990 at 440 nm. Hexane:[2] 415, 438.5, and 469 nm. $E_{1\,cm}^{1\%}$ 1920, 2990, and 3010, respectively. Carbon disulfide:[2] 439.5, 470.5, and 502.5 nm.

Nuclear Magnetic Resonance. The nuclear magnetic resonance spectrum of neurosporene has been reported.[6,7]

Mass Spectrum. The mass spectrum of neurosporene has been reported.[10,11]

Melting Point. A melting point of 124 °C has been reported.[2]

Probable Impurities: Oxidation products, *cis*-isomers, γ-carotene, and *cis*-isomers of lycopene.

Conditions of Storage: Darkness (brown vial), inert atmosphere (sealed ampoule), and low temperature (−20 °C).

References

1. T. W. Goodwin, *The Comparative Biochemistry of the Carotenoids*, Chapman and Hall, London, 1952.
2. F. Haxo, *Arch. Biochem.*, **20**, 400 (1949).
3. H. H. Trombly and J. W. Porter, *Arch. Biochem. Biophys.*, **43**, 443 (1953).
4. T. W. Goodwin, *Advan. Enzymol.*, **21**, 295 (1959).
5. S. Liaaen-Jensen, in *Bacterial Photosynthesis*, H. Gest, A. San Pietro, and L. P. Vernon, eds., Antioch Press, Yellow Springs, Ohio (1963).
6. J. B. Davis, L. M. Jackman, P. T. Siddons, and B. C. L. Weedon, *Proc. Chem. Soc.*, 261 (1961).
7. J. B. Davis, L. M. Jackman, P. T. Siddons, and B. C. L. Weedon, *J. Chem. Soc. (C)*, 2154 (1966).
8. S. C. Kushwaha, G. Suzue, C. Subbarayan, and J. W. Porter, *J. Biol. Chem.*, **245**, 4708 (1970).
9. A. Jensen, in *Carotine und Carotinoide*, K. Lang, ed., p. 119, D. Steinkopff Verlag, Darmstadt (1963).
10. B. C. L. Weedon, *Fortschr. Chem. Org. Naturstoffe*, **27**, 81 (1969).
11. O. B. Weeks, A. G. Andrewes, B. O. Brown, and B. C. L. Weedon, *Nature*, **224**, 879 (1969).

Visible Spectrum. Hexane: 449 and 478 nm. $E_{1\,cm}^{1\%}$ 1410 and 1255, respectively. Petroleum ether: 452 and 480 nm. $E_{1\,cm}^{1\%}$ 1335 and 1190, respectively.[6] Cyclohexane: 454 and 483 nm. $E_{1\,cm}^{1\%}$ 1350 and 1180, respectively. Benzene: 463 and 492 nm. $E_{1\,cm}^{1\%}$ 1250 and 1090, respectively.

Melting Point. Physalien has been reported to melt at 95–96 °C (corr., under vacuum),[6] and at 98.5–99.5 °C.[7]

Probable Impurities: Oxidation products, *cis*-isomers, and, possibly, zeaxanthin.

Conditions of Storage: Darkness (brown vial), inert atmosphere (sealed ampoule), and a low temperature (0 °C).

References

1. R. Kuhn and W. Wiegand, *Helv. Chim. Acta*, **12**, 499 (1929).
2. L. Zechmeister and L. von Cholnoky, *Ann. Chem.* **481**, 42 (1930).
3. A. Winterstein and U. Ehrenberg, *Z. Physiol. Chem.*, **207**, 25 (1932).
4. P. Karrer and H. Wehrli, *Helv. Chim. Acta*, **13**, 1104 (1930).
5. R. Kuhn, A. Winterstein, and W. Kaufmann, *Chem. Ber.*, **63**, 1489 (1930).
6. O. Isler, H. Lindlar, M. Montavon, R. Rüegg, G. Saucy, and P. Zeller, *Helv. Chim. Acta*, **39**, 2041 (1956).
7. P. Karrer and E. Jucker, *Carotenoids*, E. A. Braude (translator), Elsevier, New York (1950).

Carot-33

Physalien

(3*R*,3′*R*)β,β-Carotene-3,3′-diol Dipalmitate

(Zeaxanthin Dipalmitate)

Formula: $C_{72}H_{116}O_4$
Formula Wt.: 1045.73
Calc. %: C, 82.70; H, 11.18; O, 6.12

Sources:

Natural Sources. Physalien occurs in a wide variety of plant materials. It was first isolated from the sepals of *Physalis alkekengi* and *Physalis franchetii.*[1] Since then, it has been reported to be present in *Lycium halimifolium,*[2] *Lycium barbarum,*[3] *Solanum hendersonii,*[3] *Aspargus officinalis,*[3] and *Hippophaes rhamnoides.*[3,4]

Chemical Synthesis. Zeaxanthin dipalmitate has been synthesized from zeaxanthin.[5] The total synthesis of zeaxanthin dipalmitate has also been reported.[6]

Isolation Procedures:[5] The sepals of *Physalis alkekengi* are exhaustively extracted with benzene. The combined extracts are concentrated to a small volume, and the pigment is precipitated or crystallized by the addition of ethanol[1] or acetone.[5] The pigment may also be purified by column chromatography.

Methods of Purification:

Chromatography. Physalien may be purified by chromatography on water-deactivated aluminum oxide. Hexane–ethyl ether (19:1) is used to develop the column.

Crystallization.[7] Physalien has been crystallized from benzene–methanol and from petroleum ether–ethanol.

Methods of Assaying for Purity:

Chromatography. The purity of physalien can be determined by chromatography on a column of aluminum oxide or magnesium oxide. Chromatograms are developed with chloroform or 19:1 hexane–ethyl ether when aluminum oxide is used and with dichloromethane when magnesium oxide is the adsorbent.

Carot-34

Phytoene

15-*cis*-7,8,11,12,7′,8′,11′,12′-Octahydro-ψ,ψ-carotene

(7,8,7′,8′,11,12,11′,12′-Octahydrolycopene)

Formula: $C_{40}H_{64}$
Formula Wt.: 544.96
Calc. %: C, 88.16; H, 11.84

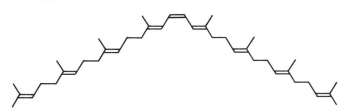

Sources:

Natural Sources. Phytoene is rather widely distributed in carotenoid-containing fruits and some other tissues not containing chlorophyll. The best sources are tomato fruits,[1–4] tomato paste,[5] and carrot oil.[5] The naturally occurring isomer has a central *cis*-configuration.[6]

Chemical Synthesis. Phytoene has been synthesized from all-*trans*-geranyllinalool.[7,8]

Isolation Procedures: Phytoene is extracted from plant materials with a solvent such as acetone, ethanol, or methanol, and then transferred into petroleum ether.[1–5] This compound is then purified by chromatography, with or without prior treatment with alcoholic potassium hydroxide.

Methods of Purification:

Chromatography. Phytoene is purified by chromatography on a column of magnesia–Supercel,[5] or alumina.[5,6]

Methods of Assaying for Purity:

Chromatography. Phytoene may be assayed for purity by chromatography on magnesia–Supercel,[5] or alumina.[5,6] It may also be assayed for purity by thin-layer chromatography on silica gel.[9]

Ultraviolet Spectrum. Hexane: 275, 285, and 297 nm.[6] $E_{1\,cm}^{1\%}$ 850 (286 nm).[5] The ultraviolet spectrum of all-*trans*-phytoene has also been reported.[6]

Infrared Spectrum. The infrared spectrum of central-*cis*-phytoene and all-*trans*-phytoene have been reported.[5,6,10]

Nuclear Magnetic Resonance Spectrum. The nuclear magnetic resonance spectrum of phytoene has been reported.[6,7,11]

Mass Spectrum. The mass spectrum of phytoene has been reported.[12,13]

Melting Point. Phytoene has not yet been crystallized. On cooling it forms a colorless glassy mass.

Probable Impurities: Waxes and *cis-trans*-isomers formed during isolation.

Conditions of Storage: In solution in petroleum ether under nitrogen at a low temperature (-20 °C).

References

1. J. W. Porter and F. P. Zscheile, *Arch. Biochem.*, **10**, 547 (1946).
2. J. W. Porter and R. E. Lincoln, *Arch. Biochem.*, **27**, 390 (1950).
3. W. J. Rabourn and F. W. Quackenbush, *Arch. Biochem. Biophys.*, **44**, 159 (1953).
4. G. Mackinney, C. M. Rick, and J. A. Jenkins, *Proc. Nat. Acad. Sci. U.S.*, **42**, 404 (1956).
5. W. J. Rabourn, F. W. Quackenbush, and J. W. Porter, *Arch. Biochem. Biophys.*, **48**, 267 (1954).
6. F. B. Jungalwala and J. W. Porter, *Arch. Biochem. Biophys.*, **110**, 291 (1965).
7. J. B. Davis, L. M. Jackman, P. T. Siddons, and B. C. L. Weedon, *Proc. Chem. Soc.*, 261 (1961).
8. J. B. Davis, L. M. Jackman, P. T. Siddons, and B. C. L. Weedon, *J. Chem. Soc. (C)*, 2154 (1966).
9. B. H. Davies, D. Jones, and T. W. Goodwin, *Biochem. J.*, **87**, 326 (1963).
10. W. J. Rabourn and F. W. Quackenbush, *Arch. Biochem. Biophys.*, **61**, 111 (1956).
11. B. C. L. Weedon, in *Chemistry and Biochemistry of Plant Pigments*, T. W. Goodwin, ed., p. 75, Academic Press, New York (1965).
12. B. C. L. Weedon, *Fortschr. Chem. Org. Naturstoffe*, **27**, 81 (1969).
13. O. B. Weeks, A. G. Andrewes, B. O. Brown, and B. C. L. Weedon, *Nature*, **224**, 879 (1969).

Carot-35
Phytofluene
15-*cis*-7,8,11,12,7′,8′-Hexahydro-ψ,ψ-carotene
(7,8,7′,8′, 11′,12′-Hexahydrolycopene)

Formula: $C_{40}H_{62}$
Formula Wt.: 542.94
Calc. %: C, 88.49; H, 11.51

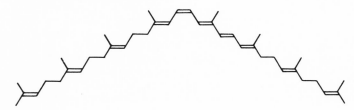

Sources:

Natural Sources. Phytofluene is rather widely distributed in carotenoid-containing fruit and in some other tissues not containing chlorophyll. It is found in persimmons,[1] red peppers,[1] carrots,[2] tomato fruits,[3,4] and tomato paste.[5,6] all-*trans*-Phytofluene also occurs naturally.[7]

Chemical Synthesis. Phytofluene has been synthesized chemically.[8,9]

Isolation Procedures: Phytofluene is extracted from plant materials with such solvents as acetone, ethanol, and methanol and is then transferred into petroleum ether.[1–6] It is then purified by chromatography, with or without a prior treatment with alcoholic potassium hydroxide.

Methods of Purification:

Chromatography. Phytofluene is purified by chromatography on calcium hydroxide–alumina,[5,6] magnesium oxide–Supercel,[10] or alumina.[11] all-*trans*-Phytofluene is readily separated from *cis*-phytofluene on partially deactivated alumina.[7]

Methods of Assaying for Purity:

Chromatography. Phytofluene may be assayed for purity by chromatography on calcium hydroxide–alumina,[5,6] magnesium oxide–Supercel,[10] or alumina.[11] Assay may also be made by chromatography on paper[12] or on a thin layer of silica gel.[13] The position of phytofluene on a chromatographic column may be readily determined by its characteristic green–white fluorescence when exposed to ultraviolet light of long wavelength.

Ultraviolet Spectrum. Petroleum ether:[3,5,14] 331, 348, and 367 nm. $E_{1\,cm}^{1\%}$ 1350 at 348 nm. Benzene:[15] 338, 355, and 374 nm. The ultraviolet spectrum of all-*trans*-phytofluene has been reported.[6,15]

Infrared Spectrum. The infrared spectrum of phytofluene has been reported.[8,16]

Nuclear Magnetic Resonance Spectrum. The nuclear magnetic resonance spectrum of phytofluene has been reported.[8,17]

Mass Spectrum. The mass spectrum of phytofluene has been published.[18,19]

Melting Point. Phytofluene has not been crystallized. On cooling, phytofluene forms a glassy mass lacking crystalline structure.

Probable Impurities: Phytoene, *trans*-phytofluene, and oxidation products.

Conditions of Storage: In solution in petroleum ether under nitrogen at a low temperature (-20 °C).

References

1. L. Zechmeister and A. Sandoval, *Arch. Biochem.*, **8**, 425 (1945).
2. H. H. Strain, *J. Biol. Chem.*, **127**, 191 (1939).
3. V. Wallace and J. W. Porter, *Arch. Biochem. Biophys.*, **36**, 468 (1952).
4. J. W. Porter and F. P. Zscheile, *Arch. Biochem.*, **10**, 547 (1946).
5. L. Zechmeister and A. Sandoval, *J. Am. Chem. Soc.*, **68**, 197 (1946).
6. F. J. Petracek and L. Zechmeister, *J. Am. Chem. Soc.*, **74**, 184 (1952).
7. S. C. Kushwaha, G. Suzue, C. Subbarayan, and J. W. Porter, *J. Biol. Chem.*, **245**, 4708 (1970).
8. J. B. Davis, L. M. Jackman, P. T. Siddons, and B. C. L. Weedon, *Proc. Chem. Soc.*, 261 (1961).
9. J. B. Davis, L. M. Jackman, P. T. Siddons, and B. C. L. Weedon, *J. Chem. Soc. (C)*, 2154 (1966).
10. J. W. Porter and R. E. Lincoln, *Arch. Biochem.*, **27**, 390 (1950).
11. D. A. Beeler and J. W. Porter, *Biochem. Biophys. Res. Commun.*, **8**, 367 (1962).
12. A. Jensen, *Acta Chem. Scand.*, **14**, 2051 (1960).
13. B. H. Davies, D. Jones, and T. W. Goodwin, *Biochem. J.*, **87**, 326 (1963).
14. B. H. Davies, in *Chemistry and Biochemistry of Plant Pigments*, T. W. Goodwin, ed., p. 489, Academic Press, New York (1965).
15. T. W. Goodwin, in *Modern Methods of Plant Analysis*, K. Paech and M. V. Tracey, eds., Vol. 3, (1955), p. 272.
16. F. B. Jungalwala and J. W. Porter, *Arch. Biochem. Biophys.*, **110**, 291 (1965).
17. B. C. L. Weedon, in *Chemistry and Biochemistry of Plant Pigments*, T. W. Goodwin, ed., p. 75, Academic Press, New York (1965).
18. O. B. Weeks, A. G. Andrewes, B. O. Brown, and B. C. L. Weedon, *Nature*, **224**, 879 (1969).
19. B. C. L. Weedon, *Fortschr. Chem. Org. Naturstoffe*, **27**, 81 (1969).

Carot-36
Prolycopene

Formula: $C_{40}H_{56}$
Formula Wt.: 536.90
Calc. %: C, 89.49; H, 10.51

Prolycopene is a poly-*cis*-lycopene of unknown stereochemical configuration. However, the suggestion has been made that it may be a symmetrical penta-*cis*-lycopene containing a central *cis*- and four other, unhindered, *cis*-double bonds.[1]

Sources:

Natural Sources. Prolycopene is found in small amounts in various fruits and flowers.[2] However, the best source of this pigment is the ripe fruit of the "tangerine" or "golden jubilee" type of tomato.[3-5]

Chemical Synthesis. The chemical synthesis of prolycopene has not yet been reported.

Isolation Procedures:[3,4] Plant tissue is extracted with a solvent such as acetone, methanol, or ethanol and the carotenes are then transferred into petroleum ether. Prolycopene in this extract is then purified by chromatography, either with or without prior saponification with alcoholic potassium hydroxide.

Methods of Purification:

Chromatography. Prolycopene may be purified by chromatography on calcium hydroxide,[3] magnesium oxide–Supercel,[5] or deactivated alumina.[6]

Crystallization. Prolycopene has been crystallized from petroleum ether.[3]

Methods of Assaying for Purity:

Chromatography. The purity of prolycopene may be determined by chromatography on calcium hydroxide,[3,7] magnesium oxide–Supercel,[5] or deactivated alumina.[6]

Visible Spectrum. Petroleum ether: 443.5 and 470 nm.[1] $E_{1\,cm}^{1\%}$ 1920.[8] Carbon disulfide: 469.5 and 500.5 nm.[3] Benzene: 454.5 and 485 nm.[3] Chloroform: 453.5 and 484 nm.[3]

Infrared Spectrum. The infrared spectrum of prolycopene has been reported.[9]

Melting Point.[3] Prolycopene melts at 111 °C.

Probable Impurities: Oxidation products, and other isomers of lycopene.[10]

Conditions of Storage: Darkness (brown vial), inert atmosphere (sealed ampoule), and low temperature (−20 °C).

References

1. L. Zechmeister, cis–trans-*Isomeric Carotenoids, Vitamins A and Arylpolyenes*, Academic Press, New York, 1962, p. 172.
2. P. Karrer and E. Jucker, *Carotenoids*, E. A. Braude (translator), Elsevier, New York, 1950, p. 125.
3. A. L. LeRosen and L. Zechmeister, *J. Am. Chem. Soc.*, **64**, 1075 (1942).
4. L. Zechmeister, A. L. LeRosen, F. W. Went, and L. Pauling, *Proc. Nat. Acad. Sci. U.S.*, **27**, 468 (1941).
5. J. W. Porter and R. E. Lincoln, *Arch. Biochem.*, **27**, 390 (1950).
6. S. C. Kushwaha, G. Suzue, C. Subbarayan, and J. W. Porter, *J. Biol. Chem.*, **245**, 4708 (1970).
7. L. Zechmeister and J. H. Pinckard, *J. Am. Chem. Soc.*, **69**, 1930 (1947).
8. L. Zechmeister, A. L. LeRosen, W. A. Schroeder, A. Polgar, and L. Pauling, *J. Am. Chem. Soc.*, **65**, 1940 (1943).
9. K. Lunde and L. Zechmeister, *J. Am. Chem. Soc.*, **77**, 1647 (1955).
10. E. F. Magoon and L. Zechmeister, *Arch. Biochem. Biophys.*, **69**, 535 (1957).

Carot-37
Proneurosporene
(Synonyms: Protetrahydrolycopene, Neoneurosporene P, Unidentified Carotene I, and Poly-*cis*-ψ-carotene)[1]

Formula: $C_{40}H_{58}$
Formula Wt.: 538.91
Calc. %: C, 89.15; H, 10.85

Proneurosporene is a poly-*cis*-neurosporene that contains a *trans*-double bond at the middle of the molecule.[2]

Sources:

Natural Sources. The principal sources of proneurosporene are the ripe berries of *Pyracantha angustifolia*[2] and fruits of the "golden jubilee" and "tangerine" varieties of tomato.[3]

Chemical Synthesis. The chemical synthesis of proneurosporene has not yet been reported.

Isolation Procedures: Berries or fruit are extracted with a solvent such as acetone, ethanol, or methanol. The carotenes are then transferred into petroleum ether and purified by chromatography, with or without prior saponification with alcoholic potassium hydroxide.

Methods of Purification:

Chromatography. Proneurosporene is purified by chromatography on 2:1 calcium hydroxide–Celite,[2] 3:1:1 magnesium oxide–calcium hydroxide–Celite,[2] 1:1 magnesium oxide–Supercel,[3] or deactivated alumina.[4]

Methods of Assaying for Purity:

Chromatography. Assays for the purity of proneurosporene may be carried out by chromatography on columns of the adsorbents just given.[2-4]

Visible Spectrum.[2] Hexane: 408, 432, and 461 nm. $E_{1\,cm}^{1\%}$ 2040 at 432 nm.

Infrared Spectrum. The infrared spectrum of proneurosporene has been reported.[5]

Nuclear Magnetic Resonance Spectrum. The nuclear magnetic resonance spectrum of proneurosporene has not yet been reported.

Melting Point. Proneurosporene has not yet been crystallized.

Probable Impurities: Prolycopene, other *cis*-isomers of neurosporene, and oxidation products.

Conditions of Storage: Darkness (brown vial), inert atmosphere (sealed ampoule), and low temperature (0 °C).

References

1. L. Zechmeister, cis–trans-*Isomeric Carotenoids, Vitamins A and Arylpolyenes*, Academic Press, New York, 1962, p. 74.
2. E. F. Magoon and L. Zechmeister, *Arch. Biochem. Biophys.*, **68**, 263 (1957).
3. H. H. Trombly and J. W. Porter, *Arch. Biochem. Biophys.*, **43**, 443 (1953).
4. S. C. Kushwaha, G. Suzue, C. Subbarayan, and J. W. Porter, *J. Biol. Chem.*, **245**, 4708 (1970).
5. K. Lunde and L. Zechmeister, *J. Am. Chem. Soc.*, **77**, 1647 (1955).

Carot-38
Retinal
(Vitamin A_1 Aldehyde)

Formula: $C_{20}H_{28}O$
Formula Wt.: 284.44
Calc. %: C, 84.45; H, 9.92; O, 5.63

Isomers: Six isomers of retinal have been reported:[1] all-*trans*; 13-*cis* (neo-a); 11-*cis* (neo-b); 9-*cis* (iso-a); 9,13-di-*cis* (iso-b); and 11,13-di-*cis* (neo-c).

Biopotency: All-*trans*-retinal and 13-*cis*-retinal have 91% of the biological activity of all-*trans*-retinyl acetate. Other isomers (*cis*) of retinal have a lower biological activity.[2]

Sources:

Natural Sources. All-*trans*-retinal is present in herring roe and in hens' eggs.[3] In the eyes of animals, marine fish, and crustacea, retinal is present as the 11-*cis* isomer.[4]

Chemical Synthesis. Retinal is formed from retinol by oxidation with activated MnO_2 in petroleum ether.[5]

Methods of Purification: Retinal may be separated from retinol and its esters by column chromatography. Retinal and its isomers may then be crystallized from petroleum ether; or their semicarbazones from ethanol; or their (2,4-dinitrophenyl)hydrazones from ethyl acetate.[6]

Methods of Assaying for Purity:

Column and Thin-Layer Chromatography. The adsorbents used for the chromatography of retinal are similar, or identical, to those used for the chromatography of retinol and its derivatives,[7,8] Retinal is eluted from water-deactivated alumina columns with 1–2% acetone. When retinal is chromatographed on thin-layer plates of silica gel G, the chromatograms are developed with ether–hexane (1:1). The isomers of retinal are partially separated in each of the above systems.

Gas–Liquid Chromatography. Retinal is more stable than retinol or retinyl acetate on gas–liquid columns. It can be recovered quantitatively under proper conditions.[9] However, isomerization occurs at high temperatures.

Ultraviolet Spectrum. The $E_{1\,cm}^{1\%}$ values and the absorption maxima of retinal and its isomers in ethanol have been reported as follows:[1,8,10] all-*trans*, 1530 (381 nm); 13-*cis*, 1250 (375 nm); 11-*cis*, 878 (376 nm); 9-*cis*, 1270 (373 nm); 9,13-di-*cis*, 1140 (368 nm); and 11,13-di-*cis*, 700 (373 nm).

Other Spectra. The infrared[6] and fluorescence[11] spectra of retinal have been reported.

Melting Points. The melting points of retinal and its isomers have been reported as follows:[10] all-*trans*, 57 and 65 °C; 13-*cis*, 77 °C; 11-*cis*, 64 °C; 9-*cis*, 64 °C; and 9,13-di-*cis*, 49 and 85 °C.

Quantitative Assays: The quantity of retinal is most frequently determined by ultraviolet absorption spectroscopy.[1,8,10] However, retinal forms a transient, highly colored complex with antimony trichloride,[3] trifluoroacetic acid,[12] and other Lewis acids. Hence, the quantity of retinal may also be determined through measurement of the quantity of light absorbed by this complex. The highest recent value of $E_{1\,cm}^{1\%}$ at 666 nm is 4150.[3] Previous values have been as low as 3340.[8,12] The presence of acetic anhydride affects the absorption maximum and the extinction coefficient. Various retinal isomers, but not retinol, also react with thiobarbituric acid. The complex has $E_{1\,cm}^{1\%}$ 2040 at 530 nm.[13]

The 9-*cis*, 11-*cis*, and 9,13-di-*cis* isomers of retinal react with opsin (isolated from the retina) to yield rhodopsin or isorhodopsin.[14,15] In 2% digitonin solution, the $E_{1\,cm}^{1\%}$ values for the opsin complexes are 9-*cis*, 1439 (487 nm); 11-*cis*, 1467 (500 nm); and 9,13-di-*cis*, 1271 (487 nm). The retinol isomers having 11-*cis* or 13-*cis* double bonds may be distinguished by the maleic anhydride test.[6,16]

Probable Impurities: Retinol, *cis*-isomers, and oxidation products.

Conditions of Storage: Darkness (brown vial), inert atmosphere (sealed ampoule), and low temperature (0 °C). Solutions of retinal in pure solvents are reasonably stable in the dark at low temperatures.

References

1. M. Kofler and S. H. Rubin, *Vitamins Hormones*, **18**, 315 (1960).
2. S. R. Ames, W. J. Swanson, and P. L. Harris, *J. Am. Chem. Soc.*, **77**, 4136 (1955).
3. P. A. Plack and S. K. Kon, *Biochem. J.*, **81**, 561 (1961).
4. G. Wald, *Vitamins Hormones*, **18**, 417 (1960).
5. H. B. Henbest, E. R. H. Jones, and T. C. Owen, *J. Chem. Soc.*, 4909 (1957).
6. C. D. Robeson, W. P. Blum, J. M. Dieterle, J. D. Cawley, and J. G. Baxter, *J. Am. Chem. Soc.*, **77**, 4120 (1955).
7. J. W. Porter and D. G. Anderson, in *Chromatography*, E. Heftmann, ed., Reinhold Publishing Corp., New York (1961), p. 465.
8. J. A. Olson, in *Newer Methods of Nutritional Biochemistry*, Vol. 2, A. A. Albanese, ed., Academic Press Inc., New York (1965), p. 345.
9. P. E. Dunagin, Jr., and J. A. Olson, *Anal. Chem.*, **36**, 756 (1964).
10. J. G. Baxter, in *Comprehensive Biochemistry*, Vol. 9, M. Florkin and E. H. Stotz, eds., Elsevier, Amsterdam (1962), p. 169.
11. W. A. Hagins and W. H. Jennings, *Discussions Faraday Soc.*, **27**, 180 (1959).
12. R. E. Dugan, N. A. Frigerio, and J. M. Siebert, *Anal. Chem.*, **36**, 114 (1964).
13. S. Futterman and J. D. Saslaw, *J. Biol. Chem.*, **236**, 1652 (1961).
14. R. Hubbard, R. I. Gregerman, and G. Wald, *J. Gen. Physiol.*, **36**, 415 (1953).
15. D. C. Herting, E. E. Drury, and P. L. Harris, *Anal. Biochem.*, **4**, 459 (1962).
16. S. R. Ames and R. W. Lehman, *J. Assoc. Offic. Agr. Chem.*, **43**, 21 (1960).

Carot-39
Retinoic Acid
(Vitamin A_1 Acid)

Formula: $C_{20}H_{28}O_2$
Formula Wt.: 300.44
Calc. %: C, 79.96; H, 9.39; O, 10.65

Isomers: The four unhindered isomers of retinoic acid (all-*trans*, 9-*cis*, 13-*cis*, and 9,13-di-*cis*) have been crystallized and characterized.[1]

Biopotency: All-*trans*-retinoic acid has 10–141% of the growth-promoting activity of retinol. This discrepancy in activity is attributable to the method of administration of retinoic acid to the animal. *cis*-Isomers of retinoic acid have less biological activity than the all-*trans*-compound. Retinoic acid does not fulfill the visual or reproductive functions of retinol or retinal.[2]

Sources:

Natural Sources. Traces of retinoic acid are found in liver and bile after the administration of retinal;[3] larger amounts are excreted in the bile as the β-D-glucosiduronic acid.[4]

Chemical Synthesis. Retinoic acid is prepared from retinal by oxidation with silver oxide, or as an intermediate in the synthesis of retinol.[1]

Methods of Purification: Retinoic acid may be separated from many other compounds by chromatography on columns of silicic acid. Complete purification of the compound is achieved by crystallization from methanol, ethanol, or isopropyl alcohol. Methyl retinoate is crystallized from methanol.[5]

Methods of Assaying for Purity:

Column and Thin-Layer Chromatography. Retinoic acid is eluted from silicic acid columns by small proportions of ethanol in hexane. This compound also migrates well on a thin-layer plate of silica gel G when a solvent system of 4:1:1 benzene–chloroform–methanol[6] is used. Retinyl β-D-glucosiduronic acid also migrates on silica gel G plates in 5:5:5:1 benzene–chloroform–methanol–acetic acid.[4]

Partition Chromatography. Retinoic acid may be separated from retinol on silicone-treated paper (reverse phase) when various polar solvents are used to develop the chromatogram.[7]

Ion-Exchange Chromatography. Retinoic acid is eluted from a DEAE-cellulose column by 0.04 M HCl in ethanol,[7] and from a Biorad AG-2-X8 anion-exchange column with 5:95 acetic acid–methanol.[3]

Gas–Liquid Chromatography. Methyl retinoate may be assayed for purity by gas–liquid chromatography. This compound may also be recovered quantitatively, since it is quite stable at elevated temperatures.[8]

Ultraviolet Spectrum. The $E_{1\,cm}^{1\%}$ values of retinoic acid and its isomers[9-11] in purified ethanol[5] are: all-*trans*, 1500 (350 nm); 13-*cis*, 1320 (354 nm); 9-*cis*, 1230 (345 nm); and 9,13-di-*cis*, 1150 (346 nm).

Melting Point. Melting points for retinoic acid and its isomers have been reported as follows: all-*trans*, 180 °C; 13-*cis*, 175 °C; 9-*cis*, 191 °C; and 9,13-di-*cis*, 136 °C.[1,10]

Quantitative Assays: Ultraviolet absorption spectroscopy is most commonly used to determine the quantity of retinoic acid.[9-11] The amount of light absorbed by the transient highly colored complex of retinoic acid with antimony trichloride,[12] trifluoroacetic acid,[13] and other Lewis acids is also a measure of the quantity of retinoic acid. An $E_{1\,cm}^{1\%}$ value of 1770 (574 nm) has been reported for this complex.[13]

Probable Impurities: *cis*-Isomers and oxidation products.

Conditions of Storage: Darkness (brown vial), inert atmosphere (sealed ampoule), and low temperature (0 °C). Solutions of the acid in pure organic solvents in the dark are reasonably stable, whereas aqueous solutions of the acid deteriorate rapidly.

References

1. O. Isler, R. Rüegg, U. Schwieter, and J. Würsch, *Vitamins Hormones*, **18**, 295 (1960).
2. J. A. Olson, *J. Lipid Res.*, **5**, 281 (1964).
3. P. E. Dunagin, Jr., and J. A. Olson, *Biochim. Biophys. Acta*, **90**, 432 (1964).
4. P. E. Dunagin, Jr., E. H. Meadows, Jr., and J. A. Olson, *Science*, **148**, 86 (1965).
5. C. D. Robeson, J. D. Cawley, L. Weisler, M. H. Stern, C. C. Eddinger, and A. J. Chechak, *J. Am. Chem. Soc.*, **77**, 4111 (1955).
6. K. Yagishita, P. R. Sundaresan, and G. Wolf, *Nature*, **203**, 410 (1964).
7. S. Futterman, *J. Biol. Chem.*, **237**, 677 (1962).
8. P. E. Dunagin, Jr., and J. A. Olson, *Anal. Chem.*, **36**, 756 (1964).
9. M. Kofler and S. H. Rubin, *Vitamins Hormones*, **18**, 315 (1960).
10. J. G. Baxter, in *Comprehensive Biochemistry*, Vol. 9, M. Florkin and E. H. Stotz, eds., Elsevier, Amsterdam (1962), p. 169.
11. J. A. Olson, in *Newer Methods of Nutritional Biochemistry*, Vol. 2, A. A. Albanese, ed., Academic Press, New York (1965), p. 345.
12. L. Jurkowitz, *Arch. Biochem. Biophys.*, **98**, 337 (1962).
13. R. E. Dugan, N. A. Frigerio, and J. M. Siebert, *Anal. Chem.*, **36**, 114 (1964).

Carot-40

Retinol
(Vitamin A₁ Alcohol)

Formula: $C_{20}H_{30}O$
Formula Wt.: 286.46
Calc. %: C, 83.86; H, 10.56; O, 5.59

Nomenclature: "Vitamin A₁ alcohol" has been designated "retinol" by the Commission on Nomenclature of Biological Chemistry,[1] and the stereochemistry of methyl groups at C-1 has been assigned by analogy with that of lanostane.

Isomers: Six isomers of vitamin A have been reported: all-*trans*; 13-*cis* (neo-a); 11-*cis* (neo-b); 9-*cis* (iso-a); 9,13-di-*cis* (iso-b); and 11,13-di-*cis* (neo-c).[2]

Biopotency: Pure all-*trans*-retinol has 3.333×10^6 I.U./g. Both the U.S.P. (United States Pharmacopeia) unit and International Unit (I.U.) are defined as the amount of all-*trans*-retinyl acetate (0.344 μg) having the biological activity of 0.300 μg of all-*trans*-retinol.

Sources:

Natural Sources. The best sources are liver oils of marine fish, where vitamin A₁ occurs mainly as retinyl esters. Free retinol is also present in the blood and tissues of vertebrates and in the eyes of crustacea.

Chemical Synthesis. Many procedures have been reported for the synthesis of retinol. These include synthesis from acetone and acetylene,[3] from β-ionone via condensation with methyl 3-methylglutaconate,[4] from β-ionone via vinyl-β-ionol by the Wittig reaction,[5] and from β-ionone via a C_{14}-aldehyde (Darzens' reaction) followed by Grignard addition of 3-methyl-2-penten-4-yn-1-ol.[6] These methods of synthesis have been reviewed.[6-8]

Methods of Purification:

Crystallization. In the past, retinol or its esters were isolated from fish-liver oils by molecular distillation. At present, however, high-potency concentrates and crystalline retinol, or its derivatives, are generally prepared by chemical synthesis. Solvated crystals of all-*trans*-retinol are obtained from methanol or ethyl formate. Solvent-free crystals are obtained from propylene oxide or petroleum ether.[9,10]

Column Chromatography. Many adsorbents have been employed, including alumina, dicalcium phosphate, calcium carbonate, magnesium oxide, magnesium carbonate, silicic acid, and bone meal.[11] Columns of water-deactivated alumina are commonly used; from these, retinol is eluted quantitatively with 3–5% acetone in hexane. Isomers of retinol may be separated on columns of dicalcium phosphate[12,13] or on thin-layer plates of silica gel G developed with petroleum ether (low boiling)—methyl-heptenone (11:2).[14] Thin-layer plates of water-deactivated alumina have also been used with various solvents.[15] Retinol may be detected by fluorescence under ultraviolet light, or by reaction with iodine vapor.

Partition Chromatography. Various adsorbents, impregnated such as with vaseline or silicone oil, may be used as a stationary phase, with relatively polar solvents as the moving phase.[2,11,12]

Gas–Liquid Chromatography. Retinol is rapidly converted into anhydroretinol under normal conditions of gas–liquid chromatography,[16] but it may be isolated with little destruction at 150 °C by use of high flow rates on columns of 1% SE-30 on siliconized 60–80 mesh Gaschrome P that has been conditioned at 250 °C and treated with an antioxidant.[17]

Methods of Assaying for Purity:

Chromatography. The above methods of chromatography may be used to assay the purity of retinol.

Ultraviolet Spectrum. Retinol and its isomers each have a single light-absorption maximum. $E_{1cm}^{1\%}$ in ethanol; all-*trans*, 1832 (325 nm); 13-*cis*, 1686 (328 nm); 11-*cis*, 1220 (319 nm) or 945 (322 nm);[9] 9-*cis*, 1480 (323 nm); 9,13-di-*cis*, 1379 (324 nm); and 11,13-di-*cis*, 908 (311 nm). Several *cis*-isomers also have small absorption maxima ($E_{1cm}^{1\%}$ about 350) between 235 and 260 nm. Spectral properties of retinol derivatives have been collated.[2,10,12]

Infrared Spectrum The infrared spectrum of retinol has been reported.[2,4,12,18]

Fluorescence Spectrum. Light of wavelength 325 nm is absorbed maximally, and emitted at 470 nm.[12,19]

Nuclear Magnetic Resonance Spectrum. The nuclear magnetic resonance spectrum of retinol has been reported.[2]

Melting Point. The melting points of retinol and some of its isomers have been reported as follows:[6] all-*trans*, 62–64 °C (solvent free), 8 °C (methanol-solvated); 13-*cis*, 58–60 °C; 9-*cis*, 82–83 °C; and 9,13-di-*cis*, 58–59 °C.

Other Properties: Polarography of retinol[20] and the x-ray powder diagram of its crystals[2] have been reported.

Quantitative Assays: The quantity of retinol is usually determined by absorption of ultraviolet light.[2,10,12] Fluorescence of retinol may be measured.[21] Assays may be made by the Carr-Price reaction. Retinol forms transient, but intensely colored, complexes with antimony trichloride, trifluoroacetic acid,[22] or other Lewis acids. The $E_{1cm}^{1\%}$ value at 620 nm of this species is 5070. All isomers of retinol give the same complex. Retinol may also be dehydrated with acid to yield anhydroretinol, which is measured spectrophotometrically. $E_{1cm}^{1\%}$ values in ethanol are 2500 at 351 nm, 3650 at 371 nm, and 3180 at 392 nm.[23]

Probable Impurities: *cis*-Isomers and oxidation products of retinol are the most common impurities.

Conditions of Storage: Darkness (brown vial), inert atmosphere (sealed ampoule), and low temperature (0 °C). Peroxide-free ethyl ether and acid-free acetone or ethyl acetate are preferable to either ethanol or petroleum ether for storage. However, ethanol is suitable as a solvent for brief periods, for spectroscopic analysis.[24]

References

1. Commission on the Nomenclature of Biological Chemistry, IUPAC, *J. Am. Chem. Soc.*, **82**, 5575 (1960); *Biochim. Biophys. Acta*, **107**, 1 (1965); *J. Biol. Chem.*, **241**, 527 (1966); *Biochemistry*, **10**, 4827 (1971).
2. M. Kofler and S. H. Rubin, *Vitamins Hormones*, **18**, 315 (1960).
3. W. Kimel, J. D. Surmatis, J. Weber, G. O. Chase, N. W. Sax, and A. Ofner, *J. Org. Chem.*, **22**, 1611 (1957).
4. C. D. Robeson, J. D. Cawley, L. Weisler, M. H. Stern, C. C. Eddinger, and A. J. Chechak, *J. Am. Chem. Soc.*, **77**, 4111 (1955).
5. H. Pommer and W. Sarnecki, German Patent 1,046,612 (Dec. 18, 1958); German Patent 1,059,900 (June 25, 1959).
6. O. Isler, R. Rüegg, U. Schwieter, and J. Würsch, *Vitamins Hormones*, **18**, 295 (1960).
7. J. G. Baxter, *Fortschr. Chem. Org. Naturstoffe*, **9**, 42 (1952).
8. N. A. Milas, in *The Vitamins*, Vol. I, W. H. Sebrell, Jr., and R. S. Harris, eds., Academic Press Inc., New York (1954), p. 4.
9. T. Moore, *Vitamin A*, Elsevier, Amsterdam (1957).
10. J. G. Baxter, in *Comprehensive Biochemistry*, Vol. 9, M. Florkin and E. H. Stotz, eds., Elsevier, Amsterdam (1962), p. 168.
11. J. W. Porter and D. G. Anderson, in *Chromatography*, E. Heftmann, ed., Reinhold Publishing Corp., New York (1961), p. 465.
12. J. A. Olson, in *Newer Methods of Nutritional Biochemistry*, Vol. 2, A. A. Albanese, ed., Academic Press Inc., New York (1965), p. 345.
13. W. Hjarde, *Acta Chem. Scand.*, **4**, 628 (1950).
14 C. v. Planta, U. Schwieter, L. Chopard-dit-Jean, R. Rüegg, and O. Isler, *Helv. Chim. Acta*, **45**, 548 (1962).

15. J. Davídek and J. Blattná, *J. Chromatog.*, **7**, 204 (1962).
16. T. Ninomiya, K. Kidokoro, M. Horiguchi, and N. Higosake, *Vitamin*, **27**, 349 (1963).
17. P. E. Dunagin, Jr., and J. A. Olson, *Anal. Chem.*, **36**, 756 (1964).
18. C. D. Robeson, W. P. Blum, J. M. Dieterle, J. D. Cawley, and J. G. Baxter, *J. Am. Chem. Soc.*, **77**, 4120 (1955).
19. D. E. Duggan, R. L. Bowman, B. B. Brodie, and S. Udenfriend, *Arch. Biochem. Biophys.*, **68**, 1 (1957).
20. E. J. Kuta, *Science*, **144**, 1130 (1964).
21. S. Futterman and J. S. Andrews, *J. Biol. Chem.*, **239**, 81 (1964).
22. R. E. Dugan, N. A. Frigerio, and J. M. Siebert, *Anal. Chem.*, **36**, 114 (1964).
23. K. Harashima, H. Okazaki, and H. Aoki, *J. Vitaminol.* (Kyoto), **7**, 150 (1961).
24. *Suggestions for the Storage and Use of Crystalline Vitamin A*, Distillation Products Industries, Rochester, N. Y. (1964).

Carot-41
Retinyl Acetate
(Vitamin A₁ Acetate)

Formula: $C_{22}H_{32}O_2$
Formula Wt : 328.50
Calc. %: C, 80.44; H, 9.82; O, 9.74

Isomers: The acetates of each of the six isomers of retinol have been synthesized.[1]

Biopotency: A U.S.P. unit or an International Unit (I.U.) of all-*trans*-retinyl acetate is 0.344 μg. Therefore, pure all-*trans*-retinyl acetate contains[2] 2.904×10^6 I.U./g.

Sources:

Natural Sources. Retinyl acetate is not found in natural materials.

Chemical Synthesis. Retinyl acetate is synthesized from retinol by treatment with acetic anhydride or acetyl chloride in pyridine, or from acetylated intermediates in the synthesis of retinol.[3]

Methods of Purification: Retinyl acetate may be separated from retinol by column chromatography. The compound may then be purified by crystallization from methanol. Purification methods were reviewed in 1960.[1]

Methods of Assaying for Purity:

Column Chromatography. Similar or identical adsorbents are used for the column chromatography of retinol and retinyl acetate.[4,5] On columns of water-deactivated alumina, retinyl acetate is eluted, after β-carotene, by hexane or 0.5% of acetone in hexane.

Partition Chromatography. Column-partition chromatography and reverse-phase paper chromatography have been used to separate retinyl acetate from retinol and other retinyl esters.[1]

Gas–Liquid Chromatography. Retinyl acetate may be assayed for purity by gas–liquid chromatography. However, retinyl acetate forms anhydroretinol during gas–liquid chromatography unless proper conditions are maintained.[6]

Ultraviolet Spectrum. The maximum for retinyl acetate differs from that of retinol in $E_{1cm}^{1\%}$ values only. The $E_{1cm}^{1\%}$ values reported for retinyl acetate in ethanol are: all-*trans*, 1560 (325–326 nm); 13-*cis*, 1430 (328 nm); 11-*cis*, 973 (320–321 nm); 9-*cis*, 1200 (323 nm); 9,13-di-*cis*, 1110 (324 nm); and 11,13-di-*cis*, 859 (310–311 nm).[1,5]

Fluorescence Spectrum. Retinyl acetate maximally absorbs light of 325 nm, and emits part of the energy at 470 nm.[1,5,7,8]

Melting Point. A value of 57–58 °C has been reported.

Other Properties: Infrared spectrum, nuclear magnetic resonance spectrum, and polarographic behavior are similar for retinol and retinyl acetate.

Quantitative Assays: Ultraviolet-light absorption and colorimetric analysis[1,5] are most commonly used in assays for quantity of retinyl acetate. Colorimetric assays may be made with the Carr-Price reagent (antimony trichloride), or with trifluoroacetic acid[9] or other Lewis acids. The $E_{1 \text{ cm}}^{1\%}$ value at 616 nm is 4420 for the colored species.

Probable Impurities: *cis*-Isomers, retinol, or oxidation products of retinyl acetate.

Conditions of Storage: Darkness (brown vial), inert atmosphere (sealed ampoule), and low temperature (0 °C). Retinyl acetate is more stable in peroxide-free ethyl ether, acid-free acetone, or acid-free ethyl acetate in the dark than it is in other solvents.[10]

References

1. M. Kofler and S. H. Rubin, *Vitamins Hormones*, **18**, 315 (1960).
2. Commission on the Nomenclature of Biological Chemistry, IUPAC, *J. Am. Chem. Soc.*, **82**, 5575 (1960).
3. N. A. Milas, in *The Vitamins*, Vol. 1, W. H. Sebrell, Jr., and R. S. Harris, eds., Academic Press Inc., New York (1954), p. 4.
4. J. W. Porter and D. G. Anderson, in *Chromatography*, E. Heftmann, ed., Reinhold Publishing Corp., New York (1961), p. 465.
5. J. A. Olson, in *Newer Methods of Nutritional Biochemistry*, Vol. 2, A. A. Albanese, ed., Academic Press Inc., New York (1965), p. 345.
6. P. E. Dunagin, Jr., and J. A. Olson, *Anal. Chem.*, **36**, 756 (1964).
7. H. Sobotka, S. Kann, and E. Loewenstein, *J. Am. Chem. Soc.*, **65**, 1959 (1943).
8. S. Futterman and J. S. Andrews, *J. Biol. Chem.*, **239**, 81 (1964).
9. R. E. Dugan, N. A. Frigerio, and J. M. Siebert, *Anal. Chem.*, **36**, 114 (1964).
10. *Suggestions for the Storage and Use of Crystalline Vitamin A*, Distillation Products Industries, Rochester, N. Y. (1964).

Carot-42
Retinyl Palmitate
(Vitamin A$_1$ Palmitate)

Formula: $C_{36}H_{60}O_2$
Formula Wt.: 524.88
Calc. %: C, 82.38; H, 11.52; O, 6.10

Isomers: Six isomers of retinyl palmitate may be formed (see Retinol, Carot-40).

Biopotency: A U.S.P. unit or an International Unit of all-*trans*-retinyl palmitate is 0.55 μg. Thus, pure all-*trans*-retinyl palmitate contains 1.817×10^6 I.U./g.

Sources:

Natural Sources. Retinyl palmitate is the major ester of retinol found in liver, intestine, and retina of many vertebrates. Smaller amounts of stearate, oleate, and other esters are also present.[1-3]

Chemical Synthesis. Retinyl palmitate is synthesized by direct esterification of retinol with palmitoyl chloride in pyridine,[4] or by reaction with methyl palmitate in the presence of sodium ethoxide.[5]

Methods of Purification: Retinyl palmitate may be separated from retinol by column chromatography. It may then be crystallized from propylene oxide.[4]

Methods of Assaying for Purity:

Column and Thin-Layer Chromatography. Similar adsorbents

are used for the column chromatography of retinol and retinyl palmitate.[6,7] Retinyl palmitate is eluted from columns of water-deactivated alumina with hexane or a very small percentage of acetone in hexane. Retinyl palmitate may also be chromatographed on a thin layer of silica gel G. Petroleum ether–isopropyl ether–acetic acid–water (180:20:2:5) or petroleum ether–acetonitrile–acetic acid–water (190:10:1:5)[2] are used to develop the chromatogram.

Partition Chromatography. Column-partition chromatography and reverse-phase paper chromatography have been used to separate retinyl palmitate from retinyl acetate and retinol.[7,8]

Gas–Liquid Chromatography. Retinyl palmitate does not emerge from gas–liquid columns at temperatures suitable for the chromatography of retinyl acetate and retinol.[9]

Ultraviolet Spectrum. Retinyl palmitate differs from retinol in its ultraviolet absorption spectrum in $E_{1 \text{ cm}}^{1\%}$ values only. In ethanol, all-*trans*-retinyl palmitate has an $E_{1 \text{ cm}}^{1\%}$ value of 1000 at 325 nm.

Fluorescence Spectrum. Light of 325 nm is absorbed maximally by retinyl palmitate. A portion of this energy is emitted[2] as light of 470 nm.

Melting Point. A value of 28–29 °C has been reported.

Quantitative Assays: The quantity of retinyl palmitate is normally determined by measurement of ultraviolet light absorbed,[7,8] or through measurement of light emitted by fluorescence.[2,3,10] Assays may also be made by measuring the absorbance of the colored complex formed with such Lewis acids as antimony trichloride and trifluoroacetic acid.[11] An $E_{1 \text{ cm}}^{1\%}$ value of 2760 is obtained at 616 nm.

Probable Impurities: Retinol, *cis*-isomers, and oxidation products.

Conditions of Storage: Darkness (brown vial), inert atmosphere (sealed ampoule), and low temperature (0 °C). Solutions of retinyl palmitate are reasonably stable in the dark at low temperatures in peroxide-free and acid-free organic solvents.[12]

References

1. S. Mahadevan and J. Ganguly, *Biochem. J.*, **81**, 53 (1961).
2. S. Futterman and J. S. Andrews, *J. Biol. Chem.*, **239**, 81 (1964).
3. S. Futterman and J. S. Andrews, *J. Biol. Chem.*, **239**, 4077 (1964).
4. J. G. Baxter and C. D. Robeson, *J. Am. Chem. Soc.*, **64**, 2407 (1942).
5. V. Mahadevan and W. O. Lundberg, *J. Lipid Res.*, **3**, 106 (1962).
6. J. W. Porter and D. G. Anderson, in *Chromatography*, E. Heftmann, ed., Reinhold Publishing Corp., New York (1961), p. 465.
7. J. A. Olson, in *Newer Methods of Nutritional Biochemistry*, Vol. 2, A. A. Albanese, ed., Academic Press, Inc., New York (1965), p. 345.
8. M. Kofler and S. H. Rubin, *Vitamins Hormones*, **18**, 315 (1960).
9. P. E. Dunagin, Jr., and J. A. Olson, *Anal. Chem.*, **36**, 756 (1964).
10. H. Sobotka, S. Kann, and E. Loewenstein, *J. Am. Chem. Soc.*, **65**, 1959 (1943).
11. R. E. Dugan, N. A. Frigerio, and J. M. Siebert, *Anal. Chem.*, **36**, 114 (1964).
12. *Suggestions for the Storage and Use of Crystalline Vitamin A*, Distillation Products Industries, Rochester, N. Y. (1964).

Carot-43
Spirilloxanthin
1,1'-Dimethoxy-3,4,3',4'-tetrahydro-1,2,1',2'-tetrahydro-
ψ,ψ-carotene
(Rhodoviolascin)

Formula: $C_{42}H_{60}O_2$
Formula Wt.: 596.95

Calc. %: C, 84.51; H, 10.12; O, 5.37; OCH₃, 10.40

Sources:

Natural Sources. Spirilloxanthin is found only in photosynthetic bacteria. *Rhodospirillum rubrum* (stationary-growth phase) is an excellent source of this compound.[1,2]

Chemical Synthesis. The total synthesis of spirilloxanthin has been reported.[3,4]

Isolation Procedure: The extraction, saponification, column chromatography, and crystallization of spirilloxanthin have been reported.[2,5–7]

Methods of Purification:

Chromatography. Column chromatography on either a calcium carbonate–calcium hydroxide mixture[6] or deactivated alumina[7] may be used to separate spirilloxanthin from other carotenoids.

Crystallization. Chloroform–petroleum ether,[6] acetone–petroleum ether,[7] benzene–petroleum ether,[7] or benzene[3,5] are used for crystallization. Spirilloxanthin is only slightly soluble in petroleum ether, moderately soluble in benzene, and readily soluble in acetone or carbon disulfide.

Methods of Assaying for Purity:

Chromatography. The purity of spirilloxanthin may be determined by chromatography on calcium hydroxide,[6] on circular filter paper having a suitable filler,[8,9] or by the thin-layer technique.[10]

Solvent Partition. The partition ratio[11] between petroleum ether and 95% methanol is 88:12.[7]

Visible Spectrum. Petroleum ether (b.p. 40–70 °C): 463, 493, and 528 nm. $E_{1\,cm}^{1\%}$ 2680 at 493 nm.[6] Spectral curve.[6,7] Acetone: 468, 498, and 534 nm. Chloroform: 479, 509, and 544 nm. Benzene: 480, 510, and 548 nm. Carbon disulfide: 495, 532, and 570 nm. Iodine-isomerized spirilloxanthin shows "*cis*-peak" absorption at 367 and 385 nm in petroleum ether.[6]

Infrared Spectrum. The infrared absorption spectrum of spirilloxanthin in a potassium bromide pellet has been reported.[12]

Nuclear Magnetic Resonance Spectrum. The nuclear magnetic resonance spectrum of spirilloxanthin has been reported.[3,13]

Mass Spectrum. The mass spectrum of spirilloxanthin has been reported.[14]

Melting Point. Spirilloxanthin melts at 216–218 °C in an evacuated tube.[3,6]

Probable Impurities: Oxidation products, *cis*-isomers, and rhodopin or anhydrorhodovibrin when spirilloxanthin is isolated from natural sources.

Conditions of Storage: Darkness (brown vial), inert atmosphere (sealed ampoule), and low temperature (−20 °C).

References

1. S. Liaaen-Jensen, in *Bacterial Photosynthesis*, H. Gest, A. San Pietro and L. P. Vernon, eds., Antioch Press, Yellow Springs, Ohio (1963), p. 19.
2. C. B. van Niel and J. H. C. Smith, *Arch. Mikrobiol.*, **6**, 219 (1935).
3. J. D. Surmatis and A. Ofner, *J. Org. Chem.*, **28**, 2735 (1963).
4. D. F. Schneider and B. C. L. Weedon, *J. Chem. Soc.*, (C), 1686 (1967).
5. P. Karrer and U. Solmssen, *Helv. Chim. Acta*, **18**, 1306 (1935).
6. A. Polgár, C. B. van Niel, and L. Zechmeister, *Arch. Biochem. Biophys.*, **5**, 243 (1944).
7. S. L. Jensen, *Kgl. Norske Vidensk. Selskabs Skrifter*, 8 (1962).
8. A. Jensen and S. Liaaen-Jensen, *Acta Chem. Scand.*, **13**, 1863 (1959).
9. A. Jensen, *Acta Chem. Scand.*, **14**, 2051 (1960).
10. H. R. Bolliger, A. König, and U. Schwieter, *Chimia*, **18**, 136 (1964).
11. F. J. Petracek and L. Zechmeister, *Anal. Chem.*, **28**, 1484 (1956).
12. S. Liaaen-Jensen, *Acta Chem. Scand.*, **17**, 500 (1963).
13. M. S. Barber, L. M. Jackman, and B. C. L. Weedon, *Proc. Chem. Soc.*, 96 (1959).

14. C. R. Enzell, G. W. Francis, and S. Liaaen-Jensen, *Acta Chem. Scand.*, **23**, 727 (1969).

Carot-44
Squalene

Formula: C₃₀H₅₀

Formula Wt.: 410.74

Calc. %: C, 87.73; H, 12.27

Sources:

Natural Sources. Squalene is found in the largest amounts in fish-liver oils, particularly those of elasmobranchs,[1,2] Squalene is also found in plants.[3,4]

Chemical Synthesis. Several chemical syntheses of squalene have been reported.[5–9] In addition, it has been shown that natural squalene synthesized with tritium at C-12, has the *R* configuration.[10] It has also been shown that natural squalene is the all-*trans*-isomer.[11]

Isolation Procedures: Squalene is removed from biological materials by extraction with such solvents as acetone, methanol, or ethanol. The mixture of compounds in the extract is then subjected to saponification, and, subsequently, squalene and other nonsaponifiable compounds are transferred into petroleum ether.

Methods of Purification:

Chromatography. Squalene is purified by chromatography on a column of alumina, on a thin-layer plate, or by the gas–liquid technique.[12,13]

Methods of Assaying for Purity:

Chromatography. An assay for squalene by chromatography on a column of alumina has been reported.[13]

Thin-Layer Chromatography. Squalene may also be assayed[12] for purity by chromatography on a thin-layer plate of silica gel G.

Gas–Liquid Chromatography. The purity of squalene may be determined[13] by gas–liquid chromatography on a column of SE-30.

Derivative Formation. The hexabromides and hexachlorides of squalene have been prepared.[2] The thiourea clathrate of squalene has also been prepared,[4] and the biochemically important squalene 2,3-oxide has been synthesized.[14,15]

Solvent Partition. Squalene is insoluble in water, soluble in ether, petroleum ether, carbon tetrachloride, or acetone, and sparingly soluble in alcohol or glacial acetic acid.[16]

Boiling Point.[16] b₂₅, 285 °C; b₄, 250 °C; b₀.₁₅, 203 °C.

Density. d_4^{20} 0.8584 g/ml; d_{20}^{20} 0.8538.

Refractive Index. n_D^{20} 1.4965.

Infrared Spectrum. The infrared spectra of natural and synthetic squalene have been reported.[7]

Nuclear Magnetic Resonance. The nuclear magnetic resonance spectrum of squalene has been reported.[17]

Probable Impurities: Oxidation products and *cis*-isomers.

Conditions of Storage: In darkness, at a low temperature (0 °C).

References

1. M. Tsujimoto, *J. Chem. Ind.* (Tokyo), **9**, 953 (1906).
2. I. M. Heilbron, E. D. Kamm, and W. M. Owens, *J. Chem. Soc.*, 1630 (1926).
3. S. Q. Alam, J. Brossard, and G. Mackinney, *Nature*, **194**, 479 (1962).

4. D. A. Beeler, D. G. Anderson, and J. W. Porter, *Arch. Biochem. Biophys.*, **102,** 26 (1963).
5. S. Trippett, *Chem. Ind.* (London), 80 (1956).
6. P. Karrer and A. Helfenstein, *Helv. Chim. Acta*, **14,** 78 (1931).
7. O. Isler, R. Rüegg, L. Chopard-dit-Jean, H. Wagner, and K. Bernhard, *Helv. Chim. Acta*, **39,** 897 (1956).
8. J. W. Cornforth, R. H. Cornforth, and K. K. Mathew, *J. Chem. Soc.*, 2539 (1959).
9. D. W. Dicker and M. C. Whiting, *Chem. Ind.* (London), 351 (1956).
10. B. Samuelson and D. S. Goodman, *J. Biol. Chem.*, **239,** 98 (1964).
11. N. Nicolaides and F. Laves, *J. Am. Chem. Soc.*, **76,** 2596 (1954).
12. E. Capstack, Jr., N. Rosin, G. A. Blondin, and W. R. Nes, *J. Biol. Chem.*, **240,** 3258 (1965).
13. G. Krishna, H. W. Whitlock, Jr., D. H. Feldbruegge, and J. W. Porter, *Arch. Biochem. Biophys.*, **114,** 200 (1966).
14. E. J. Corry, W. E. Russey, and P. R. Ortiz de Mondellano, *J. Am. Chem. Soc.*, **88,** 4750 (1966).
15. E. E. Van Tamelen, J. D. Willett, R. B. Clayton, and K. E. Lord, *J. Am. Chem. Soc.*, **88,** 4752 (1966).
16. *The Merck Index*, 7th Edition, Merck & Company, Inc., Rahway, New Jersey (1960), p. 974.
17. J. B. Davis, L. M. Jackman, P. T. Siddons, and B. C. L. Weedon, *J. Chem. Soc.* (*C*), 2154 (1966).

Carot-45

Torularhodin, Ethyl Ester

Ethyl 3′,4′-Didehydro-β,ψ-caroten-16′-oate

(β-C_{40}-Carotenoic Acid, Ethyl Ester)

Formula: $C_{42}H_{56}O_2$
Formula Wt.: 592.91
Calc. %: C, 85.08; H, 9.52; O, 5.40

Sources:

Natural Sources. Torularhodin has been isolated from microorganisms.[1-4]

Chemical Synthesis. The ethyl ester of torularhodin is prepared, in analogy to the methyl ester,[5] from β-apo-2′-carotenal (C_{37}) and [1-(ethoxycarbonyl)ethyl]triphenylphosphonium bromide.

Methods of Purification: This pigment is purified by chromatography on a column of deactivated alumina, and by crystallization from organic solvents (e.g., ethyl acetate).

Methods of Assaying for Purity:

Chromatography. This compound may be assayed for purity by chromatography on a column of deactivated alumina, or by chromatography on a thin-layer plate of secondary magnesium phosphate or silica gel G.[6] Ethyl acetate–dichloromethane (1:4), carbon disulfide, or benzene are used to develop the latter chromatograms.

Visible Spectrum. Torularhodin ethyl ester in hexane exhibits maxima at 475, 500, and 533 nm. $E_{1\,cm}^{1\%}$ (hexane) 2290, 3050, and 2430, respectively.

Nuclear Magnetic Resonance. The nuclear magnetic resonance spectrum of torularhodin ethyl ester has been reported.[7]

Melting Point. Torularhodin ethyl ester melts at 156–158 °C (uncorr.) in an evacuated tube.

Probable Impurities: *cis*-Isomers of the ethyl ester of torularhodin.

Conditions of Storage: Darkness (brown vial), inert atmosphere (sealed ampoule), and at a low temperature (−20 °C).

References

1. E. Lederer, *Compt. Rend. Acad. Sci.* **197,** 1694 (1933).
2. P. Karrer and J. Rutschmann, *Helv. Chim. Acta*, **26,** 2109 (1943).
3. P. Karrer and J. Rutschmann, *Helv. Chim. Acta*, **28,** 795 (1945).
4. P. Karrer and J. Rutschmann, *Helv. Chim. Acta*, **29,** 355 (1946).
5. O. Isler, W. Guex, R. Rüegg, G. Ryser, G. Saucy, U. Schwieter, M. Walter, and A. Winterstein, *Helv. Chim. Acta*, **42,** 864 (1959).
6. E. Stahl, H. R. Bolliger, and L. Lehnert, in *Carotine und Carotinoide*, K. Lang, ed., D. Steinkopff Verlag, Darmstadt (1963), p. 129.
7. B. C. L. Weedon, *Fortschr. Chem. Org. Naturstoffe*, **27,** 81 (1969).

Carot-46

Torularhodinaldehyde

3′,4′-Didehydro-β,ψ-caroten-16′-al

(3′,4′-Didehydro-17′-oxo-γ-carotene; β-C_{40}-Carotenal)

Formula: $C_{40}H_{52}O$
Formula Wt.: 548.86
Calc. %: C, 87.53; H, 9.55; O, 2.92

Sources:

Natural Sources. This compound has not been reported to be present in any natural sources other than *Rhodotorula* species,[1] where it is present in small amounts.

Chemical Synthesis. The synthesis of this compound from 15,15′-didehydro-β-apo-8′-carotenal by enol ether condensation has been reported.[2]

Methods of Purification: This compound may be purified by chromatography on columns of alumina, and by crystallization from organic solvents.[2]

Methods of Assaying for Purity:

Chromatography. The purity of this compound may be determined by chromatography on a column of alumina or on a thin layer of silica gel G (Merck). Ethyl ether–cyclohexane (1:4) is used to develop the latter chromatogram.[2]

Visible Spectrum. Petroleum ether (b.p. 80–105 °C): 509 and 540 nm (shoulder). Cyclohexane: 513 and 544 nm (shoulder). Benzene: 522 nm. The $E_{1\,cm}^{1\%}$ value at 508 nm is 2865 (petroleum ether, b.p. 80–105 °C).

Melting Point. A melting point of 166–168 °C has been reported.[2]

Probable Impurities: *cis*-Isomers of torularhodinaldehyde.

Conditions of Storage: Darkness (brown vial), inert atmosphere (sealed ampoule), and at a low temperature (−20 °C).

References

1. R. Bonaly and J. P. Malenge, *Biochim. Biophys. Acta*, **164,** 306 (1968).
2. R. Rüegg, M. Montavon, G. Ryser, G. Saucy, U. Schwieter, and O. Isler, *Helv. Chim. Acta*, **42,** 854 (1959).

Carot-47

Violaxanthin

(3R,5R,6S,3′S,5′R,6′S)-5,6,5′,6′-Diepoxy-5,6,5′,6′-tetrahydro-β,β-carotene-3,3′-diol

Formula: $C_{40}H_{56}O_4$
Formula Wt.: 600.89
Calc. %: C, 79.96; H, 9.39; O, 10.65

Violaxanthin has been characterized as 5,6,5′,6′-diepoxy-zeaxanthin.[1] The stereochemistry of this compound is known to be 3*S*,3′*S*,[2,3] and the configuration of each 5,6-epoxide group is thought to be 5 *R*, 6 *S*.[3]

Sources:

Natural Sources. Violaxanthin is found in many flowers,[4] fruits,[4] green leaves,[5] and algae.[6,7] Crystalline violaxanthin has been isolated from *Viola tricolor*.[8]

Chemical Synthesis. Violaxanthin has allegedly been prepared, in low yield, by oxidation of zeaxanthin with monoperoxyphthalic acid.[2] However, the main product probably differs from natural violaxanthin in the configuration of the 5,6-epoxy groups.[2]

Isolation Procedures:[8] Yellow blossoms of *Viola tricolor* are dried, and then extracted with petroleum ether. The combined extracts are concentrated, and the material in the solution is saponified. Violaxanthin is then extracted into petroleum ether, and purified by crystallization or chromatography.

Methods of Purification:

Solvent Partition. A partial purification of violaxanthin can be achieved by solvent partition.[9]

Chromatography. Violaxanthin may be purified by chromatography on a column of magnesium oxide,[5,9] zinc carbonate,[10] or calcium carbonate.[4]

Crystallization. Violaxanthin can be crystallized from methanol or carbon disulfide.[11]

Derivatives. The di-(*p*-nitrobenzoate) and the dibenzoate of violaxanthin have been reported.[10]

Methods of Assaying for Purity:

Chromatography. The homogeneity of violaxanthin can be determined by chromatography on thin-layer plates[12] or on kieselguhr paper.[13]

Solvent Partition. An observed polarity of 2.49 and an M_{50} of 66.2 have been reported[14] for this compound.

Visible Spectrum.[4] Carbon disulfide: 440, 470, and 501 nm. Chloroform: 424, 451.5, and 482 nm. Petroleum ether: 417.5, 443, and 472 nm. Ethanol: 417.5, 442.5, and 471.5 nm. $E_{1\,cm}^{1\%}$ approximately 2400 at 442.5 nm.[4,13] Spectral curves for violaxanthin have been published.[4,15]

Nuclear Magnetic Resonance. The nuclear magnetic resonance spectrum for violaxanthin has been reported.[2,16]

Optical Rotatory Dispersion. The optical rotatory dispersion curve for violaxanthin has been reported.[16]

Mass Spectrum. The mass spectrum of violaxanthin has been reported.[17]

Melting Point. Violaxanthin melts at 200 °C.[4]

Optical Rotation. An $[\alpha]_{Cd}^{20}$ of $+35°$ (chloroform) has been reported for violaxanthin.[11]

Derivatives. Auroxanthin is formed on treatment of violaxanthin with dilute acid.[12]

Color Reactions. Violaxanthin gives a persistent blue color when an ethereal solution of this compound is shaken with 20% aqueous hydrochloric acid.[13]

Probable Impurities: Antheraxanthin, auroxanthin, and zeaxanthin.
Conditions of Storage: Darkness (brown vial), inert atmosphere (sealed ampoule), and at a low temperature (-20 °C). Contact with acid vapors must be avoided.

References

1. P. Karrer and E. Jucker, *Helv. Chim. Acta*, **28**, 300 (1945).
2. L. Bartlett, W. Klyne, W. P. Mose, P. M. Scopes, G. Galasko, A. K. Mallams, B. C. L. Weedon, J. Szabolcs, and G. Tóth, *J. Chem. Soc. (C)*, 2527 (1969).
3. T. E. DeVille, M. B. Hursthouse, S. W. Russell, and B. C. L. Weedon, *J. Chem. Soc. (D)*, 1311 (1969).
4. P. Karrer and E. Jucker, *Carotenoids*, E. A. Braude (translator), Elsevier, New York (1950).
5. H. H. Strain, *Leaf Xanthophylls*, Carnegie Institution of Washington, Washington, D.C. (1938).
6. A. Jensen, in *Carotine and Carotinoide*, K. Land, ed., p. 119, D. Steinkopff Verlag, Darmstadt (1963).
7. A. Hager and H. Stransky, *Arch. Mikrobiol.*, **72**, 68 (1970).
8. L. Zechmeister and L. von Cholnoky, *Ann. Chem.*, **516**, 30 (1935).
9. A. L. Curl and G. F. Bailey, *J. Agr. Food Chem.*, **2**, 685 (1954).
10. P. Karrer and J. Rutschmann, *Helv. Chim. Acta*, **27**, 1684 (1944).
11. R. Kuhn and A. Winterstein, *Chem. Ber.*, **64**, 326 (1931).
12. H. R. Bolliger, A. König, and U. Schwieter, *Chimia*, **18**, 136 (1964).
13. A. Jensen and S. Liaaen-Jensen, *Acta Chem. Scand.*, **13**, 1863 (1959).
14. N. I. Krinsky, *Anal. Biochem.*, **6**, 293 (1963).
15. P. Karrer and E. Würgler, *Helv. Chim. Acta*, **26**, 116 (1943).
16. B. C. L. Weedon, *Fortschr. Chem. Org. Naturstoffe*, **27**, 81 (1969).
17. H. Budzikiewicz, H. Brzezinka, and B. Johannes, *Monatsh. Chem.*, **101**, 579 (1970).

Carot-48
β-Zeacarotene
7′,8′-Dihydro-β,ψ-carotene
(all-*trans*-7′,8′-Dihydro-γ-carotene)

Formula: $C_{40}H_{58}$
Formula Wt.: 538.90
Calc. %: C, 89.15; H, 10.85

Sources:

Natural Sources. β-Zeacarotene has been isolated from yellow corn grain,[1] yeast,[2] and fungi.[3]

Chemical Synthesis. The chemical synthesis of β-zeacarotene from farnesyl triphenylphosphonium bromide and 15,15′-didehydro-apo-12′-carotenal by a Wittig reaction has been reported.[4]

Isolation Procedures: β-Zeacarotene is extracted from biological materials with an organic solvent. The carotene is then transferred into petroleum ether, with or without prior saponification, and purified by chromatography.

Methods of Purification:

Chromatography. β-Zeacarotene is purified by chromatography on a column of magnesium oxide–Supercel,[1,2] alumina,[1,2] or calcium hydroxide–Celite.[1,3]

Methods of Assaying for Purity:

Chromatography. β-Zeacarotene may be assayed for purity by column chromatography as just noted. It may also be assayed for purity by chromatography on a kieselgel plate, or by thin-layer chromatography on alumina.[3]

Visible Spectrum. Petroleum ether: 406, 428, and 454 nm. $E_{1\,cm}^{1\%}$ 1660, 2520, and 2300, respectively.[4]

Infrared Spectrum. The infrared absorption spectrum of β-zeacarotene has been reported.[4]

Nuclear Magnetic Resonance. The nuclear magnetic resonance spectrum of β-zeacarotene has been reported.[5]

Mass Spectrum. The mass spectrum of β-zeacarotene has been published.[6]

Melting Point.[4] β-Zeacarotene melts at 96–97 °C.

Probable Impurities: Oxidation products, *cis*-isomers of β-zeacarotene, and, possibly, ζ-carotene.

Conditions of Storage: Darkness (brown vial), inert atmosphere (sealed ampoule), and at a low temperature (−20 °C).

References

1. E. N. Petzold, F. W. Quackenbush, and M. McQuistan, *Arch. Biochem. Biophys.*, **82**, 117 (1959).
2. K. L. Simpson, T. O. M. Nakayama, and C. O. Chichester, *J. Bacteriol.*, **88**, 1688 (1964).
3. R. J. H. Williams, B. H. Davies, and T. W. Goodwin, *Phytochemistry*, **4**, 759 (1965).
4. R. Rüegg, U. Schwieter, G. Ryser, P. Schudel, and O. Isler, *Helv. Chim. Acta*, **44**, 994 (1961).
5. B. C. L. Weedon, *Fortschr. Chem. Org. Naturstoffe*, **27**, 81 (1969).
6. H. Budzikiewicz, H. Brzezinka, and B. Johannes, *Monatsh. Chem.*, **101**, 579 (1970).

Carot-49
Zeaxanthin
(3R,3′R)β,β-Carotene-3,3′-diol
(β-Carotene-3,3′-diol)

Formula: $C_{40}H_{56}O_2$
Formula Wt.: 568.89
Calc. %: C, 84.45; H, 9.92; O, 5.62

Sources:

Natural Sources. Maize seeds,[1,2] calyx of *Physalis alkekengi* (as physalien, a zeaxanthin dipalmitate),[2–4] and in small proportions in other plant sources.[2,5,6] The absolute configuration of natural zeaxanthin from *Physalis alkekengi* and maize has been established[7,8] as 3R,3′R.

Chemical Synthesis. The total synthesis of zeaxanthin has been reported.[9,10]

Isolation Procedures: The isolation of zeaxanthin from natural sources involves extraction, saponification under mild conditions, extraction of the carotenediol with petroleum ether or ethyl ether, chromatography, and crystallization.[1–6,11]

Methods of Purification:

Solvent Partition. Zeaxanthin may be separated from carotenes by partition between petroleum ether and 95% methanol.[5,12]

Chromatography. Zeaxanthin may be purified by column chromatography on calcium carbonate, calcium hydroxide, zinc carbonate, magnesia, magnesium silicate, or deactivated alumina.[2,3,5,11]

Crystallization. Several solvent combinations may be used for crystallization. Two of these are carbon disulfide–ethyl ether–petroleum ether[5] and dichloromethane–methanol.[9] Ethanol or methanol may also be used. Zeaxanthin is almost insoluble in petroleum ether, slightly soluble in ethyl ether, and quite soluble

in chloroform or carbon disulfide. One gram of the pigment dissolves in 1.5 liters of boiling methanol.[5]

Derivatives. Diesters[5] and diethers[13] of zeaxanthin have been reported.

Methods of Assaying for Purity:

Chromatography. The purity of zeaxanthin may be determined by chromatography on columns,[3] on circular paper having a suitable filler,[14] or on thin layers of adsorbent.[15]

Solvent Partition. The partition ratio between petroleum ether and 95% methanol is 11:89; and between petroleum ether and 85% methanol is 40:60.[12]

Visible Spectrum. Petroleum ether (b.p. 40–60 °C): 423, 452, and 480 nm. $E_{1\,cm}^{1\%}$ 2350 (452 nm) and 2050 (480 nm). Spectral curve.[9] Ethanol:[5] 423 (shoulder), 451, and 483 nm. Methanol:[5] 422 (shoulder), 450, and 481 nm. Chloroform:[5] 429 (shoulder), 462, and 495 nm. Carbon disulfide: 450 (shoulder), 482, and 517 nm. Iodine-isomerized zeaxanthin shows "*cis*-peak" light absorption at 336 nm (petroleum ether).[16]

Infrared Spectrum. The infrared spectra in chloroform and bromoform have been reported.[9]

Nuclear Magnetic Resonance Spectrum. The nuclear magnetic resonance spectrum of zeaxanthin has been reported.[17]

Mass Spectrum. The mass spectrum of zeaxanthin has been reported.[18]

Optical Rotatory Dispersion. The optical rotatory dispersion curve of zeaxanthin has been reported.[8]

Melting Point. Zeaxanthin melts at 205–206 °C in an evacuated tube.[9]

Optical Rotation. An $[\alpha]_{Cd}$ value of −40° to −42° (chloroform) has been reported.[19]

Probable Impurities: Oxidation products, *cis*-isomers, and possibly lutein.

Conditions of Storage: Darkness (brown vial), inert atmosphere (sealed ampoule), and at a low temperature (−20 °C).

References

1. P. Karrer, A. Helfenstein, H. Wehrli, B. Pieper, and R. Morf, *Helv. Chim. Acta*, **14**, 619 (1931).
2. L. Zechmeister, *Carotinoide*, Julius Springer, Berlin (1934).
3. F. P. Zscheile, J. W. White, Jr., B. W. Beadle, and J. R. Roach, *Plant Physiol.*, **17**, 331 (1942).
4. R. Kuhn, A. Winterstein, and W. Kaufmann, *Chem. Ber.*, **63**, 1489 (1930).
5. P. Karrer and E. Jucker, *Carotenoids*, E. A. Braude (translator), Elsevier, New York (1950).
6. T. W. Goodwin, in *Carotine und Carotinoide*, K. Lang, ed., D. Steinkopff Verlag, Darmstadt (1963), p. 1.
7. T. E. DeVille, M. B. Hurthouse, S. W. Russell, and B. C. L. Weedon, *J. Chem. Soc. (D)*, 1311 (1969).
8. L. Bartlett, W. Klyne, W. P. Mose, P. M. Scopes, G. Galasko, A. K. Mallams, B. C. L. Weedon, J. Szabolcs, and G. Tóth, *J. Chem. Soc. (C)*, 2527 (1969).
9. O. Isler, H. Lindlar, M. Montavon, R. Rüegg, G. Saucy, and P. Zeller, *Helv. Chim. Acta* **39**, 2041 (1956).
10. D. E. Loeber, S. W. Russell, T. P. Toube, B. C. L. Weedon, and J. Diment, *J. Chem. Soc. (C)*, 404 (1971).
11. H. H. Strain, *Leaf Xanthophylls*, Carnegie Institution of Washington, Washington, D. C. (1938).
12. F. J. Petracek and L. Zechmeister, *Anal. Chem.*, **28**, 1484 (1956).
13. H. Müller and P. Karrer, *Helv. Chim. Acta*, **48**, 291 (1965).
14. A. Jensen and S. Liaaen-Jensen, *Acta Chem. Scand.*, **13**, 1863 (1959).
15. H. R. Bolliger, A. König, and U. Schwieter, *Chimia*, **18**, 136 (1964).
16. L. Zechmeister, *cis–trans Isomeric Carotenoids, Vitamins A and Arylpolyenes*, Springer-Verlag, Vienna (1962).
17. M. S. Barber, J. B. Davis, L. M. Jackman, and B. C. L. Weedon, *J. Chem. Soc.*, 2870 (1960).
18. C. R. Enzell, G. W. Francis, and S. Liaaen-Jensen, *Acta Chem. Scand.*, **23**, 727 (1969).
19. L. Zechmeister, L. v. Cholnoky, and A. Polgár, *Chem. Ber.*, **72**, 1678 (1939).

Coenzymes and Related Compounds

GENERAL REMARKS

Preparations of the coenzymes commonly employed in biochemical investigations are now available commercially. Determinations of their purity rest on such criteria as enzyme assays, chromatographic analysis, and spectral properties. It is essential that the manufacturer provide adequate information concerning the criteria of biological activity and purity used for his preparations. The criteria sheets describe material of the highest available purity. Preparations may be available that are less pure but which may be satisfactory, provided that they are adequately described and that the known impurities are stated.

The spectral reference values for nucleotide coenzymes have been selected from the literature and are considered to be the best available at present. It is recognized that some values are not known with certainty, and the Committee on Specifications and Criteria for Biochemical Compounds and its Subcommittee on Coenzymes invite suggestions for revisions, particularly with reference to precise determinations of molar absorption coefficients.

CoE-1
Acetyl Coenzyme A
Symbol: CoA-Ac

Formula of Lithium Salt: $C_{23}H_{35}Li_3N_7O_{17}P_3S$
Formula Wt.: 827.37

Usually available as lithium salt with variable amounts of water of hydration.

Enzyme Assay: Suggested Method:[1-3] Enzymic transfer of the acetyl group from acetyl-CoA to carnitine. The reaction is catalyzed by acetyl-CoA–carnitine acetyl transferase. The amount of sulfhydryl group liberated in the reaction is determined with Ellman's reagent.

Experimental Measurement: Increase in absorbancy at 412 nm.

Spectral Reference Values:

Molar absorption coefficient, $\epsilon = 15.4 \times 10^3$ l mol^{-1} cm^{-1} at 259 nm and pH 7 (calculated from value for ATP). Ratio $A_{250}/A_{260} = 0.78$ at pH 7. Ratio $A_{280}/A_{260} = 0.15$ at pH 7.

Homogeneity: Evidence of chromatographic homogeneity should be presented. The chromatographic method should be capable of detecting small amounts of structurally related compounds, such as acetyl dephosphocoenzyme A, as well as other types of impurities, such as acetylglutathione. The amount of free sulfhydryl group per unit amount of acetyl coenzyme A should also be stated.

A suitable solvent for chromatography consists of equal volumes of 95% ethanol and 0.1 M sodium acetate–acetic acid buffer, pH 4.5. A suitable paper is Whatman No. 1 or its equivalent. A suitable solvent system for diethylaminoethylcellulose column chromatography consists of a linear gradient of LiCl and HCl. The mixing vessel contains 500 ml of 10 mM tris-HCl buffer, pH 7.4, and the reservoir contains 500 ml of a solution of 0.4 M LiCl–0.04 M HCl. These volumes are suitable for column beds about 50 cm long and 1.5 cm in diameter. The column should be eluted in the cold. The fractions should be analyzed for ultraviolet absorption, as well as for concentration of thioester and free sulfhydryl.

Certain types of impurities, such as a small amount of propionyl-CoA, may be difficult to detect chromatographically. Acetyl-CoA is usually prepared by acetylation of CoA, and this type of impurity is best avoided by the use of pure acetylating agents.

References
1. J. F. A. Chase, D. J. Pearson, and P. K. Tubbs, *Biochim. Biophys. Acta*, **96**, 162 (1965).
2. G. L. Ellman, *Arch. Biochem. Biophys.*, **74**, 443 (1958).
3. G. L. Ellman, *Arch. Biochem. Biophys.*, **82**, 70 (1959).

CoE-2
3-Acetylpyridine Analog of NAD
(3-Acetylpyridine Adenine Dinucleotide; AcPyAD)

Formula: $C_{22}H_{28}N_6O_{14}P_2$
Formula Wt.: 662.45

Usually available as free acid, with water of hydration.

Enzyme Assay: Suggested Method: Reduction to 3-acetylpyridine analog of NADH, using alcohol dehydrogenase from yeast or liver.[1,2]

Experimental Measurement: Increase in absorbancy at 363 nm.

Spectral Reference Values:

Oxidized form (3-acetylpyridine analog of NAD$^+$): Molar absorption coefficient, $\epsilon = 16.5 \times 10^3$ l mol^{-1} cm^{-1} at pH 7.5 and 260 nm. Ratio $A_{250}/A_{260} = 0.81$ at pH 7.5. Ratio $A_{280}/A_{260} = 0.24$ at pH 7.5.[1]

Reduced form (3-acetylpyridine analog of NADH): Molar absorption coefficient, $\epsilon = 9.1 \times 10^3$ l mol^{-1} cm^{-1} at pH 10 and 363 nm. Cyanide adduct of 3-acetylpyridine analog of NAD$^+$: Molar absorption coefficient, $\epsilon = 8.7 \times 10^3$ l mol^{-1} cm^{-1} at pH 10 and 343 nm.

Homogeneity: Evidence for chromatography homogeneity should be presented, whenever possible, from a minimum of two solvent systems that will permit the detection of small amounts of structurally related compounds. The degree of contamination with other nucleotides, especially with adenosine diphosphoribose, NAD, and the α-isomer of NAD should be stated whenever possible. If the preparation contains residual traces of organic solvent, the amount should be stated in mols of solvent per mol of the analog.

References
1. J. M. Siegel, G. A. Montgomery, and R. M. Bock, *Arch. Biochem. Biophys.*, **82**, 288 (1959).
2. N. O. Kaplan and M. M. Ciotti, *J. Biol. Chem.*, **221**, 823 (1956).

CoE-3
3-Acetylpyridine Analog of NADP
(3-Acetylpyridine Adenine Dinucleotide Phosphate; AcPyADP)

Formula: $C_{22}H_{28}N_6NaO_{17}P_3$
Formula Wt.: 764.41

Available as sodium salt with water of hydration.

Enzyme Assay: Suggested Method: Reduction to 3-acetylpyridine analog of NADPH using isocitrate dehydrogenase (see NADP; CoE-8).

Experimental Measurement: Increase in absorbancy at 363 nm.

Spectral Reference Values:

Reduced form (3-acetylpyridine analog of NADPH): Molar absorption coefficient, $\epsilon = 9.1 \times 10^3$ l mol^{-1} cm^{-1} at pH 10 and 363 nm. Cyanide adduct of 3-acetylpyridine analog of NADP$^+$. Molar absorption coefficient, $\epsilon = 8.7 \times 10^3$ l mol^{-1} cm^{-1} at pH 10 and 343 nm (from values for acetylpyridine analog of NAD$^+$).

Homogeneity: Evidence for chromatographic homogeneity should be presented, whenever possible, from a minimum of two solvent systems that will permit detection of small amounts of structurally related compounds. The degree of contamination with other nucleotides, especially with adenosine diphosphoribose phosphate and NADP should be stated. If the preparation contains residual amounts of organic solvent, the amount should be stated in mols of solvent per mol of the analog.

CoE-4
Cobamide Coenzymes:
I. Adenylcobamide Coenzyme
II. Benzimidazolylcobamide Coenzyme
III. 5,6-Dimethylbenzimidazolylcobamide Coenzyme

Enzyme Assay: Suggested Method:[1] Conversion of glutamate to mesaconate using glutamate mutase coupled to β-methylaspartase. Maximum activities with II and III coenzymes are the same, activity with I coenzyme is very similar. The K_m values for the three coenzymes differ widely.
Experimental Measurement: Change in absorbancy at 260 nm.
Alternative Method:[2] Conversion of propanediol to propionaldehyde.
Spectral Reference Values:
II:[3] Molar absorption coefficient, $\epsilon = 34.7 \times 10^3$ l mol^{-1} cm^{-1} at 261 nm.
Homogeneity: *Paper electrophoresis.*[3] Solvent A: 0.5 M acetic acid (40 V/cm, 2 h); a single reddish-orange and ultraviolet-absorbing spot moving toward the cathode should be observed. Solvent B: 0.5 M ammonia (10 V/cm, 3 h). *Paper chromatography.*[3] A single colored and ultraviolet absorbing spot should be observed employing Whatman No. 1 paper, or equivalent, and the solvent mixture *sec*-butanol–glacial acetic acid–water (100:3:50, by volume) in the descending manner, at room temperature. R_f values are 0.16 and 0.22 for II and III coenzymes, respectively. R_f values are 0.34 and 0.09 for cyanocobalamin and I coenzyme, respectively.
Solubility test.[3] The solubility of crystalline II coenzyme in water at 24 °C is 0.0164 M. Add successive amounts of crystalline III coenzyme to water until the total exceeds the solubility of the enzyme by 60–100%. No significant increase in absorbance should occur after the solution becomes saturated with coenzyme. This test shows that the crystalline III coenzyme is free (<2%) of ultraviolet-absorbing impurities.
Stability.[3] Solutions are most stable at pH 6 to 7 (for several months at −10 °C). The dry crystalline coenzymes are moderately stable. They can be stored several months at −10 °C or several days at room temperature. Solutions of II and III coenzymes are very unstable under exposure to light or cyanide ion.[4] Either treatment causes loss of activity and a change in the absorption spectrum. No cyanide should be detectable as measured by the method described in Ref. 5.

References
1. H. A. Barker, R. D. Smyth, H. Weissbach, A. Munch-Petersen, J. I. Toohey, J. N. Ladd, B. E. Volcani, and R. M. Wilson, *J. Biol. Chem.*, **235**, 181 (1960).
2. H. A. Lee, Jr., and R. H. Abeles, *J. Biol. Chem.*, **238**, 2367 (1963); R. H. Abeles, C. Myers, and T. A. Smith, *Anal. Biochem.*, **15**, 192 (1966).
3. H. A. Barker, R. D. Smyth, H. Weissbach, J. I. Toohey, J. N. Ladd, and B. E. Volcani, *J. Biol. Chem.*, **235**, 480 (1960).
4. H. Weissbach, J. I. Toohey, and H. A. Barker, *Proc. Nat. Acad. Sci. U.S.*, **45**, 521 (1959).
5. G. E. Boxer and J. C. Rickards, *Arch. Biochem.*, **30**, 382 (1951).

CoE-5
Coenzyme A
Symbol: CoA or CoASH

Formula: $C_{21}H_{36}N_7O_{16}P_3S$
Formula Wt.: 767.54

Usually available as freeze-dried free acid containing variable proportions of water of hydration.

Enzyme Assay: Suggested Method:[1] Conversion into sorbyl CoA, catalyzed by acyl-CoA synthase.
Experimental Measurement: Increase in absorbancy at 300 nm.
Alternative Method:[2] Conversion to succinyl CoA, catalyzed by α-ketoglutarate dehydrogenase.
Experimental Measurement: Increase in absorption at 340 nm, or increase in fluorescence at 465 nm.
Spectral Reference Values:
Molar absorption coefficient, $\epsilon = 15.4 \times 10^3$ l mol^{-1} cm^{-1} at 259 nm and pH 7 (calculated from values for ATP). Ratio $A_{250}/A_{260} = 0.78$ at pH 7. Ratio $A_{280}/A_{260} = 0.15$ at pH 7.
Homogeneity: Evidence of chromatographic homogeneity should be presented. Many solvent systems for paper chromatography yield multiple spots with CoA, and this situation complicates the assessment of purity. Column chromatography on diethylaminoethylcellulose is recommended. A suitable column bed is 1.5 cm in diameter and 50 cm in height. The column is eluted with a linear gradient of salt, using 500 ml of 10 mM tris-HCl, pH 7.4, in the mixing vessel and 500 ml of 0.4 M LiCl–0.04 M HCl in the reservoir. Fractions are analyzed for absorbance at 260 nm, sulfhydryl content, and enzymically active CoA.
Air oxidation on the column may be minimized by use of oxygen-free solvents.
The degree of contamination with dephosphocoenzyme A and sulfhydryl compounds other than CoA should be stated whenever possible.

References
1. S. J. Wakil and G. Hübscher, *J. Biol. Chem.*, **235**, 1554 (1960).
2. J. R. Williamson and B. E. Corkey, *Methods Enzymol.*, **13**, (1967).

CoE-6
Nicotinamide Adenine Dinucleotide (NAD)
(Diphosphopyridine Nucleotide, DPN)

Formula: $C_{21}H_{27}N_7O_{14}P_2$
Formula Wt.: 663.44

Usually available as free acid, with variable proportions of water of hydration.

Enzyme Assay: Suggested Method:[1] Reduction to NADH using yeast alcohol dehydrogenase.
Experimental Measurement: Increase in absorbancy at 340 nm.
Spectral Reference Values:
Oxidized form (NAD$^+$):[2,3] Molar absorption coefficient, $\epsilon = 18.0 \times 10^3$ l mol^{-1} cm^{-1} at pH 7 and 260 nm. Ratio $A_{250}/A_{260} = 0.83$ at pH 7. Ratio $A_{280}/A_{260} = 0.22$ at pH 7.
Reduced form (NADH): Molar absorption coefficient, $\epsilon = 6.2 \times 10^3$ l mol^{-1} cm^{-1} at pH 10 and 340 nm.

NAD–cyanide adduct:[4,5] Molar absorption coefficient, $\epsilon = 5.9 \times 10^3$ l mol^{-1} cm^{-1} at pH 10 and 327 nm. Ratio $A_{250}/A_{260} = 0.82$ at pH 10.

Homogeneity: Evidence for chromatographic homogeneity should be presented, whenever possible, from a minimum of two solvent systems that will permit the detection of small amounts of structurally related compounds. The degree of contamination with other nucleotides, especially with adenosine diphosphoribose, NADP, and the α-isomer of NAD, should be stated whenever possible. NAD purified by ion-exchange chromatography is usually isolated by precipitation with organic solvents. When this method is employed, the amount of residual organic solvent in the final product should be stated, preferably in mols of solvent per mol of NAD.

References

1. M. M. Ciotti and N. O. Kaplan, *Methods Enzymol.*, 3, 890 (1957).
2. A. Kornberg and W. E. Pricer, *Biochem. Prep.*, 3, 20 (1953).
3. J. M. Siegel, G. A. Montgomery, and R. M. Bock, *Arch. Biochem. Biophys.*, 82, 288 (1959).
4. B. L. Horecker and A. Kornberg, *J. Biol. Chem.*, 175, 385 (1948).
5. S. P. Colowick, N. O. Kaplan, and M. M. Ciotti, *J. Biol. Chem.*, 191, 447 (1951).

CoE-7

Nicotinamide Adenine Dinucleotide Phosphate (NADP)
(Triphosphopyridine Nucleotide, TPN)

Formula: $C_{21}H_{27}N_7NaO_{17}P_3$
Formula Wt.: 765.40

Usually available as sodium salt, with variable proportions of water of hydration.

Enzyme Assay: Suggested Method:[1] Reduction to NADPH by isocitrate dehydrogenase.

Experimental Measurement: Increase in absorbancy at 340 nm.

Spectral Reference Values:

Oxidized form (NADP$^+$): Molar absorption coefficient, $\epsilon = 18.0 \times 10^3$ l mol^{-1} cm^{-1} at pH 7 and 260 nm.[2] Ratio $A_{250}/A_{260} = 0.83$ at pH 7;[3] ratio $A_{280}/A_{260} = 0.21$ at pH 7.[3]

Reduced form (NADPH): Molar absorption coefficient, $\epsilon = 6.2 \times 10^3$ l mol^{-1} cm^{-1} at pH 10 and 340 nm.

NADP–cyanide adduct: Molar absorption coefficient, $\epsilon = 5.9 \times 10^3$ l mol^{-1} cm^{-1} at pH 10 and 327 nm.[4]

Homogeneity: Evidence for chromatographic homogeneity should be presented, whenever possible, from a minimum of two solvent systems that will permit the detection of small amounts of structurally related compounds. The degree of contamination by other nucleotides should be stated whenever possible.

References

1. M. M. Ciotti and N. O. Kaplan, *Methods Enzymol.*, 3, 892 (1957).
2. A. Kornberg and W. E. Pricer, *Biochem. Prep.*, 3, 28 (1953).
3. *U.V. Spectra of 5'-Ribonucleotides*, Circular OR-10, Pabst Laboratories, Milwaukee (1956), p. 19.
4. B. L. Horecker and A. Kornberg, *J. Biol. Chem.*, 175, 385 (1948).

CoE-8

Nicotinamide Hypoxanthine Dinucleotide (NHD)
(Deamino Analog of NAD)

Formula: $C_{21}H_{25}N_6NaO_5P_2$
Formula Wt.: 686.40

Available as sodium salt, with water of hydration.

Enzyme Assay: Suggested Method:[1] Reduction to deamino analog of NADH using alcohol dehydrogenase.

Experimental Measurement: Increase in absorbancy at 340 nm.

Spectral Reference Values:

Oxidized form (deamino analog of NAD$^+$):[1,2] Molar absorption coefficient, $\epsilon = 14.7 \times 10^3$ l mol^{-1} cm^{-1} at pH 6 and 249 nm. Ratio $A_{250}/A_{260} = 1.36$ at pH 6. Ratio $A_{280}/A_{260} = 0.28$ at pH 6.

Reduced form (deamino analog of NADH): Molar absorption coefficient, $\epsilon = 6.2 \times 10^3$ l mol^{-1} cm^{-1} at pH 10 and 340 nm (from value for NADH).

Homogeneity: Evidence for chromatographic homogeneity should be presented, whenever possible, from a minimum of two solvent systems that will permit the detection of small amounts of structurally related compounds. The degree of contamination with other nucleotides, especially with NAD, the deamino analog of the α-isomer of NAD, inosine diphosphoribose, and adenosine diphosphoribose should be stated whenever possible. If the preparation contains residual traces of organic solvent, the amount should be stated in mols of solvent per mol of the analog.

References

1. M. E. Pullman, S. P. Colowick, and N. O. Kaplan, *J. Biol. Chem.*, 194, 593 (1952).
2. J. M. Siegel, G. A. Montgomery, and R. M. Bock, *Arch. Biochem. Biophys.*, 82, 288 (1959).

CoE-9

Reduced Nicotinamide Adenine Dinucleotide (NADH)
(Reduced Diphosphopyridine Nucleotide, DPNH)

Formula: $C_{21}H_{27}N_7Na_2O_{14}P_2$
Formula Wt.: 709.42

Usually available as sodium salt, with variable proportions of water of hydration.

Enzyme Assay: Suggested Method:[1] Oxidation to NAD$^+$ using alcohol dehydrogenase and acetaldehyde.

Experimental Measurement: Decrease in absorbancy at 340 nm.

Spectral Reference Values:

NADH:[2] Molar absorption coefficient, $\epsilon = 6.22 \times 10^3$ l mol^{-1} cm^{-1} at pH 10 and 338 nm; 14.4×10^3 l mol^{-1} cm^{-1} at pH 10 and 259 nm.

Homogeneity: Evidence for chromatographic homogeneity should be presented, whenever possible, from a minimum of two solvent systems that will permit the detection of small amounts of structurally related compounds. The degree of contamination with other nucleotides should be stated, including NAD and the α-isomer of NAD. NADH is usually isolated by precipitation with organic solvents. When this method is used, the amount of residual organic solvent in the final product should be stated, preferably in mols of solvent per mol of NADH.

Prolonged storage, or unsuitable storage conditions, lead to the conversion of NADH into a substance that inhibits some dehydrogenases. Assays for the inhibitor produced by storage have not yet been standardized. The date of preparation, method of storage, and details of transportation of NADH should be stated where possible.

References

1. M. M. Ciotti and N. O. Kaplan, *Methods Enzymol.*, **3**, 893 (1957).
2. J. M. Siegel, G. A. Montgomery, and R. M. Bock, *Arch. Biochem. Biophys.*, **82**, 288 (1959).

CoE-10
Reduced Nicotinamide Adenine Dinucleotide Phosphate (NADPH)
(Reduced Triphosphopyridine Nucleotide, TPNH)

Formula: $C_{21}H_{26}N_7Na_4O_{17}P_3$
Formula Wt.: 833.36

Usually available as sodium salt, with variable proportions of water of hydration.

Enzyme Assay: Suggested Method:[1] Oxidation of NADPH using NADPH-specific glutathione reductase.
  Experimental Measurement: Decrease in absorbancy at 340 nm.
Spectral Reference Values:
  Molar absorption coefficient, $\epsilon = 18.0 \times 10^3$ l mol^{-1} cm^{-1} at 260 nm and pH 7; 6.2×10^3 l mol^{-1} cm^{-1} at 340 nm and pH 7.
Homogeneity: Evidence for chromatographic homogeneity should be presented, whenever possible, from a minimum of two solvent systems that will permit detection of small amounts of structurally related compounds. The degree of contamination with other nucleotides should be stated. If the preparation contains residual amounts of organic solvent, the amount should be stated in mols of solvent per mol of NADPH.

References

1. M. M. Ciotti and N. O. Kaplan, *Methods Enzymol.*, **3**, 894 (1957).
2. B. L. Horecker and A. Kornberg, *J. Biol. Chem.*, **175**, 385 (1948).

CoE-11
Uridine Diphosphoglucose (UDPG or UDP-Glc)

Formula: $C_{15}H_{22}N_2Na_2O_{17}P_2$
Formula Wt.: 610.27

Usually available as sodium salt, with water of hydration.

Enzyme Assay: Suggested Method:[1] Oxidation to UDP–glucuronic acid using UDPG dehydrogenase and NAD$^+$. (Theory requires 2 mol of NAD$^+$ per mol of UDPG).
  Experimental Measurement: Increase in absorbancy at 340 nm.
Spectral Reference Values:
  Molar absorption coefficient, $\epsilon = 10.0 \times 10^3$ l mol^{-1} cm^{-1} at 262 nm and pH 7.[2,3] Ratio $A_{250}/A_{260} = 0.74$ at pH 7; ratio $A_{280}/A_{260} = 0.38$ at pH 7. The average constants for UMP, UDP, and UTP are employed.[2,3]
Homogeneity: Evidence for chromatographic homogeneity should be presented, whenever possible, from a minimum of two solvent systems that will permit the detection of small amounts of structurally related compounds.

References

1. J. L. Strominger, E. S. Maxwell, J. Axelrod, and H. M. Kalckar, *J. Biol. Chem.*, **224**, 79 (1957).
2. J. M. Ploeser and H. S. Loring, *J. Biol. Chem.*, **178**, 431 (1949).
3. R. M. Bock, N. S. Ling, S. A. Morell, and S. H. Lipton, *Arch. Biochem. Biophys.*, **62**, 253 (1956).

Enzymes

GENERAL REMARKS

The Subcommittee on Enzymes recognizes that chemists who purchase commercial enzyme preparations will be primarily concerned with the activity, stability, and purity of an enzyme and are likely to be the most frequent users of these specifications and criteria for enzymes.

The application of an enzyme determines how rigorously it must conform to ideal specifications. Frequently, a preparation containing multiple activities may be useful for specific purposes as a consequence of, or perhaps in spite of, the presence of the other enzymes. In other cases, a minute trace of a second activity may render the preparation useless. The present specifications and criteria have been compiled with these considerations in mind. Furthermore, with increasing use of enzymic techniques for convenient and specific preparations (e.g., asymmetric syntheses of radioactive compounds or resolution of optical isomers), greater attention will be paid to recommended standards for commercial enzyme preparations.

FORMAT FOR CRITERIA SHEETS

The following plan has been used in the preparation of the Enzymes Section. *Specifications* apply to analytical-grade enzymes obtainable at a high level of activity and purity (no such materials are listed in this edition); maximum permissible activities of contaminating enzymes should be specified. *Criteria* apply to all other enzymes. As additional data become available, criteria will be replaced by specifications in future editions.

Tentative Designations

Each set of data in the present edition should be regarded as "Tentative" until general acceptance indicates that deletion of "Tentative" is justified. Future specifications or criteria not designated "Tentative" will also be subject to amendment, as research progress indicates changes desirable to reflect the state of the art.

Nomenclature

An enzyme is designated by its most generally acceptable common name. Configurational relationships in stereospecific reactions, abbreviations, and symbols are indicated according to accepted rules.[1,2] To facilitate cross-referencing to classification of enzymes by the IUB,[3] the number and name assigned to each enzyme by the IUB appear in a footnote and are, in some cases, followed by other common names.

Equations

Whenever possible, an equation is given for the reaction of principal, uncomplexed, ionic species at the pH specified in the assay. Structural formulas are not given. However, stereochemical configurations[1,2] are indicated.

Methods of Isolation

Appropriate references are listed.

Physical Constants

A reported molecular weight (or range of weights), with references, is given. The Subcommittee emphasizes that these values should not necessarily be accepted as authoritative, but rather as a guide to the relative molecular sizes of various enzymes.

Procedures for Handling

Procedures believed (at the time of publication) to be the most reliable for retaining activity during storage, shipment, dialysis, and dilution are given.

Commercially Available Substrates

Pertinent commercially available compounds are listed. Sources of supply are not indicated.

Activity Unit

Whenever possible, and unless specifically stated otherwise, the activity unit, U, is defined as that amount of activity catalyzing the transformation of 1 μmol of substrate per minute under the conditions specified.[3] Where this is neither practical nor desirable, the most reasonable alternative has been adopted and is described under "Assay Procedure" for each such enzyme.

Specific Activity

The specific activity is defined as activity divided by mass of protein and is expressed in the unit, U/mg. Reported values for the specific activity of an enzyme preparation believed to be of the highest purity (at the time of publication) are cited on each sheet. Whenever possible, these values are expressed in terms of activity obtained, with the reference assay procedure.

Protein Assay

Absorbance in the ultraviolet has generally been specified for protein assays. Wavelengths are given in nanometers (nm). The following symbols are used[4]: A, absorbance; a, specific absorption coefficient (absorption coefficient for 1 mg of protein/ml of solution); ϵ, molar absorption coefficient; l, internal cell-length; ρ, mass concentration; $\rho = A/al$. Unless otherwise stated, $l = 1$ cm, and ρ is expressed in mg (of protein)/ml (of solution). Data are usually given in the form: $A \times y =$ milligrams of protein/milliliter of solution, where $y = 1/al$. In most cases, an appropriate warning is given when absorbance measurements are unsuitable for grossly impure preparations or when contaminants having absorbance at the measured wavelength are likely to be present.

Activity Assay

Suitable procedures for reference (or referee) activity are given in detail. The given procedure may not necessarily be the best for other purposes, but has been chosen because of its precision, simplicity, and general utility. These procedures are primarily given to permit comparison by a manufacturer and a consumer of a commercially available enzyme preparation.

In some instances, references to alternative procedures that have definite advantages for specific purposes are listed.

In general, it is convenient to determine the number of enzyme units present in the total volume of reaction mixture used in the assay measurement. The specific activity of the enzyme is readily calculated from the number of units present and the number of milligrams of protein added to this volume.

Impurities

Probable contaminants in crude preparations are listed. For analytical-grade enzymes, and those of lesser purity likely to be used for specific purposes, a recommended maximum for troublesome impurities is given. Ultimately, these activities will be expressed in U/mg as determined by the referee assay procedure reported on the parent sheet for the contaminating enzyme. Publication of *specifications* with recommended minimum activity values for each enzyme, and maximum activities for contaminants, is planned for the next edition of this volume.

Isoenzymes

The Subcommittee on Enzymes does not believe that treatment of isoenzymes (isozymes) is desirable at this time. Where appropriate, suitable references to the problems of isoenzymes have been included.

Recommendations to Enzyme Manufacturers

The Subcommittee suggests that the following information should be made available to users of commercially available enzymes:

1. Source, method of isolation, number of recrystallizations (if applicable), solvents or media used for precipitation or crystallization, and additives present at the time of shipment, e.g., ammonium sulfate or EDTA.

2. A minimum specific activity at the time of packaging and the date of packaging.

3. A list of contaminating activities known to be present and, in *specifications* (for analytical-grade en-

zymes), the actual activities present at the time of preparation.

4. If assays have been conducted by a procedure other than the method described herein, details of the procedure should be published in the manufacturer's brochure, accompanied by a factor permitting conversion into the equivalent unit obtained by the NRC reference procedure.

Commercially Available Enzymes for Which Adequate Criteria Cannot Be Written at Present

Criteria for the following enzymes have been studied by the Subcommittee on Enzymes. It has been agreed that, for the reasons stated, adequate criteria cannot be formulated at present.

1. Glutamate Decarboxylase (*Escherichia coli* ATC No. 11246)

Of those commercially available, this is the only amino acid decarboxylase for which the Subcommittee has attempted to formulate criteria. After careful consideration, the Committee reached the conclusion that more research is needed. Publication of criteria at this time is inappropriate for an enzyme that frequently contains troublesome impurities leading to misleading observations.

2. Deoxyribonuclease (Bovine Pancreas)

In this instance, enzyme preparations may be of higher quality than the substrates. Since no suitable reproducible substrate is available for activity comparisons, criteria for this enzyme are not submitted at this time.

3. Lysozyme (Hen Egg White)

A suitable substrate is not available; the activity varies with the particular lot of substrate. Attempts are being made to make samples of a reference enzyme available for intercomparison by interested individuals.

4. D-Glucose Oxidase

Several fungal sources are used. Increases in D-glucosidase activity during purification and lack of specificity for D-glucose have been reported[5] for both crude and "purified" preparations. The Subcommittee believes that further study of this enzyme is needed.

ACKNOWLEDGMENTS

The aid of a considerable number of consultants who participated in the writing of Criteria for this section is greatly appreciated. The names of these individuals follow the criteria to which they contributed. The dedicated help of Mrs. Nona Kelley, who did most of the secretarial work for the Enzymes Section, is gratefully acknowledged.

REFERENCES

1. "Abbreviations and Symbols for Chemical Substances of Interest in Biological Chemistry." *J. Biol. Chem.,* **241,** 527 (1966) and elsewhere.
2. "Tentative Rules for Carbohydrate Nomenclature Part 1, 1969." *J. Biol. Chem.,* **247,** 613 (1972) and elsewhere.
3. *Enzyme Nomenclature, Second Edition, Recommendations (1964) of the International Union of Biochemistry,* Elsevier Publishing Co., New York (1965). (New edition in press.)
4. International Union of Pure and Applied Chemistry, "Manual of Symbols and Terminology for Physiochemical Quantities and Units." *Pure Appl. Chem.,* **21,** 1 (1970).
5. R. S. Crowne and K. R. L. Mansford, *Analyst,* **87,** 294 (1962).

E-1
Acetylcholinesterase*
(*Electrophorus electricus*)

Reaction: Acetylcholine$^+$ + H_2O → acetate$^-$ + choline$^+$ + H^+

Method of Isolation: Crystallized from ammonium sulfate solution.[1-3]

CRITERIA (Tentative)

Physical Constant: Reported molecular weight:[1,4-6] 230,000 to 300,000.

Procedure for Handling:

Storage: Suspensions of crystals in ammonium sulfate solution, dry powders, or solutions (>0.1% of protein) of the enzyme in dilute neutral phosphate buffers are stable for at least 6 months when refrigerated. Purified preparations are extremely labile below pH 5.5 and above pH 9.0.

Dilution: Solubilization and dilution of the enzyme may be made in 0.02 M sodium phosphate, pH 7.0. Because of the acidic isoionic point (pH approximately[2] 5.3) and acid lability, solutions of this enzyme should be buffered near neutrality, particularly when conducting dialysis, gel filtration, etc.

Commercially Available Substrates: Acetyl- and propionyl-choline (as halides).

Specific Activity: One unit of activity for this enzyme is defined as the amount that catalyzes the hydrolysis of 1 μmol of acetylcholine per min at 25 °C and pH 7.4. The crystalline, electrophoretically homogeneous enzyme has an activity of 12,500 U/mg (calculated from Leuzinger *et al*.[2]) when assayed as indicated below.

Protein Assay: A_{280} × 0.62 = mg of protein per ml (in 0.02 M ammonium carbonate).[2]

Activity Assay:

Method:[3] The rate of release of acetic acid from acetylcholine is measured titrimetrically, at pH 7.4 and 25 °C.

Reagents: Prepare, with CO_2-free distilled water: 200 mM NaCl containing 40 mM $MgCl_2$ and 0.02% crystalline bovine serum albumin; 2.16 × 10^{-2} M acetylcholine chloride. Acetylcholine esterase: 10 to 50 μg per ml in 0.02% albumin solution.

Procedure: Standardize an autotitrator at 25 °C. Pipette into a 12-ml reaction vessel 4.0 ml NaCl–$MgCl_2$–albumin solution, 1.0 ml substrate, and 3.0 ml water. Adjust the pH to 7.4 with freshly prepared, standard 0.010 M NaOH. Initiate reaction by adding 10 μl of enzyme solution; record the volume of standard NaOH required to maintain the pH at 7.4 for several minutes.

Calculations: From the linear portion of the curve, determine the volume of standard NaOH required per minute. No. of units (μmol/min) = [volume NaOH (ml)/time (min)] × molar conc. (mol/l) × 10^3.

Impurities: None has been reported for the crystalline enzyme.

References

1. W. Leuzinger and A. L. Baker, *Proc. Nat. Acad. Sci. U.S.*, **57**, 446 (1967).
2. W. Leuzinger, A. L. Baker, and E. Cauvin, *Proc. Nat. Acad. Sci. U.S.*, **59**, 620 (1968).
3. L. T. Kremzner and I. B. Wilson, *J. Biol. Chem.*, **238**, 1714 (1963).
4. L. T. Kremzner and I. B. Wilson, *Biochemistry*, **3**, 1902 (1964).
5. W. Leuzinger, personal communication (1968).
6. W. Leuzinger, M. Goldberg, and E. Cauvin, *J. Mol. Biol.*, **40**, 217 (1964).

Acknowledgments

Original compilation of Criteria: A. L. Baker. Contributions by: W. Leuzinger and I. B. Wilson.

*IUB Classification: 3.1.1.7, Acetylcholine hydrolase (1964 Report) [not to be confused with 3.1.1.8 acylcholine acylhydrolase (cholinesterase or pseudocholinesterase)].

E-2
Alcohol Dehydrogenase*
(Horse Liver)

Equation: R—CH_2OH + NAD$^+$ $\rightleftharpoons$ R—CHO + NADH + H^+

A variety of primary and secondary alcohols, but not methanol, isopropanol, or tertiary alcohols, are oxidized, some with a rate exceeding that of ethanol.[1,2] Modification of the amino groups of the enzyme significantly increases the rate of oxidation of ethanol.[3] Liver alcohol dehydrogenase exists as isozymes formed by hybridization of dissimilar subunits, one of which oxidizes steroids;[4,5] in commerical preparations, the ethanol-active isozymes preponderate.[5] NADP$^+$ is 1% as active as NAD$^+$ as a coenzyme.[6]

Method of Isolation: Crystallized from horse liver.[7-10]

CRITERIA (Tentative)

Physical Constants: Reported molecular weight: 79,100, based on $s^o_{20,w}$ = 5.08 S, $D_{20,w}$ = 6.23 × 10^{-7} cm^2/s (extrapolated to zero concentration), $\bar{V}$ = 0.750 ml/g.[11] There are four bound zinc atoms per molecule of enzyme.[12] The absorbance ratio[13] at 280 and 260 nm is 1.7. A crystalline complex is formed in the presence of pyrazole.[14]

Procedures for Handling: Usually stored as crystals, in 25–30% aqueous ethanol at −15 to −20 °C. Storage in ammonium sulfate solution is not recommended, because of relatively poor stability.[15] Lyophilized commercial preparations have been reported to have good stability.[16] The enzyme cannot be dialyzed against ion-free water without extensive loss in activity, but may be dialyzed against cold 0.01 M phosphate buffer (pH 7.3) and diluted in this buffer just prior to analysis.

Commercially Available Substrates: Ethanol, acetaldehyde.

Specific Activity: One unit (U) is the amount of activity oxidizing 1 μmol ethanol/min. A specific activity of 3.1 U/mg was obtained in the assay given below (recalculated from Dalziel).[17]

Protein Assay:[8] Spectrophotometrically at 280 nm. A_{280} × 2.2 = mg of protein/ml.

Activity Assay:[15]

Reagents: 0.18 M ethanol in 0.1 M glycine buffer, pH 10.0; 1.26 mM NAD$^+$.

Procedure: To 1.85 ml glycine buffer, 0.15 ml ethanol solution, and 1.0 ml NAD$^+$ solution in a 1.0-cm cuvette is added 1 to 10 μl of enzyme solution. The increment in absorbance at 340 nm per minute is recorded.

Calculations: Number of units (U) in reaction vessel = 0.483 × [$\Delta A_{340}/\Delta t$(min)].

References

1. A. D. Merritt and G. M. Tompkins, *J. Biol. Chem.*, **234**, 2778 (1959).
2. A. D. Winer, *Acta Chem. Scand.*, **12**, 1695 (1958).
3. B. V. Plapp, *J. Biol. Chem.*, **245**, 1727 (1970).
4. R. Pietruszko, A. Clark, J. M. H. Graves, and H. J. Ringold, *Biochem. Biophys. Res. Commun.*, **23**, 526 (1966).
5. R. Pietruszko, H. J. Ringold, T.-K. Li, B. L. Vallee, Å. Åkeson, and H. Theorell, *Nature*, **221**, 440 (1969).
6. M. E. Pullman, S. P. Colowick, and N. O. Kaplan, *J. Biol. Chem.*, **194**, 593 (1952).
7. R. K. Bonnichsen and A. Wassen, *Arch. Biochem. Biophys.*, **18**, 361 (1948).
8. R. K. Bonnichsen and N. G. Brink, *Methods Enzymol.*, **1**, 495 (1955).
9. K. Dalziel, *Acta Chem. Scand.*, **12**, 459 (1958).
10. K. Dalziel, *Biochem. J.*, **80**, 440 (1961).
11. R. W. Green and R. H. McKay, *J. Biol. Chem.*, **244**, 5034 (1969).
12. Å. Åkeson, *Biochem. Biophys. Res. Commun.*, **17**, 211 (1964).
13. R. K. Bonnichsen, *Acta Chem. Scand.*, **4**, 715 (1950).
14. T. Yonetani and H. Theorell, *Arch. Biochem. Biophys.*, **100**, 554 (1963).
15. Å. Åkeson, personal communication.

16. R. Flora, personal communication.
17. K. Dalziel, *Acta Chem. Scand.*, 11, 397 (1957).

Acknowledgments

Original compilation of Criteria: J. B. Neilands. Contributions by Å. Åkeson and R. Flora. Revision by: J. B. Neilands.

*IUB Classification: 1.1.1.1, Alcohol:NAD oxidoreductase.

E-3
Alcohol Dehydrogenase*
(Yeast)

Equation: $R—CH_2OH + NAD^+ \rightleftharpoons R—CHO + NADH + H^+$

Ethanol and other primary, straight-chain alcohols are oxidized at various rates. Reactivity with secondary and branched-chain alcohols is generally low. The specificity is narrower than that of the liver enzyme.[1] NADP is not a coenzyme for yeast alcohol dehydrogenase.[2]

Methods of Isolation: The enzyme has been crystallized from brewers' yeast,[3] but the usual source is bakers' yeast.[4,5] The two preparations are very similar.[6]

CRITERIA (Tentative)

Physical Constants: Reported molecular weight: 141,000, based on $s^0_{20,w} = 7.61$ S, $D_{20,w} = 5.08 \times 10^{-7}$ cm^2/s (independent of protein concentration), $\bar{V} = 0.743$ ml/g.[7] The enzyme consists of four subunits, each of molecular weight 35,000.[7] There are four tightly bound zinc atoms per molecule.[8] The absorbance ratio[9] at 280 and 260 nm is 1.82.

Procedures for Handling: The enzyme is unstable at pH values below 6.0 and above 8.2 (Ref. 2) and is sensitive to oxidizing agents (including air) and to chelating agents.[5] To avoid inactivation, the use of an ultrarapid device for dialysis has been recommended.[10] The enzyme may be stored in 50% saturated ammonium sulfate at −20 °C and diluted, prior to use, with 10 mM potassium phosphate buffer (pH 7.5) containing 0.1% of gelatin or bovine serum albumin.[11]

Commercially Available Substrates: Ethanol, acetaldehyde.

Specific Activity: One unit (U) is the amount of activity oxidizing 1 μmol ethanol/min. A specific activity of 500 U/mg was obtained in the assay given below.[12] The pH optimum is close to 8.6 in either tris or pyrophosphate buffers.[2]

Protein Assay: Protein is determined spectrophotometrically at 280 nm. $A_{280} \times 0.796 = $ mg of protein/ml (Ref. 9). The biuret procedure should be used for impure preparations.[13]

Activity Assay:[14] (Modified to yield increased concentrations of NAD$^+$ and ethanol.)

Reagents: 2 M ethanol; 45 mM pyrophosphate, pH 8.8; 25 mM NAD$^+$.

Stock solution of enzyme, containing approximately 1 mg/ml in 100 mM phosphate buffer (pH 7.5). Dilute, just before use, in 10 mM phosphate buffer (pH 7.5) containing 0.1% gelatin, to give a concentration of about 5 μg/ml.

Procedure: To 1.0 ml of pyrophosphate buffer, 0.5 ml of ethanol solution, and 1.0 ml NAD$^+$ solution in a 1-cm cuvette at 25 °C is added sufficient distilled water to give, after introduction of the enzyme solution, a volume of 3.0 ml. The enzyme is added in approximately 0.2 ml of solution containing about 1 μg of protein. A blank cuvette is charged with all the reagents except the

enzyme. The increment in absorbance over a 30-s period during the initial (or linear) phase of the reaction is used to calculate activity units.

Calculations: Number of enzyme units (U) in reaction vessel = $0.483 \times [\Delta A_{340}/\Delta t(\text{min})]$.

References

1. F. M. Dickinson and K. Dalziel, *Nature*, 214, 31 (1967).
2. H. Sund and H. Theorell, *Enzymes*, 7A, 25 (1963).
3. E. Negelein and H. J. Wulff, *Biochem. Z.*, 293, 351 (1937).
4. E. Racker, *J. Biol. Chem.*, 184, 313 (1950).
5. K. Wallenfels and H. Sund, *Biochem. Z.*, 329, 17 (1957).
6. T. Keleti, *Acta Physiol. Acad. Sci. Hung.*, 13, 309 (1958).
7. M. Bühner and H. Sund, *Eur. J. Biochem.*, 11, 73 (1969).
8. K. Wallenfels, H. Sund, A. Faessler, and W. Burchard, *Biochem. Z.*, 329, 31 (1957).
9. J. E. Hayes and S. F. Velick, *J. Biol. Chem.*, 207, 225 (1954).
10. A. Larrson, personal communication.
11. E. Racker, *Methods Enzymol.*, 1, 500 (1955).
12. C. Worthington, personal communication.
13. H. W. Robinson and C. G. Hogden, *J. Biol. Chem.*, 135, 727 (1940).
14. B. L. Vallee and F. L. Hoch, *Proc. Nat. Acad. Sci. U.S.*, 41, 327 (1955).

Acknowledgments

Original compilation of Criteria: J. B. Neilands. Contributions by: A. Larrson and C. Worthington. Revision by: J. B. Neilands.

*IUB Classification: 1.1.1.1, Alcohol:NAD oxidoreductase.

E-4
Aldolase*
(Rabbit Skeletal Muscle)

Equation: $C_6H_{10}O_6(PO_3)_2^{4-} \rightleftharpoons C_3H_5O_3PO_3^{2-} + C_3H_5O_3PO_3^{2-}$

D-fructose 1,6-bisphosphate   D-glyceraldehyde 3-phosphate   dihydroxyacetone phosphate

Methods of Isolation: Taylor *et al.*;[1] Beisenherz *et al.*[2]

CRITERIA (Tentative)

Physical Constant: Reported molecular weight:[3,4] 158,000.

Procedures for Handling:

Storage: As a crystalline suspension in 2 M ammonium sulfate at 0–4 °C.

Dilution: For assay, dilution of the ammonium sulfate suspension with H$_2$O is satisfactory. Enzyme preparations that are more concentrated and contain less salt may be obtained by centrifuging and dissolving the crystals in buffer or water. The salt may be removed by dialysis or gel filtration. Aldolase is unstable below pH 4 and above pH 10.

Commercially Available Substrates: D-Fructose 1,6-bisphosphate (Ba, Ca, Mg, Na, and tricyclohexylammonium salts); DL-glyceraldehyde diethyl acetal 3-phosphate Ba salt; dihydroxyacetone dimethyl acetal† phosphate cyclohexylammonium salt monohydrate; and many other aldehydes.

Specific Activity: One unit U of activity corresponds to 1 μmol D-fructose 1,6-bisphosphate (I) cleaved per minute at 25 °C. The purest enzyme has a specific activity of 16 U/mg, assayed as described below, but values as high as 20 U/mg have been reported.[5]

Protein Assay: $A_{280} \times 1.06 = $ mg of enzyme/ml, 1-cm path length.[6]

Activity Assay: The dihydroxyacetone phosphate formed in the aldolase reaction is reduced by NADH with glycerophosphate dehydrogenase. The reaction is followed spectrophotometrically[7] at 340 nm.

Reagents: (I), 10 mM sodium salt of D-fructose 1,6-bisphosphate. NADH, 3 mM (2.2 mg/ml), α-Glycerophosphate dehydrogenase, 1:10 dilution (in water) of the commercial ammonium sulfate suspension, 1 mg/ml (8 U/ml). (See E-22.)

Procedure: To 0.3 ml of 100 mM tris-HCl buffer (pH 7.5–8.0) in a cuvette, add 0.15 ml NADH, 0.3 ml (I), 0.1 ml α-glycerophosphate dehydrogenase, and water to give a volume of 3.0 ml. Record any blank reaction that may occur. At zero time, add 0.1 ml or less of aldolase, mix, and record the decrease in optical density at 340 nm.

Calculation of Activity: Number of units (U) in the cuvette (3 ml) = $[\Delta A_{340}/\Delta t(\min)] \times 0.483$.

Impurities: For the purest enzyme, the following activities should not be present in more than trace quantities: lactate dehydrogenase; glyceraldehyde 3-phosphate dehydrogenase; pyruvate kinase; α-glycerophosphate dehydrogenase; triose phosphate isomerase; and phosphofructokinase. Crude aldolase preparations may contain troublesome amounts of these impurities. Chromatography on O-[2-(diethylamino)ethyl]cellulose may be used to eliminate triose phosphate isomerase.[5]

Notes: Aldolase is not significantly affected by pH in the range from 7 to 9. All salts act as competitive inhibitors as a function of their ionic strength.[8] The substrate concentration given is approximately 100 times the K_m, and, therefore, need not be adjusted precisely. In the presence of triose phosphate isomerase, the rate of oxidation of NADH is double.[‡3] The same assay may be used with all other substrates of aldolase, such as D-fructose 1-phosphate.

References

1. J. F. Taylor, A. A. Green, and G. T. Cori, *J. Biol. Chem.*, **173**, 591 (1948).
2. G. Beisenherz, H. J. Boltze, T. Bücher, R. Czok, K. H. Garbade, E. Meyer-Arendt, and G. Pfleiderer, *Z. Naturforsch.*, **8b**, 555 (1953).
3. K. Kawahara and C. Tanford, *Biochemistry*, **5**, 1578 (1966).
4. F. J. Castellino and R. Barker, *Biochemistry*, **7**, 2207 (1968).
5. O. C. Richards and W. J. Rutter, *J. Biol. Chem.*, **236**, 3185 (1961).
6. J. W. Donovan, *Biochemistry*, **3**, 67 (1964).
7. E. Racker, *J. Biol. Chem.*, **167**, 843 (1947).
8. A. H. Mehler, *J. Biol. Chem.*, **238**, 100 (1963).

Acknowledgments

Original compilation of Criteria: A. H. Mehler. Contributions by: W. J. Rutter. Revision by: F. Wold.

* IUB Classification: 4.1.2.13, D-Fructose 1,6-bisphosphate D-glyceraldehyde-3-phosphate-lyase. (Previously, bisphosphate was called diphosphate.)

† Previously termed a ketal, "acetal" is now recommended.

‡ The presence of triose phosphate isomerase may be detected by using glyceraldehyde 3-phosphate as a substrate in place of (I) in the assay system described above.

CRITERIA (Tentative)

Physical Constants: Reported molecular weight: 80,000 (Ref. 4), 86,000 (Ref. 5). The enzyme is composed of 2 identical subunits,[6] and contains from 2 to 4 zinc atoms per molecule.[7,8]

Procedure for Handling: Crystal suspensions are stable at room temperature for months and can be stored at 0 °C, but are not stable to freezing.[2] Inorganic phosphate is a powerful inhibitor (K_i = 5 μM).[5]

Commercially Available Substrates: p-Nitrophenyl phosphate, and many other monoester phosphates.

Specific Activity: One unit (U) of this enzyme is the activity that hydrolyzes 1 μmol of p-nitrophenyl phosphate per minute under the conditions described in "Activity Assay." The crystalline enzyme has been reported to have a specific activity of 48 U/mg.†

Protein Assay:[9] To 2.4 ml of 0.06 M ammonium sulfate, add 0.7 to 3 mg of protein, mix, and add 0.1 ml of 2.5 M trichloroacetic acid. Dilute to 3.0 ml, mix, and, after several minutes, read A_{278}. $A_{278} \times 1.39$ = mg enzyme/ml (Ref. 2, 3).

Activity Assay:[2]

Method: Measurement at 420 nm of p-nitrophenolate released at pH 8 and 27 °C.

Reagents: p-Nitrophenyl phosphate, 1 mM in 1 M tris (hydroxymethyl)aminomethane hydrochloride buffer, pH 8.

Procedure: To 2.9 ml of buffered p-nitrophenyl phosphate at 27 °C, add 0.1 ml of enzyme. Read change in ΔA_{420} per minute. The molar absorption coefficient is 1.32×10^4 for the mixture of p-nitrophenol and p-nitrophenolate liberated at this pH.

Calculation: Number of enzyme units (U) in reaction vessel (3 ml) = $[\Delta A_{420}/\Delta t (\min)] \times 0.227$

Impurities: Nucleases may be present as impurities.

References

1. W. B. Anderson and R. C. Nordlie, *J. Biol. Chem.*, **242**, 114 (1967).
2. M. Malamy and B. L. Horecker, *Methods Enzymol.*, **9**, 639 (1966).
3. M. Malamy and B. L. Horecker, *Biochemistry*, **3**, 1893 (1964).
4. A. Garen and C. Levinthal, *Biochim. Biophys. Acta*, **38**, 470 (1960).
5. M. J. Schlesinger and K. Barrett, *J. Biol. Chem.*, **240**, 4284 (1965).
6. F. Rothman and R. Byrne, *J. Mol. Biol.*, **6**, 330 (1963).
7. D. J. Plocke, C. Levinthal, and B. L. Vallee, *Biochemistry*, **1**, 373 (1962).
8. J. A. Reynolds and M. J. Schlesinger, *Biochemistry*, **6**, 3552 (1967).
9. T. Bücher, *Biochim. Biophys. Acta*, **1**, 292 (1947).

Acknowledgments

Original compilation of Criteria: J. Larner. Contributions by: M. J. Schlesinger and B. L. Horecker.

* IUB Classification: 3.1.3.1, Orthophosphoric-monoester phosphohydrolase.

† Calculated from data given by Malamy and Horecker.[2]

E-5
Alkaline Phosphatase*
(*Escherichia coli*)

Equation: $ROPO_3^{2-} + H_2O \rightarrow HPO_4^{2-} + ROH$ (R = alkyl or aryl groups)

Inorganic pyrophosphate (PP$_i$) is hydrolyzed:[1] $P_2O_7^{3-} + H_2O \rightarrow 2HPO_4^{2-} + H^+$ Transfer of phosphate from PP$_i$ to D-glucose has also been reported.[1] D-Glucose + $P_2O_7^{3-} \rightarrow$ D-glucose 6-phosphate^{2-} + $HPO_4^{2-} + H^+$

Method of Isolation: Crystallized from ammonium sulfate solution.[2,3]

E-6
D-Amino Acid Oxidase*
(Pig Kidney)

Reaction: $RCHNH_3^+COO^- + O_2 + H_2O \rightarrow RCOCOO^- + NH_4^+ + H_2O_2$

Catalyzes the oxidative deamination of D-amino acids. For notes on specificity, see Burton[1] and Boulanger and Osteux.[2]

Methods of Isolation: Crystallized from pig kidney.[3,4] (The crystalline enzyme is a combination of holoenzyme and benzoate, which, as a substrate substitute, acts as a stabilizer.)

CRITERIA (Tentative)

Physical Constants: Reported molecular weight: For monomer, by the molecular-sieve technique, 35,000–40,000 (Ref. 5) and 50,000 (Ref. 6). The molar ratio of apoenzyme monomer to FAD has been reported to be 1:1.[5] At low concentrations (0.3 mg/ml), the benzoate complex contains two fractions, one of Mol. Wt. 48,200, and the other of Mol. Wt. 95,300 (Ref. 6). At 1–5 mg/ml, the benzoate complex has[6] a Mol. Wt. of 95,300.

Procedure for Handling: The purified enzyme, lyophilized in 13.3 mM phosphate buffer (pH 6.2), is stable at room temperature for several weeks. The enzyme is inactivated by reagents that react with sulfhydryl groups.

Commercially Available Substrates: Alanine (D- and DL-) and other D-amino acids.

Specific Activity: One unit (U) represents the amount of activity that catalyzes the deamination of 1 μmol of D-alanine per minute at 37 °C under the specified conditions, with air as the gas phase. Massey et al.[4] reported that the crystalline enzyme has a specific activity of 25 U/mg at 37 °C, pH 8.3 (280 μl O_2/min per milligram of protein).

Protein Assay:[4] In $Na_4P_2O_7$ buffer (20 mM), pH 8.3, $A_{280} \times 0.625 =$ mg enzyme/ml.

Activity Assay: Excess catalase is added to ensure complete decomposition of the H_2O_2 formed in the oxidase reaction; no interference is then introduced by the various levels of catalase activity in the samples being assayed. With excess of catalase, the consumption of 1 molecule (2 atoms) of oxygen corresponds to the oxidation of 2 molecules of D-amino acid.

Method: The rate of uptake of O_2 is measured.[1]

Enzyme: 0.5 mg of enzyme/ml in 20 mM $Na_4P_2O_7$, pH 8.3.

Substrate: 300 mM DL-alanine in 20 mM $Na_4P_2O_7$, pH 8.3.

Procedure: In single-arm Warburg flasks, place pyrophosphate buffer (20 mM, pH 8.3, containing ~2.5 μg of crystalline catalase/ml), 1.8 ml into a test flask, and 2.3 ml into the control. Add 0.5 ml of enzyme to the test flask. Into the side arm of each flask, add 0.5 ml of substrate, equilibrate 10 min, tip in the enzyme, and read at 5-min intervals for 30 min.† Number of units (U) = [volume of O_2 taken up (μl)/t (min)]/11.2.

Impurities: Contaminating impurities have not been determined.

References

1. K. Burton, *Methods Enzymol.*, **2**, 199 (1965).
2. P. Boulanger and P. Osteux, *Methods of Enzymatic Analysis*, H. U. Bergmeyer, ed., Academic Press, New York (1963), p. 367.
3. H. Kubo, T. Yamano, M. Iwatsubo, H. Watari, T. Shiga, and A. Isomoto, *Bull. Soc. Chim. Biol.*, **42**, 569 (1960).
4. V. Massey, G. Palmer, and R. Bennett, *Biochim. Biophys. Acta*, **48**, 1 (1961).
5. S. E. Henn and G. M. Ackers, *J. Biol. Chem.*, **244**, 465 (1969).
6. M. L. Fonda and B. M. Anderson, *J. Biol. Chem.*, **243**, 5635 (1968).

Acknowledgments

Original compilation and revision of Criteria: C. Worthington. Contributions by: A. Meister.

* IUB Classification: 1.4.3.3, D-Amino acid: oxygen oxidoreductase (deaminating).

† If aspartate is the substrate, 0.2 ml 10% KOH should be added to the center well to trap CO_2 formed by decarboxylation of the oxaloacetate that is formed.

E-7

L-Amino Acid Oxidase*
(*Crotalus adamanteus* Venom)

Reaction: $RCHNH_3^+COO^- + O_2 + H_2O \rightarrow RCOCOO^- + NH_4^+ + H_2O_2$

Catalyzes the oxidative deamination of L-amino acids. For notes on specificity, see Ratner.[1]

Methods of Isolation: Crystallized by the method of Wellner and Meister.[2]

CRITERIA (Tentative)

Physical Constants: Reported molecular weight: 130,000–140,000 with two molecules of FAD per molecule of enzyme.[3]

Procedure for Handling: The crystalline enzyme is soluble in 100 mM KCl. Solutions containing ammonium sulfate are stable for up to a week at room temperature, and for several months at 5 °C. Dialysis against cold water causes the enzyme to crystallize.[4] The enzyme may be inactivated[5] if it is stored at −5 to −60 °C. The optimum pH depends on the amino acid used as the substrate.

Commercially Available Substrate: L-Leucine and many other L-amino acids.

Specific Activity: One unit (U) for this enzyme is the activity that catalyzes the oxidative deamination of 1 μmol of L-leucine per minute under the specified conditions at 37 °C, with air as the gas phase. Highest specific activity reported:† 5.13 U/mg.

Protein Assay:[2] In 100 mM KCl, $A_{275} \times 0.56 =$ mg enzyme/ml.

Activity Assay: Excess catalase is added to ensure complete decomposition of the H_2O_2 formed in the oxidase reaction; no interference is then introduced by the various levels of catalase activity in the samples being assayed. With excess catalase, the consumption of 1 molecule (2 atoms) of oxygen corresponds to the oxidation of 2 molecules of L-amino acid.

Method: The rate of uptake of O_2 is measured.[2]

Procedure: In Warburg flasks, place 1 ml of tris-HCl buffer (200 mM, pH 7.8) containing 2–3 μg of crystalline catalase, 0.1 to 0.2 ml of enzyme (0.25–0.50 mg/ml in 100 mM KCl), and 1.2 to 1.1 ml of 100 mM KCl. Add 0.5 ml of L-leucine (100 mM) to the side arm. Equilibrate for 10 min, tip in the substrate, and read at 5-min intervals for 30 min.‡ Number of units (U) = [volume of O_2 taken up (μl)/t(min)]/11.2.

Impurities: Contaminating activities have not been reported.

References

1. S. Ratner, *Methods Enzymol.*, **2**, 204 (1955).
2. D. Wellner and A. Meister, *J. Biol. Chem.*, **235**, 2013 (1960).
3. A. deKok and A. B. Rawitch, *Biochemistry*, **8**, 1405 (1968).
4. W. K. Paik and S. Kim, *Biochim. Biophys. Acta*, **96**, 66 (1965).
5. B. Curti, V. Massey, and M. Zmudka, *J. Biol. Chem.*, **243**, 2306 (1968).

Acknowledgments

Original compilation and revision of Criteria: C. Worthington. Contributions by: A. Meister.

* IUB Classification: 1.4.3.2, L-Amino acid: oxygen oxidoreductase (deaminating).

† Calculated from the data of Wellner and Meister.[2]

‡ If aspartate is the substrate, 0.2 ml 10% KOH should be added to the center well to trap CO_2 formed by decarboxylation of the oxaloacetate that is formed.

E-8
Aminoacylase*
(Pig Kidney)

Equation: $RCONHCHR'COO^- \rightarrow RCOO^- + {}^+H_3NCHR'COO^-$ where R is F_3C-, $ClCH_2-$, CH_3CH_2-, CH_3-, H−, etc., and R′ is the side chain of many amino acids.

Reaction: Catalyse the hydrolysis of amide bonds between carboxyl-terminal L-amino acids and N-acyl groups; shows little activity toward carboxyl-terminal L-aspartic acid and carboxyl-terminal aromatic L-amino acids.

Method of Isolation: Purified from fresh or fresh-frozen hog kidney.[1,2]

CRITERIA (Tentative)

Physical Constant: Reported molecular weight: 76,500.[1]

Procedures for Handling:

Storage: The dry powder of partially purified enzyme is stable for months at 5 °C.

Dilution: In 0.1 M phosphate buffer, pH 7.0.

Dialysis: Of concentrated solutions (more than 50 mg/ml) against distilled water or 5×10^{-3} M NaCl.

Commercially Available Substrates: Acetyl-L-alanine. N-(chloroacetyl)-L-alanine, acetyl-L-methionine (AM), and numerous other N-acylated L-amino acids.

Specific Activity: One unit of activity, U, is the amount that catalyzes the hydrolysis of 1 μmol of AM per minute under the conditions specified below. Highest specific activity reported: About 150 U/mg (calculated from data of Bruns and Schulze[1]).

Protein Assay:[3] $A_{280} \times 1.0$ = mg of enzyme/ml.

Activity Assay:

Method: The rate of hydrolysis[3] of AM is measured by the decrease in absorbance at 238 nm.

Procedure:[3] The spectrophotometer is nulled against 0.025 M L-methionine in 0.1 M phosphate buffer (pH 7.0) at 238 nm with a slit width of ~0.3 mm. The methionine solution is replaced by 3 ml of 0.025 M solution of AM in the same buffer. The absorbance difference is 0.53. A 10-μl portion of enzyme solution, containing 0.05 to 0.15 mg of enzyme, is added, and the decrease in A_{238} is recorded at 30-s intervals. The hydrolysis of 1 μmol of AM causes a decrease in A_{238} of 0.00707. Rate measurements should not be made below an absorbance of 0.31 (60% of the total initial A).

Calculation of Activity: Number of units (U) = $[\Delta A_{238}/\Delta t$ (min)] $\times 141.5$.

Impurities: Appreciable D-amino acid oxidase activity is present in samples of aminoacylase I current available.[2]

References

1. F. H. Bruns and C. Schulze, *Biochem. Z.*, **336**, 162 (1962).
2. S. M. Birnbaum, L. Levintow, R. B. Kingsley, and J. P. Greenstein, *J. Biol. Chem.*, **194**, 455 (1952).
3. M. A. Mitz and R. J. Schlueter, *Biochim. Biophys. Acta*, **27**, 168 (1958).

Acknowledgments

Original compilation of Criteria: J. E. Folk. Contributions by: G. E. Perlmann.

* IUB Classification: 3.5.1.4, Aminoacylase; formerly called Acylase I.

E-9
α-Amylase*
(Pig Pancreas)

Reaction: Randomly hydrolyzes α-D-(1 → 4) bonds in polysaccharides containing three or more adjacent α-D-(1 → 4)-linked D-glucose residues.

Method of Isolation: Crystallized from pig pancreas.[1-3]

CRITERIA (Tentative)

Physical Constants: Reported molecular weight:[4] 45,000. The enzyme contains at least 1 atom of Ca per molecule.[5]

Procedures for Handling: Preparation may be available as a crystalline suspension in distilled water. At 5 °C, such suspensions are stable for several months. Solutions (2 mg/ml) in 10 mM phosphate (pH 7.0–7.2) and 20 mM NaCl are stable for 10 weeks at 5 °C. Freeze-drying partially inactivates the enzyme. Freezing also results in a loss of activity.[3] Protease impurities are removed during purification, resulting in increased stability of the enzyme.

Commercially Available Substrates: Soluble starch, glycogen, and potato amylopectin.

Specific Activity: One unit (U) for this enzyme is the activity that liberates 1 μmol of reducing groups per minute at 25 °C. Specific activity as high as 1400 U/mg has been reported[6] in assay conditions similar to, but not identical with, those described under "Activity Assay."

Protein Assay: Measure A_{280} in water. $A_{280} \times 0.41$ = mg of enzyme/ml, calculated from Caldwell *et al.*[3] and Hsiu *et al.*[6]

Activity Assay:

Method: Measurement of the rate at which reducing groups are formed from the substrate, estimated (as maltose) by reaction with 3,5-dinitrosalicyclic acid. Cl⁻ (or NaCl) is required for maximum activity.[7]

Reagents: Substrate—1% of potato amylopectin in 20 mM sodium phosphate, pH 6.9, containing 6.7 mM NaCl. Color Reagent—1 g of 3,5-dinitrosalicylic acid is dissolved in 70 ml of H_2O containing 1.60 g of NaOH, 30 g of potassium sodium tartrate is added, and the solution is diluted to 100 ml; protect from CO_2.

Procedure: Dilute the enzyme with H_2O to 1–2 μg/ml. Add 0.5 ml of enzyme solution to 0.5 ml of substrate at 25 °C. Incubate for 3 min. Add 1 ml of color reagent. Heat in a boiling-water bath for 5 min and cool. Add 10 ml of H_2O and measure the absorbance at 540 nm against a blank containing buffer without enzyme.

Calculations: The number of micromoles of reducing groups is estimated by comparing the absorbance with a standard curve, established with maltose (0.1–1 mg/ml of H_2O).

Impurities: Protease activities may be present in samples that have not been highly purified.[8]

References

1. K. H. Meyer, E. H. Fischer, and P. Bernfeld, *Experientia*, **3**, 106 (1947).
2. E. H. Fischer and P. Bernfeld, *Helv. Chim. Acta*, **31**, 1831 (1948).
3. M. L. Caldwell, M. Adams, J. T. Kung, and G. C. Toralballa, *J. Am. Chem. Soc.*, **74**, 4033 (1952).
4. C. E. Danielsson, *Nature*, **160**, 899 (1947).
5. B. L. Vallee, E. A. Stein, W. N. Sumerwell, and E. H. Fischer, *J. Biol. Chem.*, **234**, 2901 (1959).
6. J. Hsiu, E. H. Fischer, and E. A. Stein, *Biochemistry*, **3**, 61 (1964).
7. P. Bernfeld, *Methods Enzymol.*, **1**, 149 (1955).
8. E. H. Fischer and E. A. Stein, *Enzymes*, **4**, 313 (1960).

Acknowledgments

Original compilation of Criteria: C. Worthington. Contributions by: P. Bernfeld, M. Adams, M. L. Caldwell, and F. Smith.† Revision: G. Gorin, P. Bernfeld, and R. Flora.

* IUB Classification: 3.2.1.1, 1,4-α-Glucan glucanohydrolase. Committee Note: In the IUB report, the configuration is not indicated. The same should be α-D-(1 → 4)-glucan 4-glucanohydrolase.
† Deceased 1965.

E-10
β-Amylase*
(Sweet Potato)

Reaction: Hydrolyzes α-D-(1 → 4) bonds in α-D-(1 → 4)-glucans, removing successive maltose residues from the nonreducing end of the chain.

Method of Isolation: Crystallized from sweet potato.[1,2]

CRITERIA (Tentative)

Physical Constants: Reported molecular weight:[3] 152,000 ±15,000. This is probably[4] an aggregate of active subunits of weight 50,000.

Procedures for Handling: Store as a crystalline suspension in 2.3 M ammonium sulfate at 4 °C.

Commercially Available Substrates: Soluble starch, potato amylopectin, amylose, and glycogen.

Specific Activity: One unit (U) for this enzyme is the activity that liberates 1 μmol of maltose per minute at 25 °C. A specific activity of about 1160 U/mg has been found[5] for preparations of the highest purity by the procedure described under "Activity Assay."

Protein Assay: Determine by the biuret procedure,[6] with bovine serum albumin as the standard. For highly purified preparations ($\geq$1000 U/mg), A_{280} may be used. Measure A_{280} in acetate buffer (16 mM, pH 4.8). $A_{280} \times 0.59$ = mg of enzyme/ml (Ref. 3).

Activity Assay:

Method: Measurement of the rate at which maltose is formed from the substrate, estimated by reaction with 3,5-dinitrosalicylic acid.[7]

Reagents: Substrate—1% of potato amylopectin in 16 mM sodium acetate, pH 4.8. Color Reagent—1 g of 3,5-dinitrosalicylic acid is dissolved in 70 ml of H_2O containing 1.60 g of NaOH, 30 g of potassium sodium tartrate is added, and the solution is diluted to 100 ml. Protect from CO_2.

Procedure: Dilute enzyme to 1 μg/ml with H_2O. Add 0.5 ml of enzyme solution to 0.5 of substrate at 25 °C. Incubate for 3 min. Add 1 ml of color reagent. Heat in a boiling-water bath for 5 min and cool. Add 10 ml of H_2O and measure absorbance at 540 nm against a blank containing buffer without enzyme.

Calculations: The number of micromoles of maltose is estimated by comparing absorbance with a standard curve, established with maltose (0.1–1.0 mg/ml of H_2O).

Impurities: α-Amylase and acid phosphatase may occur in some samples, but can be removed by purification.

References

1. A. K. Balls, M. K. Walden, and R. R. Thompson, *J. Biol. Chem.*, **173**, 9 (1948).
2. S. Nakayama and S. Amagase, *J. Biochem.*, **54**, 375 (1963).
3. S. England and T. P. Singer, *J. Biol. Chem.*, **187**, 213 (1950).
4. J. A. Thoma, D. E. Koshland, Jr., J. Ruscica, and R. Baldwin, *Biochem. Biophys. Res. Commun.*, **12**, 184 (1963).
5. P. Bernfeld, unpublished data.
6. A. G. Gornall, C. J. Bardawill, and M. M. David, *J. Biol. Chem.*, **177**, 751 (1949).
7. P. Bernfeld, *Methods Enzymol.*, **1**, 149 (1955).

Acknowledgments

Original compilation of Criteria: C. Worthington. Contributions by: P. Bernfeld, M. Adams, M. L. Caldwell, and F. Smith.† Revision: G. Gorin, P. Bernfeld, and R. Flora.

* IUB Classification: 3.2.1.2, 1,4-α-Glucan maltohydrolase. Committee Note: In the IUB report, the configuration is not indicated. The name should be α-D-(1 → 4)-glucan maltohydrolase.

† Deceased 1965.

E-11

ATP–Creatine Phosphotransferase*
(Rabbit Skeletal Muscle)

Equation: ATP^{4-} + creatine $\overset{Mg^{2+}}{\rightleftharpoons}$ ADP^{3-} + phosphocreatine^{2-} + H^+

Method of Isolation: Crystallized from aqueous ethanol.[1-3]

CRITERIA (Tentative)

Physical Constant: Reported molecular weight: 81,000[4], 82,600[5].

Procedures for Handling: Limits of stability range, pH 5–10. Stable for months as a crystalline suspension (at −10 °C), or when frozen (5–6% solution). May be lyophilized in the presence of glycine, pH 9, and stored indefinitely, if kept dry and cold (−10 °C).[1] May be dialyzed at 3 °C under a variety of conditions, provided that an atmosphere of nitrogen is used to prevent oxidation of reactive –SH groups.[3,7]

Specific Activity: One unit of activity is equivalent to 1 μmol of phosphocreatine formed per minute at 30 °C and pH 8.8. The enzyme, recrystallized 2–3 times, has a specific activity of 130–200 U/mg.[3,7]

Protein Assay: $A_{280} \times 1.13$ = mg of protein per ml (50 mM phosphate, pH 7.0).[8]

Activity Assay:

Principle: At pH 8.8, the reaction yields one equivalent of H^+ per mol of ATP consumed.[3,9]

Reagents: Prepare with CO_2-free distilled water: 32 mM ATP (neutralized with NaOH); 32 mM $MgSO_4$; 80 mM creatine; 4% (w/v) crystalline bovine serum albumin; 10.0 mM NaOH (freshly diluted from standardized NaOH).

ATP-creatine transphosphorylase: Just prior to measurement, dilute with ice-cold 1 mM glycine, pH 9, to about 100 to 500 μg/ml.

Procedure: Standardize a pH-stat (a commercial autotitrator) at 30 °C with a standard buffer. Pipet into a 12-ml reaction vessel (water-jacketed at 30 °C and equipped with a stirrer) 4.00 ml of creatine, 1.00 ml of ATP, 1.00 ml of $MgSO_4$, and 0.20 ml of albumin. Pass a slow stream of N_2 over the liquid (to exclude interference by CO_2 during titration), and start stirring. Adjust to pH 8.8 with 0.01 M NaOH; add distilled water to a total volume of 8.0 ml and readjust to pH 8.8. After temperature equilibration (3–5 min), with pH-stat set to record at pH 8.8, initiate the enzymic reaction by adding 10 μl of ATP-creatine transphosphorylase (1 to 5 μg). Repeat, but substitute water for creatine, to determine the relatively low "blank" rate.

Calculations: The initial rate is calculated from the linear portion of the curve. The blank rate is subtracted from the overall rate to obtain the enzymic rate, which is expressed in activity units (microequivalents of NaOH consumed per minute).

Impurities: May contain ATPase activity to the extent of 10^{-5} of the ATP : creatine transphosphorylase activity.[10] Significant impurities have not been reported for the thrice-recrystallized enzyme. Traces of glycolytic enzymes (e.g., triose phosphate isomerase) may contaminate amorphous preparations.

References

1. S. A. Kuby, L. Noda, and H. A. Lardy, *J. Biol. Chem.*, **209**, 191 (1954).
2. L. Noda, T. Nihei, and M. F. Morales, *J. Biol. Chem.*, **235**, 2830 (1960).
3. T. A. Mahowald, E. A. Noltmann, and S. A. Kuby, *J. Biol. Chem.*, **237**, 1535 (1962).
4. L. Noda, S. A. Kuby, and H. A. Lardy, *J. Biol. Chem.*, **209**, 203 (1954).
5. R. H. Yue, R. H. Palmieri, O. E. Olson, and S. A. Kuby, *Biochemistry*, **6**, 3204 (1967).
6. E. A. Noltmann, T. A. Mahowald, and S. A. Kuby, *J. Biol. Chem.*, **237**, 1146 (1962).
7. M. L. Tanzer and C. Gilvarg, *J. Biol. Chem.*, **234**, 3201 (1959).
8. S. A. Kuby, T. A. Mahowald, and E. A. Noltmann, *Biochemistry*, **1**, 748 (1962).
9. S. A. Kuby and E. A. Noltmann, *Enzymes*, **6**, 515 (1962).
10. T. Sasa and L. Noda, *Biochim. Biophys. Acta*, **81**, 270 (1964).

Acknowledgments

Original compilation of Criteria: S. A. Kuby. Contributions by: O. E. Olson and E. A. Noltmann.

* IUB Classification: 2.7.3.2, ATP : creatine phosphotransferase; commonly used name: creatine kinase.

E-12
Carbonic Anhydrase*
(Bovine Erythrocytes)

Equations:
$$CO_2 + H_2O \rightleftharpoons HCO_3^- + H^+ \qquad (1)$$
$$CH_3—CHO + H_2O \rightleftharpoons CH_3—CH(OH)_2 \qquad (2)$$
$$RCOOR + H_2O \rightleftharpoons RCOO^- + H^+ + ROH \qquad (3)$$

Equation (1) represents the physiological function of the enzyme. The hydration of acetaldehyde was first demonstrated by Pocker and Meany.[1] The hydrolysis of p-nitrophenyl acetate and certain other esters has been studied by Pocker and Stone,[2] Thorslund and Lindskog,[3] and Kaiser and Lo.[4]

Methods of Isolation: From bovine erythrocytes by DEAE chromatography and zone electrophoresis on cellulose columns (Lindskog[5] and Liefländer[6]). Some commercial preparations are made by using the procedure of Keilin and Mann.[7]

CRITERIA (Tentative)

Physical Constants: Reported molecular weight: 31,000 ±1,000 (Ref. 5), 30,000 (Ref. 8). The enzyme contains one atom of zinc per molecule of enzyme.

Procedures for Handling: Concentrated solutions (1–10 mg/ml) can be stored for several days at room temperature between pH 5.5 and 12 (ionic strength, 0.15) and for months in the refrigerator (pH 6–10). Lyophilization may cause a loss of crystallizability.[9] The enzyme may be stored in 2 M ammonium sulfate at −20 °C. Very dilute solutions of the enzyme (∼1 μg/ml) are sensitive to agitation and heavy-metal impurities but can be stabilized by 0.05% peptone.[10] The enzyme is inhibited by most monovalent anions.[2] Aromatic and heterocyclic sulfonamides are potent inhibitors of carbonic anhydrase. Sulfate and phosphate are practically noninhibitory.

Commercially Available Substrates: Carbon dioxide, sodium hydrogen carbonate, acetaldehyde, p-nitrophenyl acetate.

Specific Activity: One unit (U) is the activity that hydrolyzes 1 μmol of p-nitrophenyl acetate/min under conditions given in the "Activity Assay".† Highly purified preparations have a specific activity of 0.73 U/mg.

Protein Assay:[3] $A_{280} \times 0.526$ = mg of enzyme/ml.

Activity Assay:[3]

Reagents: 10 mM p-nitrophenyl acetate in acetone; tris-H_2SO_4 buffer, pH 7.6, ionic strength, 0.1. (Dissolve 1.0 g of tris(hydroxymethyl)aminomethane in 75 ml of H_2O, neutralize to pH 7.6 with 0.5 M H_2SO_4, and dilute to 100 ml.)

Procedure: Buffer and enzyme (total volume, 2.4 ml; final enzyme concentration ∼50 μg/ml) are mixed in a 1-cm cuvette. Substrate (0.1 ml) is added, and the solution is rapidly mixed. The reaction is followed at 348 nm. A blank without enzyme is determined, and the rate of the uncatalyzed reaction is subtracted from the total rate, to give the enzyme-catalyzed rate. Activity is calculated from the initial rate, $(dA_{348}/dt)_{t=0}$, of the reaction. (If possible, measure from the initial slope of the curve on a recording of the reaction.)

Calculations: Number of units (U) in reaction vessel = 0.486 × $[dA_{348}/dt]_{t=0}$ (t in min).

Impurities: Owing to the presence of other esterases in erythrocytes, an impure sample of enzyme may yield an esterase activity higher inhibitor, such as acetazolamide, permits non-carbonic anhydrase esterase activity to be measured.[13]

References

1. Y. Pocker and J. E. Meany, *Biochemistry*, **4**, 2535 (1965).
2. Y. Pocker and J. T. Stone, *Biochemistry*, **6**, 668 (1967).
3. A. Thorslund and S. Lindskog, *Europ. J. Biochem.*, **3**, 117 (1967).
4. E. T. Kaiser and K.-W. Lo, *J. Am. Chem. Soc.*, **91**, 4912 (1969).
5. S. Lindskog, *Biochim. Biophys. Acta*, **39**, 218 (1960).
6. M. Liefländer, *Z. Physiol. Chem.*, **335**, 125 (1964).
7. D. Keilin and T. Mann, *Biochem. J.*, **34**, 1163 (1940).
8. P. O. Nyman and S. Lindskog, *Biochim. Biophys. Acta*, **85**, 141 (1964).
9. B. Strandberg, B. Tilander, K. Fridborg, S. Lindskog, and P. O. Nyman, *J. Mol. Biol.*, **5**, 583 (1962).
10. A. M. Clark and D. D. Perrin, *Biochem. J.*, **48**, 495 (1951).
11. E. E. Rickli, S. A. S. Ghazanfer, B. H. Gibbons, and J. T. Edsall, *J. Biol. Chem.*, **239**, 1065 (1964).
12. S. Lindskog, unpublished results.
13. J. A. Verpoorte, S. Mehta, and J. T. Edsall, *J. Biol. Chem.*, **242**, 4221 (1967).

Acknowledgments

Original compilation of Criteria: P. O. Nyman and S. Lindskog. Contributions by: J. T. Edsall and G. Gorin.

* IUB Classification: 4.2.1.1, Carbonate hydro-lyase; carbonate dehydratase.
† In the carbon dioxide hydration reaction using the procedure described by Rickli et al.,[11] a unit is defined as $10(t_b/t_c) − 1$, where t_b and t_c are times in seconds required for a given change in pH for blank and enzyme-catalyzed reactions, respectively. A value of about 200,000 unit/mg for the bovine enzyme[12] is obtained by this procedure. The unit is arbitrary, and not convertible to μmol/min without uncertainties.

E-13
Carboxypeptidase A*
(Bovine Pancreas)

Equation:
$$R'CONHCH(R'')COO^- \rightarrow R'COO^- + {}^+H_3NCH(R'')-COO^-$$
or
$$R'COOCH(R'')COO^- \rightarrow R'COO^- + HOCH(R'')COO^-$$
where
R' = acyl side chain, acylamino acid side chain, or peptide side chain; and
R'' = specific amino or hydroxy acid side chain, preferentially that of an aromatic acid or branched-chain aliphatic acid.

Reaction: Catalyzes the hydrolysis of peptide or ester bonds of certain aromatic and aliphatic, carboxyl-terminal, amino or hydroxy acids; pH optimum 7–9.

Method of Isolation: The crystalline enzyme is routinely isolated from bovine pancreas,[1,2]

CRITERIA (Tentative)

Physical Constant: Reported molecular weight:[3] 32,000; calculated:[4] 34,440.

Procedures for Handling:

Storage: Suspensions of crystals in water in the presence of a trace of toluene are stable at 4 °C for at least 6 months. Solutions (above 0.1%) in 10% LiCl may be stored at 4 °C for approximately 1 week without significant loss in activity. Significant losses in activity result from freezing either suspensions of crystals or solutions of the enzyme.

Dilution: Crystals of the enzyme may be washed with large volumes of cold water. The crystals dissolve slowly in cold 10% LiCl, to form 0.6 to 0.8% solutions. Even at low concentrations, the enzyme may crystallize rapidly from solutions that are less than 0.2 M in salt.

Dialysis: Against 10% LiCl or 1 M NaCl.

Commercially Available Substrates: (Benzyloxycarbonyl)glycyl-L-phenylalanine†, benzoylglycyl-L-phenylalanine (hippuryl-L-phenylalanine, I), (benzyloxycarbonyl)glycyl-L-tryptophan, and (benzyloxycarbonyl)glycyl-L-leucine. Many other protein and peptide substrates are available.

Specific Activity: One unit catalyzes the hydrolysis of 1 μmol I/min at 25 °C and pH 7.5. Preparations having the highest purity have a specific activity of about 88 U/mg by the method following.

Protein Assay: $A_{278} \times 0.515$ = mg of enzyme/ml; solvent, 10% LiCl.[5]

Activity Assay:

Method: The rate of hydrolysis of I is measured by the increase in absorbance at 254 nm (based upon the method of Folk and Schirmer[6]).

Procedure: The assay is conducted as follows:

Take I (10 mM, in H_2O), adjusted to pH 6–8 with about one equivalent of NaOH, 0.3 ml; tris buffer 50 mM (pH 7.5) 1 M in NaCl, 1.5 ml; H_2O, 1.2 ml. Add the enzyme, 5 to 20 μl containing 0.5 to 2.5 μg of protein, to the substrate mixture in a 1-cm quartz cuvette, and record A_{254} at 30-s intervals.

The hydrolysis of 1 μmol of I causes an increase in A_{254} of 0.12. Number of units = $[\Delta A_{254}/\Delta t(\min)] \times 8.33$.

Impurities: Recrystallized samples may have slight activity toward trypsin and chymotrypsin substrates. These activities can be eliminated by treatment with diisopropyl phosphorofluoridate.[7] Recrystallized samples may also be slightly contaminated with carboxypeptidase B, as indicated by release of lysine or arginine from certain substrates at high levels of enzyme.

References

1. M. L. Anson, *J. Gen. Physiol.*, 20, 663, 777 (1937).
2. H. Neurath, *Methods Enzymol.*, 2, 77 (1955).
3. F. W. Putnam and H. Neurath, *J. Biol. Chem.*, 166, 603 (1946).
4. E. L. Smith and A. Stockell, *J. Biol. Chem.*, 207, 501 (1954).
5. B. L. Vallee, J. A. Rupley, T. L. Coombs, and H. Neurath, *J. Biol. Chem.*, 235, 64 (1960).
6. J. E. Folk and E. W. Schirmer, *J. Biol. Chem.*, 238, 3884 (1963).
7. H. Fraenkel-Conrat, J. I. Harris, and A. L. Levy, *Methods Biochem. Anal.*, 2, 397 (1955).

Acknowledgments

Original compilation of Criteria: J. E. Folk. Contributions by: F. W. Putnam, C. H. W. Hirs, and R. L. Hill.

* IUB Classification: 3.4.12.2, Peptidyl-L-amino acid hydrolase.

† (Benzyloxycarbonyl)glycyl-L-phenylalanine = (carbobenzoxy)glycyl-L-phenylalanine.

E-14
Carboxypeptidase B*
(Pig Pancreas)

Equation: $R'CONHCH(R'')COO^- \rightarrow R'COO^- + {}^+H_3NR''COO^-$
   or
   $R'COOCH(R'')COO^- \rightarrow R''COO^- + HOCH(R'')COO^-$
where
   R' = acyl side chain, acylamino acid side chain, or peptide side chain; and
   R'' = side chain of lysine, arginine, ornithine, or homoarginine.

Reaction: Catalyzes the hydrolysis of peptide or ester bonds of carboxyl-terminal basic, amino, or hydroxy acids; pH optimum 7–9.

Method of Isolation: The enzyme is isolated from aqueous extracts of an acetone powder of autolyzed, porcine pancreas.[1]

Physical Constant: Reported molecular weight,[1] 34,300.

Procedures for Handling:

Storage: Concentrated aqueous solutions (10 mg/ml and above), stored frozen at −10 °C, are stable for at least 6 months.

Dilution: In cold water; should be made up daily.

Dialysis: Against water at 4 °C.

Commercially Available Substrates: α-*N*-Benzoylglycyl-L-arginine (hippuryl-L-arginine, I), α-*N*-benzoylglycyl-L-lysine (hippuryl-L-lysine).

Specific Activity: One unit represents the hydrolysis of 1 μmol I/min at 25 °C and pH 8. Preparations having the highest purity have an activity of about 275 U/mg by the following method:

Protein Assay: $A_{278} \times 0.468$ = mg of enzyme/ml; solvent, 5 mM tris buffer, pH 8.0.[1]

Activity Assay:

Method: The rate of hydrolysis of I is measured by the increase in absorbance at 254 nm.[1,2]

Procedure: The assay is conducted as follows:

Take I, 10 mM in H_2O, 0.3 ml; tris buffer, 50 mM (pH 8.0), 1.5 ml; and H_2O, 1.2 ml. Add enzyme, 5 to 20 μl containing 0.1 to 0.5 μg of protein, to the substrate mixture in a 1-cm quartz cuvette, and record A_{254} at 30-s intervals.

The hydrolysis of 1 μmol of I causes an increase in A_{254} of 0.12. Rates must be calculated from ΔA values below 0.14. Number of units = $[\Delta A_{254}/\Delta t(\min)] \times 8.33$.

Impurities: The purest samples may have slight activity toward trypsin and chymotrypsin substrates. These activities can be eliminated by treatment of solutions with diisopropyl phosphorofluoridate in a manner similar to that outlined for carboxypeptidase A.[3] The purest samples show less than 0.1% of carboxypeptidase A activity when assayed by the procedure described for carboxypeptidase A (E–13).

References

1. J. E. Folk, K. A. Piez, W. R. Carroll, and J. A. Gladner, *J. Biol. Chem.*, 235, 2272 (1960).
2. E. C. Wolff, E. W. Schirmer, and J. E. Folk, *J. Biol. Chem.*, 237, 3097 (1962).
3. H. Fraenkel-Conrat, J. I. Harris, and A. L. Levy, *Methods Biochem. Anal.*, 2, 397 (1955).

Acknowledgments

Original compilation of Criteria: J. E. Folk. Contributions by: F. W. Putnam, C. H. W. Hirs, and R. L. Hill.

* IUB Classification: 3.4.12.3, Peptidyl-L-lysine hydrolase.

E-15
Chymotrypsin A*
(Bovine Pancreas)

Equation: $R'R''CHCOR''' \rightarrow R'R''CHCOO^- + {}^+H_3NR$ or HOR
where
   R' = (usually) an acylamido group;
   R'' = a specific amino acid side chain, preferentially that of an aromatic amino acid; and
   R''' = $-NHR$ or $-OR$.

Reaction: Catalyzes the hydrolysis of peptide (amide) or ester bonds at the carboxyl linkage of aromatic and certain long-chain aliphatic amino acids; pH optimum, 7–9.

Method of Isolation: The active enzyme is routinely prepared following slow tryptic activation of the zymogen, chymotrypsinogen A.[1]

CRITERIA (Tentative)

Physical Constant: Reported molecular weights:[2] 21,600 to 27,000; calculated:[2] 24,500.

Procedures for Handling:

Storage: Salt-free lyophilized samples are stable when stored at 4 °C.

Dilution: In cold 1 mM HCl, 1 to 10 mg/ml, stable for several weeks at 4 °C.

Dialysis: Against 1 mM HCl at 4 °C.

Commercially Available Substrates: N-Benzoyl-L-tyrosine ethyl ester (I), N-benzoyl-L-tyrosinamide, N-acetyl-L-tyrosine ethyl ester, N-acetyl-L-tyrosinamide, hemoglobin, and casein. Many other related protein and synthetic substrates are available.

Specific Activity: One unit (U) catalyzes the hydrolysis in 25% methanol of 1 μmol I/min at 25 °C and pH 7.8. Preparations having the highest purity have a specific activity of about 35 U/mg by the following method:

Protein Assay: $A_{280} \times 0.495$ = mg of enzyme/ml; solvent, 1 mM HCl.[3]

Activity Assay:

Method: The rate of hydrolysis of I is measured by the increase in absorbance at 256 nm (based on the methods of Hummel[4]).

Procedure: The assay is conducted as follows:

Take I, 5 mM in 50% methanol (w/w), 0.3 ml; tris buffer, 80 mM (pH 7.8), 1.5 ml; 50% methanol (w/w), 1.2 ml. Add enzyme, 5 to 20 μl containing 1 to 5 μg of protein, to the substrate mixture in a 1-cm quartz cuvette, and record A_{256} at 30-s intervals.

The hydrolysis of 1 μmol of I causes an increase in A_{256} of 0.32. The rate must be calculated from ΔA values below 0.14. Number of units (U) = $[\Delta A_{256}/\Delta t(\text{min})] \times 3.13$

References

1. M. Kunitz and J. H. Northrop, *J. Gen. Physiol.*, **19**, 991 (1936).
2. P. Desnuelle, *Enzymes*, **4**, 93 (1960).
3. M. Laskowski, *Methods Enzymol.*, **2**, 8 (1955).
4. B. C. W. Hummel, *Can. J. Biochem. Physiol.*, **37**, 1393 (1959).

Acknowledgments

Original compilation of Criteria: J. E. Folk. Contributions by: F. W. Putnam C. H. W. Hirs, and R. L. Hill.

* IUB Classification: 3.4.16.1, Chymotrypsin A. Also designated α-chymotrypsin. The active enzyme is herein designated chymotrypsin A to distinguish it from chymotrypsin B, an enzyme of similar specificity that occurs in bovine pancreas as the zymogen, chymotrypsinogen B.[3]

least 0.1 mg/ml in 20–100 mM phosphate buffer, pH 7.0, retain their activity for many months at -20 °C. Ovalbumin (0.1%) can be added to stabilize more-dilute solutions for temporary storage at -20 °C.

Dilution: Dilute solutions for assay may be prepared in either phosphate or tris-HCl buffers, pH 7.4–8.1.

Commercially Available Substrates: Acetyl-CoA, oxaloacetate, CoASH, and citrate (malate dehydrogenase, NAD, and L-malate are available for a coupled, assay system[5]).

Specific Activity: One unit of activity for this enzyme catalyzes the formation of 1 μmol of free coenzyme A from acetyl-CoA per minute at 25 °C and pH 8.1. Recrystallized pig-heart citrate synthase has a specific activity of approximately 180 U/mg, calculated from Srere *et al.*[4] The specific activity in the malate–NAD–malate dehydrogenase assay system is 64 U/mg.[2,4]

Protein Assay:[4] $A_{280} \times 0.56$ = mg enzyme/ml.

Activity Assay:

Method: The rate of formation of free coenzyme A is measured with 5,5'-dithiobis(2-nitrobenzoate) (Nbs$_2$).[5–7]

Reagents: Nbs$_2$, 1 mM in 1.0 M tris-HCl, pH 8.1; acetyl-CoA, 13 mM in water; oxaloacetate, 10 mM in 100 mM tris-HCl, pH 8.1. Citrate synthetase, 1 to 5 μg protein/ml of 0.1% bovine serum albumin.

Procedure: Into a 1-ml cuvette, pipette the following: 0.1 ml of Nbs$_2$ solution, 0.02 ml of acetyl-CoA, a suitable aliquot of the enzyme, and H$_2$O to make a volume of 0.97 ml. Initiate the reaction by adding 0.03 ml of oxaloacetate, and follow the increase in absorbance at 412 nm.

Calculation: Number of units (U) = $[\Delta A_{412}/\Delta t(\text{min})] \times 0.0735$.

Impurities: Crystalline preparations of enzyme may contain traces of malate dehydrogenase and isocitrate dehydrogenase.

References

1. J. R. Stern, C. S. Hegre, and G. Bambers, *Biochemistry*, **5**, 1119 (1966).
2. P. A. Srere and G. W. Kosicki, *J. Biol. Chem.*, **236**, 2557 (1961).
3. S. Ochoa, J. R. Stern, and M. C. Schneider, *J. Biol. Chem.*, **193**, 691 (1951).
4. P. A. Srere, M. Singh, and G. C. Brooks, personal communication.
5. P. A. Srere, *Methods Enzymol.*, **13**, 3 (1969).
6. P. A. Srere, H. Brazil, and L. Gonen, *Acta Chem. Scand.*, **17**, 129 (1963).
7. N. O. Jangaard, J. Unkeless, and D. E. Atkinson, *Biochim. Biophys. Acta*, **151**, 225 (1968).

Acknowledgments

Original compilation of Criteria: A. L. Baker. Contributions by: P. A. Srere and J. R. Stern.

* IUB Classification: 4.1.3.7, Citrate oxaloacetate-lyase (CoA-acetylating). Other common name, citrate-condensing enzyme.

E-16
Citrate Synthase*
(Pig Heart)

Reaction: Acetyl-CoA + H$_2$O + oxaloacetate^{2-} $\rightleftharpoons$ citrate^{3-} + CoASH + H$^+$. The pig-heart enzyme is 100% stereospecific and forms the (S)-citrate enantiomer exclusively.[1]

Method of Isolation: Crystallized from ammonium sulfate solution.[2,3]

CRITERIA (Tentative)

Physical Constant: Reported molecular weight:[4] $\sim$94,000.

Procedures for Handling:

Storage: The enzyme is relatively stable from pH 5.5 to 10.0. Solutions of the enzyme in dilute phosphate buffers, pH 7.4, are stable for several weeks when maintained at 4 °C. Solutions of at

E-17
Enolase*
(Rabbit Muscle)

Equation: D-Glycerate 2-phosphate^{3-} $\rightleftharpoons$ phospho*enol*pyruvate^{3-} (PEP) + H$_2$O

Method of Isolation: Purified from fresh or frozen rabbit muscle. Crystallized from ammonium sulfate.[1]

CRITERIA (Tentative)

Physical Constants: Reported molecular weight,[2,3] 82,000. Dissociation into subunits has been reported.[4]

Procedure for Handling: The enzyme is stable as a crystalline suspension in ammonium sulfate.

Commercially Available Substrates: Sodium and barium salts of D-glycerate 2-phosphate. If the barium salt is used, barium may be removed by using the procedure given by Winstead and Wold.[1]

Specific Activity: One unit of activity for this enzyme will convert 1 μmol of D-glycerate 2-phosphate into PEP per minute at 25 °C by using the conditions given under "Activity Assay." Pure rabbit-muscle enolase has been reported[1] to have a specific activity of 90 U/mg at 25 °C.

Protein Assay:[2] $A_{280} \times 1.11 =$ mg enzyme/ml (50 mM imidazole buffer, pH 7).

Activity Assay:

Method: The rate of increase of absorbance at 240 nm at pH 6.9 and 25 °C due to PEP formation is measured.

Reaction Medium:

D-Glycerate 2-phosphate	10 mM	0.3 ml
Magnesium sulfate	10 mM	0.3 ml
KCl	800 mM	1.5 ml
Imidazole-HCl, pH 6.7 (500 mM in imidazole)		0.3 ml
H$_2$O		0.5 ml

Procedure: To 2.9 ml of reaction medium at 25 °C in a silica cell, in a spectrophotometer with wavelength set at 240 nm, add 0.1 ml of enzyme containing 10–100 μg of enzyme/ml. Plot ΔA_{240} vs. time, and draw a tangent to the curve. From the slope of the curve, calculate the activity. Number of units (U) in 3 ml of reaction mixture = $2.3 \times [\Delta A_{240}/\Delta t(\text{min})]$.

Impurities: Likely contaminating activities are other glycolytic enzymes. The recrystallized enzyme gives a single band upon electrophoresis on poly(acrylamide) gel.

References

1. J. A. Winstead and F. Wold, *Biochem. Prep.*, **11**, 31 (1965).
2. A. Holt and F. Wold, *J. Biol. Chem.*, **236**, 3227 (1961).
3. B. G. Malmstrom, *Arch. Biochem. Biophys.*, **Suppl. 1**, 247 (1962).
4. J. A. Winstead and F. Wold, *Biochemistry*, **4**, 2145 (1965).

Acknowledgments

Original compilation of Criteria: F. Wold. Contributions by: J. Larner.

* IUB Classification: 4.2.1.11, 2-Phospho-D-glycerate hydro-lyase.

E-18
Enolase*
(Yeast)

Equation: D-Glycerate 2-phosphate^{3-} $\rightleftharpoons$ phospho*enol*pyruvate^{3-} (PEP) + H$_2$O

Method of Isolation: Purified from yeast extracts by fractionation with acetone and alcohol.[1,2]

CRITERIA (Tentative)

Physical Constants: Reported molecular weight 88,000 (Ref. 3), 67,000 (Ref. 4). Dissociation into subunits has been reported.[5,6]

Procedures for Handling: May be stored at pH 8–8.5 in 0.1 mM Mg^{2+} in the presence of a drop of toluene. Solutions of the enzyme may be lyophilized if frozen rapidly.[2] (The lyophilized powder must be redissolved in buffer at 0 °C.)

Commercially Available Substrates: Sodium and barium salts of D-glycerate 2-phosphate. If the barium salt is used, barium may be removed by using the procedure given by Westhead.[2]

Specific Activity: One unit of activity (U) for this enzyme converts 1 μmol of D-glycerate 2-phosphate into PEP per minute at 30 °C under the conditions stated in "Activity Assay." The specific activity of pure yeast enolase is reported[2,3] to be 200–220 U/mg.

Protein Assay:[2] $A_{280} \times 1.12 =$ mg enzyme/ml (in water).[1]

Activity Assay:[2]

Method: The rate of increase of absorbance due to PEP formation is measured at 230 nm, at pH 7.8 and 30 °C.

Reaction Medium:

Sodium D-glycerate 2-phosphate	20 mM	0.3 ml
Magnesium acetate	10 mM	0.3 ml
Tris acetate buffer,† pH 7.8	100 mM	1.5 ml
EDTA	100 μM	0.3 ml
H$_2$O		0.5 ml

Procedure: To 2.9 ml of reaction medium at 30 °C in a silica cell, add 0.1 ml of enzyme containing 3–30 μg enzyme/ml. Read, plot ΔA_{230} vs. time, and draw a tangent to the curve. From the slope, calculate the activity. Number of units (U) in 3 ml of reaction mixture = $0.968 \times [\Delta A_{230}/\Delta t(\text{min})]$ (calculated from Ref. 2).

Impurities: Enolase purified with acetone–ethanol has 25% of the maximum specific activity and may be used in the phosphoglycerate mutase assay after chromatography on Dowex 1 to remove phosphoglycerate mutase and pyruvate kinase.[7] Pure enolase is obtained by further fractionation with DEAE-cellulose and phosphocellulose.[2] The enzyme consists of at least three active electrophoretically distinct components.

References

1. O. Warburg and W. Christian, *Biochem. Z.*, **210**, 384 (1942).
2. E. W. Westhead, *Methods Enzymol.*, **9**, 670 (1966).
3. K. G. Mann, F. J. Castellino, and P. A. Hargrave, *Biochemistry*, **9**, 4002 (1970).
4. B. G. Malmstrom, *Enzymes*, **5**, 471 (1961).
5. J. M. Brewer and G. Weber, *Proc. Nat. Acad. Sci. U.S.*, **59**, 216 (1968).
6. T. H. Gawronski and E. W. Westhead, *Biochemistry*, **8**, 4261 (1969).
7. S. Grisolia, D. B. Bogart, and A. Torralba, *Biochim. Biophys. Acta*, **151**, 298 (1968).

Acknowledgments

Original compilation of Criteria: F. Wold. Contributions by: J. Larner and E. Westhead.

* IUB Classification: 4.2.1.11, 2-Phospho-D-glycerate hydro-lyase.
† Neutralize 50 ml of 200 mM acetic acid with 1 M tris(hydroxymethyl)aminomethane to pH 7.8 and dilute to 100 ml.

E-19
Extracellular Nuclease*
(*Staphylococcus aureus*)

Reaction: Endonucleolytic and exonucleolytic cleavage of 5'-O-phospho ester bonds in ribonucleates or 2'-deoxyribonucleates, producing 3'-O-phosphonucleosides and dinucleotides.[1,2] Also cleaves a number of *p*-nitrophenyl ester derivatives of 2'-deoxythymidine 5'-phosphate, releasing *p*-nitrophenyl phosphate.[3]

Method of Isolation: The enzyme is obtained by ammonium sulfate fractionation of an extracellular culture medium of growing bacteria, followed by ion-exchange column chromatography.[4–6]

CRITERIA (Tentative)

Physical Constants: Reported molecular weight:[7] 16,800.

Procedure for Handling:

Storage: Lyophilized preparations are stable when kept at −10 °C.

Handling: Enzyme dilutions of less than 10 μg/ml should be made in 0.1% albumin solution.[2] The enzyme is not retained by some commercial dialysis membranes, and dialysis should be done in heated, or boiled, dialysis membranes.[7]

Commercially Available Substrates: Thymidine 3'-phosphate 5'-(p-nitrophenyl phosphate) tri-Li salt·4H$_2$O, Mol. Wt. = 613.2; λ_{max} = 271 nm, ϵ_{271} = 15,400 mol^{-1} l cm^{-1}. K_m for nuclease = 1.0×10^{-5} M.[3]

Specific Activity:[3] 1 mg of purified enzyme hydrolyzes 0.62 μmol of substrate per minute. The specific activity is 0.62 U/mg under the conditions described below. (See "Activity Assay.")

Protein Assay:[5] $A_{277} \times 1.03$ = mg of enzyme/ml, in 50 mM tris buffer, pH 7.5.

Activity Assay:[3]

Method: The assay is based on the release of p-nitrophenyl phosphate, which is accompanied by a shift of its ultraviolet spectrum (to higher wavelengths) resulting from conversion of the mono- into the dianionic phosphate species.

Procedure: The assay mixture contains 100 μM thymidine 3'-phosphate 5'(p-nitrophenyl phosphate) in a total volume of 1.0 ml of 50 mM tris-HCl buffer, pH 8.8, 10 mM CaCl$_2$. A 10 mM stock substrate solution (in distilled water) can be kept frozen for many weeks. The assay mixture should be prepared daily. The mixture is placed on a 1.0-ml quartz cuvette, and 5–30 μl of enzyme, containing 1–50 μg of protein, is added. The increase in absorbance at 330 nm is continuously recorded, with the temperature of the cuvette chamber at 24 °C.† Number of units (U) in cuvette = $0.235[\Delta A_{330}/\Delta t(\text{min})]$.

Impurities: Phosphatases and other protein contaminants may be present in all but highly purified preparations.[4,5]

References

1. M. Alexander, L. A. Heppel, and J. Hurwitz, *J. Biol. Chem.*, **236**, 3014 (1961).
2. P. Cuatrecasas, S. Fuchs, and C. B. Anfinsen, *J. Biol. Chem.*, **242**, 1541 (1967).
3. P. Cuatrecasas, M. Wilchek, and C. B. Anfinsen, *Biochemistry*, **8**, 2277 (1969).
4. E. Sulkowski and M. Laskowski, Sr., *J. Biol. Chem.*, **241**, 4386 (1966).
5. S. Fuchs, P. Cuatrecasas, and C. B. Anfinsen, *J. Biol. Chem.*, **242**, 4768 (1967).
6. L. Moravek, C. B. Anfinsen, J. L. Cone, and H. Taniuchi, *J. Biol. Chem.*, **244**, 497 (1969).
7. H. Taniuchi, C. B. Anfinsen, and A. Sodja, *J. Biol. Chem.*, **242**, 4752 (1967).

Acknowledgments

Original compilation of Criteria: S. Fuchs. Contributions by: P. Cuatrecasas and C. B. Anfinsen.

* IUB Classification: 3.1.4.7, Ribonucleate(deoxyribonucleate) 3'-nucleotidohydrolase; micrococcal nuclease.

† The number of μmol of p-nitrophenyl phosphate formed in 1 ml per minute = $4.26[\Delta A_{330}/\Delta t(\text{min})]$. $\Delta \epsilon_{330}$ for p-nitrophenyl phosphate = 4,260 mol^{-1} l cm^{-1}.

Commercially Available Substrates: D-Glucose 6-phosphate, NADP$^+$.

Specific Activity: One unit of activity (U) is equivalent to 1 μmol of NADPH formed per minute at pH 8.0 and 30 °C. Four-times recrystallized enzyme has a specific activity of about 680 U/mg.[1]

Protein Assay: Biuret procedure:[3] $A_{540} \times 32$ = mg of protein per 10 ml.[1]

Activity Assay:[1]

Reagents: 100 mM glycylglycine, pH 8.0; 30 mM D-glucose 6-phosphate; 10 mM NADP$^+$, pH 7.0; 150 mM MgSO$_4$; 50 mM EDTA (pH 8.0) containing crystalline bovine serum albumin (1 mg/ml).

D-Glucose-6-phosphate dehydrogenase: Just prior to measurement, dilute with ice-cold 50 mM EDTA, pH 8.0, containing crystalline bovine serum albumin (1 mg/ml) to a concentration of about 0.2 to 1.0 μg/ml.

Procedure: A recording spectrophotometer, equipped with thermospacers through which water at 30 °C is circulated, is used to follow the change in absorbance at 340 nm. Into a 1.0-cm cuvette, pipette the following: glycylglycine (100 mM, pH 8.0), 2.5 ml; D-glucose 6-phosphate (30 mM), 0.1 ml; NADP$^+$ (10 mM), 0.1 ml; MgSO$_4$ (150 mM), 0.2 ml (measurements are made relative to a blank cuvette containing the above reaction mixture + 0.1 ml of H$_2$O). After temperature equilibration (about 5 min), the reaction is initiated by the addition of 0.1 ml of D-glucose 6-phosphate dehydrogenase.

Calculations: Calculations of the activity are made from the initial slope of the curve of the absorbance vs. time: Number of units (U) per cuvette volume = $0.483 [\Delta A_{340}/\Delta t (\text{min})]$.

Impurities: In amorphous preparations: hexokinase,[4] myokinase, phosphoglucomutase,[4] 6-phosphogluconate dehydrogenase, glutathione reductase, phosphohexose isomerase.[4] In preparations recrystallized four or five times, none of the contaminating activities should exceed 0.05% of the D-glucose-6-phosphate dehydrogenase activity.

References

1. E. A. Noltmann, C. J. Gubler, and S. A. Kuby, *J. Biol. Chem.*, **236**, 1225 (1961).
2. R. H. Yue, E. A. Noltmann, and S. A. Kuby, *Biochemistry*, **6**, 1174 (1967).
3. A. G. Gornall, C. J. Bardawill, and M. M. David, *J. Biol. Chem.* **177**, 751 (1949).
4. L. Glaser and D. H. Brown, *J. Biol. Chem.*, **216**, 67 (1955).

Acknowledgments

Original compilation of Criteria: S. A. Kuby. Contributions by: E. A. Noltmann.

* IUB Classification: 1.1.1.49, D-Glucose 6-phosphate:NADP oxidoreductase. Commonly used name: Zwischenferment.

E-20

D-Glucose-6-phosphate Dehydrogenase*
(Brewers' Yeast)

Equation: D-Glucose 6-phosphate^{2-} + NADP$^+$ $\rightleftharpoons$ 6-O-phospho-D-glucono-1,5-lactone^{2-} + NADPH + H$^+$

Method of Isolation: Crystallized from aqueous (NH$_4$)$_2$SO$_4$ solutions.[1]

CRITERIA (Tentative)

Physical Constant: Reported molecular weight:[2] 102,000.

Procedure for Handling: Stable as crystalline suspension in (NH$_4$)$_2$SO$_4$ solution.[1] At concentrations of micrograms per milliliter, the enzyme is unstable below pH 4 and above pH 11.

E-21

D-Glyceraldehyde-3-phosphate Dehydrogenase*
(Rabbit Muscle)

Equation: D-Glyceraldehyde 3-phosphate^{2-} + NAD$^+$ + HPO$_4$$^{2-}$ $\rightleftharpoons$ 1,3-diphosphoglycerate^{4-} + NADH + H$^+$

Method of Isolation: Crystallized from ammonium sulfate solutions.[1–3]

CRITERIA (Tentative)

Physical Constant: Reported molecular weight:[3–7] 118,000–150,000; 140,000 (based on 4 subunits of 35,000 each).[6]

Procedure for Handling: Stability properties have been summarized.[4] Solutions of the enzyme are stabilized by the presence of EDTA, reducing agents, and NAD$^+$. For storage over prolonged periods of time, the enzyme may be kept at 0–4 °C as a crystalline suspension in ammonium sulfate solution (2.4–2.8 M, pH 7.5–8.3). Enzyme solutions are unstable[3] below pH 6.

Commercially Available Substrates: DL-Glyceraldehyde 3-phosphoric acid (aq. solution), slight decomposition occurs in solutions after several weeks in the frozen state; DL-glyceraldehyde 3-phosphate diethyl acetal monobarium salt dihydrate; NAD$^+$.

Specific Activity: One unit of activity for this enzyme reduces 1 μmol of NAD$^+$ per minute at pH 8.4 and 25 °C. The specific activity of once-crystallized preparations (e.g., commercial preparations) ranges from 25 to 30 U/mg at 25 °C by the assay procedure described below. An activity of 85 U/mg has been reported[3] at 27 °C for the best preparation, with substrate concentrations of 480 μM (NAD$^+$ and D-glyceraldehyde 3-phosphate).

Protein Assay: When isolated by the procedures described,[1-3] the enzyme contains bound NAD$^+$ and has an absorption maximum at 276 nm: $A_{276} \times 0.95$ = mg/ml (in 0.03 M pyrophosphate, pH 8.4).

Activity Assay: Modified from procedure of Cori et al.[1] (based on Warburg and Christian[8]) as summarized by Velick,[3] with arsenate replacing phosphate, and under conditions that permit measurements of *initial velocities.*

Reagents: Sodium pyrophosphate, 30 mM, pH 8.4; NAD$^+$, 7 mM; D-glyceraldehyde 3-phosphate, 8.8 mM†; disodium arsenate, 170 mM; cysteine, 40 mM, pH 8.4 (freshly prepared) in pyrophosphate, 30 mM, pH 8.4. D-glyceraldehyde-3-phosphate dehydrogenase: Prior to measurement ($\sim$5–10 min), dilute with ice-cold 30 mM pyrophosphate solution (containing 4 mM cysteine, pH 8.4) to a concentration of 10–20 μg/ml.

Procedure: Into a cuvette, pipette the following:

Pyrophosphate	1.7 ml
Arsenate	0.30 ml
Cysteine	0.30 ml
NAD$^+$	0.10 ml

Allow 5 min for temperature equilibration in a recording spectrophotometer having water jackets at 25 °C. Add 0.50 ml of D-glyceraldehyde 3-phosphate‡ (8.8 mM); 2 min later, initiate the reaction with the addition of 0.1 ml of properly diluted D-glyceraldehyde-3-phosphate dehydrogenase. (Measurements are made relative to a blank cuvette containing the above reaction mixture, added in the above order, and containing 0.1 ml of H$_2$O instead of enzyme.)

Follow ΔA_{340} to determine the initial rate.

Calculations: Calculation of the activity is made from the initial slope of the curve of the absorbance vs. time. Number of units (U) per cuvette volume = $0.483 \times [\Delta A_{340}/\Delta t \text{ (min)}]$.

Impurities: Once-crystallized preparations may contain the following contaminants: Nucleoside diphosphokinase,[10] myokinase, phosphoglycerate kinase,[10] and traces of other glycolytic enzymes (e.g., triose phosphate isomerase, aldolase).

References

1. G. T. Cori, M. W. Slein, and C. F. Cori, *J. Biol. Chem.,* **173**, 605 (1948).
2. G. Beisenherz, H. J. Boltze, T. Bücher, R. Czok, K. G. Garbade, F. Meyer-Arendt, and G. Pfleiderer, *Z. Naturforsch.,* **8b**, 555 (1953).
3. S. F. Velick, *Methods Enzymol.,* **1**, 401 (1955).
4. S. F. Velick and C. Furfine, *Enzymes,* **7**, 243 (1963).
5. J. B. Fox and W. Dandliker, *J. Biol. Chem.,* **218**, 53 (1956).
6. J. I. Harris, in *Structure and Activity of Enzymes, FEBS Symp. No. 1,* T. W. Goodwin, J. I. Harris, and B. S. Hartley, eds., Academic Press, London (1964), p. 97.
7. B. Chance and J. H. Park, *J. Biol. Chem.,* **242**, 5093 (1967).
8. O. Warburg and W. Christian, *Biochem. Z.,* **303**, 40 (1939).
9. C. E. Ballou and H. O. L. Fischer, *J. Am. Chem. Soc.,* **77**, 3329 (1955).
10. R. L. Ratliff, R. H. Weaver, H. A. Lardy, and S. A. Kuby, *J. Biol. Chem.,* **239**, 301 (1964).

Acknowledgemnts

Original compilation of Criteria: S. A. Kuby. Contributions by: E. A. Noltmann.

* IUB Classification: 1.2.1.12, D-Glyceraldehyde 3-phosphate: NAD oxidoreductase (phosphorylating). Commonly used names: Triose phosphate dehydrogenase; glyceraldehyde-3-phosphate dehydrogenase; phosphoglyceraldehyde dehydrogenase.

† Prepared from the diethyl acetal barium salt, by removal of barium with Dowex 50 ion-exchange resin, and conversion into the free acid as described by Ballou and Fischer.[9] Adjust to pH 7.0; assay enzymically by above test system, with limiting concentrations of substrate, at least ten times the amount of enzyme, and omission of the cysteine. Dilute to 8.8 mM in D-isomer, and store at −20° C.

‡ D-Glyceraldehyde 3-phosphate is added just before the final addition of the enzyme, to minimize formation of an addition compound with cysteine; cysteine should not be added to the stock aldehyde solution.

E-22
L-Glycerol-3-phosphate Dehydrogenase*
(Rabbit Skeletal Muscle)

Equation: Dihydroxyacetone 3-phosphate^{2-} + NADH + H$^+$ $\rightleftharpoons$ L-glycerol 3-phosphate^{2-} + NAD$^+$

Method of Isolation: Crystallized from ammonium sulfate solution.[1,2]

CRITERIA (Tentative)

Physical Constants: Reported molecular weight:[3] 78,000. One molecule of NADH bound[4,5] per 39,000.

Procedure for Handling:

Storage: A suspension of crystals in 2 M (NH$_4$)$_2$SO$_4$ solution is stable for several weeks at 0–4 °C.

Dilution: Solubilization of enzyme, centrifuged to remove most of the (NH$_4$)$_2$SO$_4$, and dilution may be made in water containing 1 mM EDTA, pH 7.5. Stability is maintained even at high dilutions. Freezing is not recommended.[6] Inactivation on freezing is dilution-dependent. Partial restoration of activity occurs slowly after thawing.[6]

Commercially Available Substrates: Dihydroxyacetone 3-phosphate dimethyl acetal dicyclohexylammonium salt monohydrate, DL-α-glycerophosphate.

Specific Activity: One unit of activity for this enzyme catalyzes the reduction of 1 μmol of dihydroxyacetone phosphate per minute at 25 °C, pH 7.4. Crystalline enzyme has been reported[7] to have a specific activity of 83 U/mg.

Protein Assay: The enzyme contains bound adenosine diphosphate ribose or a closely related compound, removal of which does not affect the crystalline form or the catalytic activity of the enzyme but does affect the electrophoretic mobility.[3,6,7] Native nucleotide-containing enzyme has A_{280}/A_{260} = 1.1, and α_{280} = 0.63 mg^{-1} ml cm^{-1} at 280 nm; nucleotide-free enzyme has A_{280}/A_{260} = 1.6, and α_{280} = 0.53 mg^{-1} ml cm^{-1} at[7] 280 nm. Protein may also be determined by the biuret method, by use of bovine serum albumin as the standard.[8] $A_{280} \times 1.59$ = mg protein/ml.

Activity:[1]

Method: The decrease in absorbance of NADH at 340 nm is monitored at 25 °C as a function of time. Near-optimum conditions are 140 μM NADH, 100 μM dihydroxyacetone phosphate, pH 7.5.

Reaction Medium:

NADH	2.7 mM	0.15 ml
Dihydroxyacetone phosphate	10 mM	0.10 ml
Triethanolamine† · HCl	150 mM	1.0 ml
H$_2$O		1.65 ml

Procedure: To 2.9 ml of reaction medium at 25 °C in a silica cell, add 0.1 ml of enzyme containing about 1 μg of enzyme.

Calculation: Calculation of the activity is made from the initial slope of the curve of abosrbance at 340 nm vs. time. Number of units (U) per cuvette volume = $0.483 \times [\Delta A_{340}/\Delta t(\text{min})]$.

Impurities: Probable contaminating enzymes are lactate dehydrogenase, glyceraldehyde-3-phosphate dehydrogenase, aldolase, and pyruvate kinase.

References

1. G. Beisenherz, H. J. Boltze, T. Bücher, R. Czok, K. H. Garbade, E. Meyer-Arendt, and G. Pfleiderer, *Z. Naturforsch.*, **8b**, 555 (1953).
2. G. Beisenherz, T. Bücher, and K. H. Garbade, *Methods Enzymol.*, **1**, 391 (1955).
3. J. van Eys, B. J. Nuenke, and M. K. Patterson, Jr., *J. Biol. Chem.*, **234**, 2308 (1959).
4. G. Pfleiderer and F. Auricchio, *Biochem. Biophys. Res. Commun.*, **16**, 53 (1964).
5. S. J. Kim and B. M. Anderson, *J. Biol. Chem.*, **244**, 1547 (1969).
6. J. van Eys, J. Judd, J. Ford, and W. B. Womack, *Biochemistry*, **3**, 1755 (1964).
7. H. Ankel, T. Bücher, and R. Czok, *Biochem. Z.*, **332**, 315 (1960).
8. T. E. Weichselbaum, *Am. J. Clin. Pathol.*, *Tech. Suppl.*, **10**, 40 (1946).
9. H. J. Hohorst, in *Methods of Enzymatic Analysis*, H. U. Bergmeyer, ed., Academic Press, New York (1963), p. 215.
10. E. Baer and H. O. Fischer, *J. Biol. Chem.*, **128**, 491 (1939).

Acknowledgments

Original compilation of Criteria: R. W. Von Korff. Contributions by: J. van Eys. Revision by: C. H. Suelter.

* IUB Classification: 1.1.1.8, L-Glycerol-3-phosphate:NAD oxidoreductase.

The enzyme has been incorrectly designated L(−)glycerol 1-phosphate dehydrogenase by Hohorst.[9] The configuration of naturally occurring α-glycerophosphate has been established as the mirror image of that of D-glyceraldehyde 3-phosphate; it is L-glycerol 3-phosphate or D-glycerol 1-phosphate.[10] The enzyme is commonly known as α-glycerophosphate dehydrogenase.

† 2,2′,2″-Nitrilotriethanol.

E-23
Hexokinase*
(Bakers' Yeast)

$$(\text{Mg}^{2+})$$

Equation: $\text{ATP}^{4-} + \text{D-glucose} \rightleftharpoons \text{ADP}^{3-} + \text{D-glucose 6-phosphate}^{2-} + \text{H}^+$

Method of Isolation: Crystallized from aqueous $(\text{NH}_4)_2\text{SO}_4$ solutions.[1–3]

CRITERIA (Tentative)

Physical Constant: Reported molecular weight: 96,600,[1] 102,000,[4] 111,000.[5] Dissociation into subunits (half- and quarter-molecules) has been reported.[6]

Procedures for Handling: Unstable below pH 4.[3] For storage over long periods of time, the enzyme may be kept[3] as a crystalline suspension in ammonium sulfate solution containing 2 mM EDTA, pH 7 or as a crystalline filter cake at 0 °C.[1] The enzyme is unstable at high dilution (micrograms of protein/milliliter) and can be stabilized by D-glucose.[1]

Commercially Available Substrates: D-Glucose, D-glucose 6-phosphate.

Specific Activity: One unit of activity (U) is equivalent to 1 μmol of ADP formed per minute at 30 °C and pH 8. Enzyme that has been recrystallized 4–6 times has a specific activity of about 600 U/mg.[3,7]

Protein Assay: $A_{280} \times 0.77 = $ mg of protein per ml.[2]

Activity Assay: According to Mahowald et al.[8]

Principle: The hexokinase reaction at pH 8.0 yields one equivalent of H^+ per mol of ATP consumed.[1,9–11]

Reagents: Prepare with CO_2-free distilled water. 32 mM ATP (neutralized with NaOH); 32 mM MgSO_4; 200 mM D-glucose; 4% (w/v) crystalline bovine serum albumin; 10.0 mM NaOH (freshly diluted from standardized NaOH).

Hexokinase: Just prior to measurement, dilute with ice-cold 1 mM EDTA, pH 8.0, to about 50–250 μg of protein/ml.

Procedure: Standardize a pH-stat (a commercial autotitrator) at 30 °C with standard buffer. Into a 12-ml reaction vessel (water-jacketed at 30 °C and equipped with a stirrer) pipet 2.00 ml of D-glucose solution, 1.00 ml of distilled water, 1.00 ml of ATP, 1.00 ml of MgSO_4, and 0.20 ml of albumin. Pass a slow stream of nitrogen over the surface (to exclude interference by carbon dioxide during titration) and begin stirring. Adjust the pH to 8.0, and after temperature equilibration (3–5 min), with the pH-stat set to record at pH 8.00, initiate the enzymic reaction by addition of 10 μl of hexokinase (0.5–2.5 μg). Repeat, but substitute distilled water for the D-glucose solution, to determine the relatively low "blank" rate.

Calculations: Calculations of the initial rate are made from the linear portion of the curve; the blank rate is subtracted from the over-all rate to yield the enzymic rate, and this is expressed in microequivalents of NaOH consumed per minute. One unit (U) = one μmol of NaOH per minute.

Impurities: After six recrystallizations, the ATPase should be diminished to 5×10^{-6} of the activity of hexokinase,[7] and the following contaminating impurities should be negligible or less than 3×10^{-4} of the hexokinase activity: adenylate kinase, 6-phosphogluconate dehydrogenase, glucose 6-phosphate dehydrogenase, pyrophosphatase, phosphohexose isomerase, and triose phosphate dehydrogenases.[3] The crystalline enzyme prepared by Darrow and Colowick[3] contained traces of proteolytic enzymes. The proteolytic activity can be removed by adsorption on diethylaminoethyl-cellulose.[6]

References

1. M. Kunitz and M. R. McDonald, *J. Gen. Physiol.*, **29**, 393 (1946).
2. M. R. McDonald, *Methods Enzymol.*, **1**, 269 (1955).
3. R. A. Darrow and S. P. Colowick, *Methods Enzymol.*, **5**, 226 (1962).
4. N. R. Lazarus, M. Derechin, and E. A. Barhard, *Biochemistry*, **7**, 2390 (1968).
5. J. S. Easterby and M. A. Rosemeyer, *FEBS Lett.*, **4**, 84 (1969).
6. U. W. Kenkare and S. P. Colowick, *J. Biol. Chem.*, **240**, 4570 (1965).
7. K. A. Trayser and S. P. Colowick, *Arch. Biochem. Biophys.*, **94**, 161 (1961).
8. T. A. Mahowald, E. A. Noltmann, and S. A. Kuby, *J. Biol. Chem.*, **237**, 1535 (1962).
9. S. P. Colowick and H. M. Kalckar, *J. Biol. Chem.*, **137**, 789 (1941).
10. S. P. Colowick and H. M. Kalckar, *J. Biol. Chem.*, **148**, 117 (1943).
11. J. Wajzer, *Compt. Rend.*, **229**, 1270 (1949).

Acknowledgments

Original compilation of Criteria: S. A. Kuby. Contributions by: O. E. Olson, E. A. Noltmann, and S. P. Colowick.

* IUB Classification: 2.7.1.1, ATP:D-Hexose 6-phosphotransferase.

E-24
D(—)3-Hydroxybutyrate Dehydrogenase*
(Bovine Heart)

Equation: D(—)-3-Hydroxybutyrate$^-$ + NAD$^+$ $\rightleftharpoons$ acetoacetate$^-$ + NADH + H$^+$

The enzyme is specific for the D(—) enantiomer and NAD$^+$, with only slight reactivity for the NAD analogs, 3-acetylpyridine and pyridine-3-aldehyde.[1] The purified apodehydrogenase reacts specifically with lecithin[1–3] in the presence of thiols to form the active enzyme complex.

Method of Isolation: The enzyme is found exclusively in mammalian mitochondria, and is isolated from bovine heart mitochondria by solubilization with potassium cholate.[3,4] The highly purified apodehydrogenase is obtained by ammonium sulfate fractionation of the active lecithinoprotein complex.[1]

CRITERIA (Tentative)

Physical Constants: The highly purified apodehydrogenase is insoluble, and requires 1.8 mg of pure micellar lecithin per milligram of apoenzyme and 50 mM monothiol (or 5 mM dithiol) for solubilization and formation of the active enzyme complex.[1] Reported[5] molecular weight, 87,000, based on one gram-atom of Zn/mol.

Procedure for Handling: The enzyme is firmly bound to the electron-transport particle[6–8] and is stable when attached to respiratory membranes.[3,7] Further protection is afforded by the addition of a thiol, NAD$^+$, and reducible substrate.[9] Lyophilized preparations of the apodehydrogenase are stable,[1] but the active lecithinoprotein complex dissociates[1] in the absence of a thiol and NAD$^+$.

Commercially Available Substrates: Sodium DL-3-hydroxybutyrate; the concentration of the D(—) monomer may be determined enzymically.[10]

Specific Activity: One unit of activity (U) for this enzyme represents 1 μmol of NAD$^+$ reduced per minute by 3-hydroxybutyrate at pH 8.1 at 37 °C. The specific activity reported for the most highly purified preparation is 100 U/mg when assayed as described below.[1]

Protein Assay: Determined by the biuret method,[11] with crystalline bovine serum albumin as the standard.

Activity Assay:[1] A prewarming is required, to allow the active lecithinoprotein complex to be formed. Tris-chloride buffer, pH 8.1, 1.0 M, 0.15 ml; bovine serum albumin 2% (w/v), 0.06 ml; EDTA, 10 mM, 0.15 ml; 95% ethanol, 0.06 ml; cysteine (a freshly prepared neutral solution), 1.0 M, 0.15 ml; NAD, 20 mM, 0.3 ml; micellar lecithin, 10 mg/ml, 0.06–0.12 ml; apodehydrogenase, 10 mg/ml, 0.03 ml; and deionized H$_2$O to give a volume of 2.9 ml. After incubation for 15 min at 37 °C, DL-3-hydroxybutyrate, 600 mM, 0.1 ml, is added to initiate the reaction. Follow ΔA_{340}. Number of units (U) in cuvette = [$\Delta A_{340}/\Delta t$(min)] × 0.483.

Impurities: Flavoproteins, cytochrome, and phospholipid components released from the submitochondrial particles upon solubilization by cholate.

References

1. P. Jurtshuk, I. Sekuzu, and D. E. Green, *J. Biol. Chem.*, **238**, 3595 (1963).
2. P. Jurtshuk, I. Sekuzu, and D. E. Green, *Biochem. Biophys. Res. Commun.*, **6**, 76 (1961).
3. I. Sekuzu, P. Jurtshuk, and D. E. Green, *J. Biol. Chem.*, **238**, 975 (1963).
4. I. Sekuzu, P. Jurtshuk, and D. E. Green, *Biochem. Biophys. Res. Commun.*, **6**, 71 (1961).
5. P. Jurtshuk, I. Sekuzu, and D. E. Green, unpublished data (1963).
6. D. E. Green, J. G. Dewan, and L. F. Leloir, *Biochem. J.*, **31**, 934 (1937).
7. D. M. Ziegler and A. W. Linnane, *Biochim. Biophys. Acta*, **30**, 53 (1958).
8. A. L. Lehninger, H. C. Sudduth, and J. B. Wise, *J. Biol. Chem.*, **235**, 2450 (1960).
9. J. B. Wise and A. L. Lehninger, *J. Biol. Chem.*, **237**, 1363 (1962).
10. D. H. Williamson, J. Mellanby, and H. A. Krebs, *Biochem. J.*, **82**, 90 (1962).
11. A. G. Gornall, C. J. Bardawill, and M. M. David, *J. Biol. Chem.*, **177**, 751 (1959).

Acknowledgments

Original compilation of Criteria: Peter Jurtshuk. Contributions by: Dan M. Ziegler and David E. Green.

* IUB Classification: 1.1.1.30, D-3-Hydroxybutyrate:NAD oxidoreductase.

E-25
D(—)-3-Hydroxybutyrate Dehydrogenase*
(*Rhodopseudomonas spheroides*)

Equation: D(—)3-Hydroxybutyrate$^-$ + NAD$^+$ $\rightleftharpoons$ acetoacetate$^-$ + NADH + H$^+$

The enzyme is specific for the D(—) enantiomer of 3-hydroxybutyrate, but is slightly reactive with 3-hydroxypentanoate and 3-hydroxyhexanoate. The enzyme is specific for NAD$^+$, with slight reactivity for the NAD analogs, 3-acetylpyridine and thionicotinamide adenine dinucleotides.[1]

Method of Isolation:[1] Isolated from extracts of *Rhodopseudomonas spheroides*, and purified by precipitation with protamine sulfate, ammonium sulfate fractionation, and adsorption on DEAE-Sephadex.

CRITERIA (Tentative)

Physical Constants: Reported molecular weight:[1] 85,000.

Procedures for Handling: The enzyme is stabilized by the presence of divalent cations. Diluted solutions of the *R. spheroides* enzyme are inactivated by prewarming to 37 °C but are protected[1] by Ca^{2+} or NADH.

Commercially Available Substrates: Sodium DL-3-hydroxybutyrate; the concentration of the D(—) enantiomer may be determined enzymically.[2]

Specific Activity: One unit of activity (U) for this enzyme reduces 1 μmol of NAD$^+$ per minute with D(—)-3-hydroxybutyrate at pH 8.4 and 25 °C. The specific activity reported for the most highly purified preparation is approximately 17 U/mg when assayed as described below.[1]

Protein Assay: Determined by the biuret method,[3] with crystalline bovine serum albumin as the standard.

Activity Assay:[1] Tris-chloride buffer, pH 8.4, 330 mM, 0.3 ml; NAD$^+$, 18 mM, 0.3 ml; D-3-hydroxybutyrate, 440 mM, 0.15 ml (or 0.3 ml of the DL-salt); and deionized H$_2$O to give a final volume of 3.0 ml. (CN$^-$, 15 mM, 0.1 ml should be added when assaying extracts that contain NADH-oxidizing enzymes.[4]) Follow ΔA_{340} with time after adding 10–50 μl of enzyme containing 5 mg of protein/ml. Number of units (U) in cuvette = [$\Delta A_{340}/\Delta t$ (min)] × 0.483.

Impurities: NADH oxidases[2,4] are present in crude preparations that have not been centrifuged at 144,000 g for at least 60 min. Malic dehydrogenase is present in partially purified preparations.[2]

References

1. H. U. Bergmeyer, K. Gawehn, H. Klotzsch, H. A. Krebs, and D. H. Williamson, *Biochem. J.*, **102**, 423 (1967).
2. D. H. Williamson, J. Mellanby, and H. A. Krebs, *Biochem. J.*, **82**, 90 (1962).
3. A. G. Gornall, C. J. Bardawill, and M. M. David, *J. Biol. Chem.*, **177**, 751 (1959).
4. P. Jurtshuk, S. Manning, and C. R. Barrera, *Can. J. Microbiol.*, **14**, 775 (1968).

Acknowledgments

Original compilation of Criteria: Peter Jurtshuk. Contributions by: M. Doudoroff.

* IUB Classification: 1.1.1.30, D-3-Hydroxybutyrate:NAD oxidoreductase.

E-26
Inorganic Pyrophosphatase*
(Yeast)

$$(Mg^{2+})$$

Equation: $HP_2O_7^{3-} + H_2O \rightarrow 2\ HPO_4^{2-} + H^+$

Method of Isolation: Isolated from bakers' yeast as crystals by fractionation with alcohol[1,2] or by chromatography on DEAE-Sephadex.[3]

CRITERIA (Tentative)

Physical Constant: Reported molecular weight:[4] 71,000. Two subunits, apparently identical.[3–5]

Procedure for Handling: Dry preparations are stable for many months at 5 °C. The enzyme is soluble in water, and stock solutions (1 mg/ml) at pH 6.7–7.0 are stable for several months under refrigeration. Mg ions are necessary to activate the enzyme. The optimum conditions for activity are 40 °C, pH 7.0, 3–4 mM pyrophosphate, and an equivalent concentration of Mg^{2+}.

Commercially Available Substrate: Sodium pyrophosphate.

Specific Activity: One unit of activity (U) liberates 2 μmol of orthophosphate (P_i) per minute at 30 °C under the conditions specified below.† Specific activities of 600–650 U/mg at 30 °C have been reported for 3–5 times recrystallized enzyme.[6]

Protein Assay:[6] $A_{280} \times 0.69$ = mg of enzyme/ml (in either H_2O or 100 mM HCl).

Activity Assay:

Method: The rate of formation of inorganic orthophosphate from inorganic pyrophosphate (PP_i) is measured.[7]

Enzyme: 0.1–1.0 μg in cold 100 mM tris buffer, pH 7.2.

Procedure: Into a series of tubes maintained at 30 °C are placed the following: tris buffer, 50 mM, 4.0 ml; sodium pyrophosphate, 10 mM, 1.0 ml; magnesium chloride, 10 mM, 1.0 ml; and enzyme (at zero time), 1.0 ml. Include a blank containing 20 mM tris buffer (pH 7.2) in place of the enzyme. After 15 min at 30 °C, remove a 1-ml aliquot and add to mixture of 1 ml of 2.5 M H_2SO_4 and 5.0 ml of H_2O for orthophosphate determination by the method of Fiske and SubbaRow.[8,9] From a standard curve for inorganic phosphate, the number of micromoles of orthophosphate formed may be determined. Number of units (U) = [amount of phosphate (μmol) liberated in total reaction mixture]/2 × 15 min.

Impurities: The preparation should be free of ATPase. Kunitz[6] reported that twice-recrystallized enzyme shows some contaminating, inorganic polyphosphatase activity.

It has been reported by Schlesinger and Coon,[10] and confirmed by Kunitz,[11] that yeast pyrophosphatase (even five times recrystallized) exerts a slow, phosphohydrolytic action on ATP and ADP in the presence of Zn ions.

Avaeva and co-workers[12] reported that, in the presence of Zn^{2+}, yeast pyrophosphatase will hydrolyze O-pyrophosphoserine and related compounds.

References

1. M. Kunitz, *Arch. Biochem. Biophys.*, **92**, 270 (1961).
2. L. A. Heppel and R. J. Hilmoe, *J. Biol. Chem.*, **192**, 87 (1951).
3. B. S. Cooperman, manuscript in preparation.
4. L. G. Butler, *Enzymes*, **4**, 529 (1971).
5. S. M. Avaeva and G. I. Achmedor, *Biokhimiya*, **35**, 31 (1970).
6. M. Kunitz, *J. Gen. Physiol.*, **35**, 423 (1952).
7. L. A. Heppel, *Methods Enzymol.*, **2**, 570 (1955).
8. C. H. Fiske and Y. SubbaRow, *J. Biol. Chem.*, **66**, 375 (1925).
9. C. H. Fiske and Y. SubbaRow, *J. Biol. Chem.*, **81**, 629 (1929).
10. M. J. Schlesinger and M. J. Coon, *Biochim. Biophys. Acta*, **41**, 30 (1960).
11. M. Kunitz, *J. Gen. Physiol.*, **45**, Suppl., 31 (1962).
12. S. M. Avaeva, S. N. Kara-Murza, and M. M. Botvinik, *Biokhimiya*, **32**, 205 (1967).

Acknowledgments

Original compilation of Criteria: C. Worthington. Contributions by: M. Kunitz. Revised in 1970 by B. Cooperman.

* IUB Classification: 3.6.1.1, Pyrophosphate phosphohydrolase.

† In previous editions, the unit was defined as the activity liberating 1 μmol of orthophosphate per minute. The IUB unit is expressed in terms of one cycle of reaction, or one bond broken. Therefore, one International Unit is equal to 1 μmol of pyrophosphate cleaved or 2 μmol of orthophosphate liberated per minute.

E-27
L(+)-Lactate Dehydrogenase*
(Bovine Heart)

Equation: $L(+)$-Lactate$^-$ + NAD$^+ \rightarrow$ pyruvate$^-$ + NADH + H$^+$

Method of Isolation: Crystallized from ammonium sulfate solution. As generally prepared, the crystalline enzyme is a mixture[1] of 80% isozyme H_4 and 20% H_3M (H = heart; M = muscle). The pure H_4 form is more acidic than the H_3M form from which it can be separated by DEAE-cellulose chromatography.[2]

CRITERIA (Tentative)

Physical Constants: Molecular weight reported from sedimentation and diffusion measurements is 123,000 to 131,000 for the H_4 tetramer and 136,000 to 153,000 for the M_4 tetramer.[2]

Procedure for Handling: Store at 5 °C as a suspension in 2.25 M ammonium sulfate. Prior to assay, the suspension is centrifuged, and the pellet is dissolved in phosphate buffer to give a concentration of about 1 mg/ml; this solution serves as a stock from which further dilutions can be made.

Commercially Available Substrates: Sodium pyruvate; sodium, calcium and lithium salts of $L(+)$-lactic acid; NAD; NADH.

Specific Activity: One unit of activity (U) for this enzyme causes an initial rate of oxidation of 1 μmol of NADH per minute at 25 °C in 100 mM phosphate buffer, pH 7.5. The reported turnover number of 49,400 obtained at the optimum pyruvate concentration of 600 μM, when used in conjunction with a molecular weight of 130,000, yields a specific activity of 380 U/mg for the best preparations. The turnover number of the M_4 form is approximately twice that of the H_4 form.[2]

Protein Assay: $A_{280} \times 0.66$ = mg enzyme/ml.

Activity Assay: The decrease in A_{340} of NADH is monitored as a function of time at 25 °C. A 3-ml reaction mixture is prepared to contain 1.5 ml of 200 mM potassium phosphate buffer, pH 7.5, 0.5 ml of 2 mM sodium pyruvate, sufficient lactate dehydrogenase to give $\Delta A_{340} = 0.1$–0.2 per minute, water, and 0.5 ml of 700 μM NADH (added last).

Calculation: The activity is calculated from the initial slope of the curve of A_{340} vs. time. Number of units (U) per cuvette volume = $0.483 \times [\Delta A_{340}/\Delta t (min)]$.

References

1. D. B. Millar, V. Frattali, and G. E. Willick, *Biochemistry*, **8**, 2416 (1969).
2. A. Pesce, R. H. McKay, F. Stolzenbach, R. D. Cahn, and N. O. Kaplan, *J. Biol. Chem.*, **239**, 1753 (1964).

Acknowledgments

Original compilation of Criteria: J. B. Neilands. Contributions by: A. C. Wilson and N. O. Kaplan.

*IUB Classification: 1.1.1.27, L-Lactate:NAD oxidoreductase.

E-28
Lipoyl Dehydrogenase*
(Pig Heart)

Equation:

S—S

$$\boxed{}\!\!>\!\!—(CH_2)_4COR\dagger + \text{NADH} + H^+ \rightleftharpoons$$
(+)-1,2-Dithiolane-3-valerate
[(+)-Lipoate]

$$\underset{|}{\overset{SH}{}} \quad \underset{|}{\overset{SH}{}}$$
$$CH_2—CH_2—CH—(CH_2)_4—COR + \text{NAD}^+$$
(−)-6,8-dimercaptooctanoate
[(−)-dihydrolipoate or (−)-H₂lipoate]

where R = OH, NH₂, or various amino acid residues.

Reaction: The enzyme catalyzes the interconversion of reduced and oxidized lipoic acid and derivatives by the NAD–NADH system.[1-3] Menadione, methylene blue, 2,6-dichlorophenolindophenol, and potassium ferricyanide can also serve as hydrogen acceptors.[2]

Method of Isolation: Fractionation with ammonium sulfate, and purification on calcium phosphate gel.[2]

CRITERIA (Tentative)

Physical Constants: Reported molecular weight:[2] ∼100,000. The enzyme contains 2 mol FAD/mol. $A_{280}/A_{455} = 5.35$ (Ref. 2); $A_{455}/A_{360} = 1.30$ (Ref. 3).

Procedure for Handling: There is no appreciable loss of activity in several months when an enzyme suspension in 3 M ammonium sulfate, pH 6, is stored at 4 °C[4] or when a solution in 30 mM phosphate, pH 7, containing 0.3 mM EDTA is frozen.[2]

Commercially Available Substrates: (±)-Lipoic acid, (±)-lipoamide.

Specific Activity: One unit of activity (U) catalyzes the oxidation of 1 μmol of NADH per minute at 25 °C, in the assay system described under "Activity Assay." Pure enzyme has been reported to have a specific activity of 25 U/mg.[1] The activity with lipoamide is 80-fold greater.[2]

Protein Assay: $A_{280} \times 0.94$ = mg enzyme/ml. (The factor is based on a biuret procedure and, owing to various uncertainties, may be in error[5-7] by as much as 10%.)

Activity Assay:[2]

Method: The change in absorbance of NADH at 340 nm is measured spectrophotometrically.

Reagents:

Sodium citrate buffer, 1 M, pH 5.65

Crystalline bovine serum albumin 2% in 30 mM EDTA

(±)-Lipoic acid, 20 mM

NADH, 10 mM

NAD, 10 mM

The activity is dependent on the ionic strength and on the presence of NAD.

Procedure: Prepare two cuvettes as follows (all quantities in ml):

	Blank	Assay
Citrate buffer	2.6	2.5
Albumin–EDTA	0.1	0.1
(±)-Lipoic acid	0.0	0.1
NADH	0.03	0.03
NAD	0.03	0.03
H₂O	0.14	0.14
Enzyme solution	0.10	0.10

Read the sample cuvette against the blank cuvette at 340 nm. Number of units (U) in cuvette = $0.483 \times [\Delta A_{340}/\Delta t(min)]$. The blank corrects for NADH oxidase activity of the enzyme.

Impurities: Impurities have not been reported.

References

1. V. Massey, *Biochim. Biophys. Acta*, **37**, 314 (1960).
2. V. Massey, *Methods Enzymol.*, **9**, 272 (1966).
3. V. Massey, *Enzymes*, **7A**, 275 (1963).
4. C. F. Boehringer und Sohn Corp., *Biochemica Catalogue*, New York, N. Y. (1968).
5. L. Casola, P. E. Brumby, and V. Massey, *J. Biol. Chem.*, **241**, 4977 (1966).
6. V. Massey, T. Hofmann, and G. Palmer, *J. Biol. Chem.*, **237**, 3820 (1962).
7. V. Massey, personal communication.
8. K. Mislow and W. C. Meluch, *J. Am. Chem. Soc.*, **78**, 5920 (1956).

Acknowledgments

Original compilation of Criteria: G. Gorin. Contributions by: V. Massey and L. T. Reed.

* IUB Classification: 1.6.4.3, NADH:Lipoamide oxidoreductase; recommended name: lipoamide dehydrogenase.

† The stereospecificity of the enzyme has been studied by Massey.[1] The unnatural (−) enantiomer reacts only 7–12% as rapidly as the (±) mixture. The absolute configuration of the substrate has been established.[8]

E-29
Myokinase*
(Rabbit Skeletal Muscle)

$$(Mg^{2+})$$

Equation: $2\text{ADP}^{3-} \rightleftharpoons \text{ATP}^{4-} + \text{AMP}^{2-}$

Method of Isolation: Crystallized from aqueous $(NH_4)_2SO_4$ solutions.[1-3]

CRITERIA (Tentative)

Physical Constant: Reported molecular weight:[4] 21,000.

Procedures for Handling:

Storage: The enzyme is stable for several months at 1 °C as a crystalline suspension in $(NH_4)_2SO_4$ solutions containing 10 mM EDTA at about pH 6.[3,5] Solutions in the microgram/milliliter range of concentration are stabilized by reducing agents, EDTA, and bovine serum albumin.[1,6] One to three percent aqueous solutions are stable for several weeks at pH 6 to 7 under N₂.

Lyophilization: May be lyophilized in the presence of succinate (10 mM) at about pH 6. If kept desiccated, the dry powder may be stored indefinitely.

Dialysis: May be dialyzed against aqueous solutions saturated with N₂ to prevent oxidation of reactive SH groups.[2,5]

Commercially Available Substrates: ATP, ADP, and AMP.

Specific Activity: One unit of activity (U) is equivalent to 1 μmol of ATP formed† per minute at pH 8.0 and 30 °C. Enzyme recrystallized 2–3 times has a specific activity of 1,800–2,200 U/mg at pH 8.0 and 30 °C.[3,5]

Protein Assay: Biuret procedure:[7] $A_{540} \times 31$ = mg of protein per 10 ml total volume.[8]

Activity Assay: According to Olson and Kuby:[9]

$$2\,\text{ADP}^{3-} \xrightarrow{\text{adenylate kinase}} \text{ATP}^{4-} + \text{AMP}^{2-}$$

Principle:

$$\text{ATP}^{4-} + \text{D-Glc} \xrightarrow{\text{hexokinase}} \text{ADP}^{3-} + \text{D-Glc-6-P}^{2-} + H^+$$

Overall reaction: $\text{ADP}^{3-} + \text{D-Glc} \rightarrow \text{AMP}^{2-} + \text{D-Glc-6-P}^{2-} + H^+$

The rate of liberation of H+ at pH 8.0 is a measure of the rate of production of ATP^{4-} in the myokinase reaction.

Reagents: Prepare with CO_2-free distilled water. 32 mM ADP (neutralized with NaOH); 16 mM $MgSO_4$; 200 mM D-glucose; 4% (w/v) crystalline bovine serum albumin; 0.0100 M NaOH (freshly diluted from standardized NaOH). Hexokinase[10,11] is stored at 1 °C as a crystalline suspension (about 40–45 mg of protein/ml) in $(NH_4)_2SO_4$ (about 2.2–2.4 M) containing 2 mM EDTA, pH 7. The hexokinase should be recrystallized until low in adenylate kinase activity, and the hexokinase activity should be at least 140 U/mg (preferably, 500 U/mg).

Myokinase: Just prior to measurement, dilute with ice-cold 1 mM EDTA–1 mM 1,4-dithioerythritol, pH 8.0 (adjusted with NaOH) to about 10-50 μg of protein/ml.

Procedure: Standardize a pH-stat (a commercial autotitrator), while stirring at 30 °C with a standard buffer (e.g., of pH 7.0). Pipet into a 12-ml reaction vessel (water-jacketed at 30 °C and equipped with a stirrer) 2.00 ml of D-glucose, 1.00 ml of distilled H_2O, 1.00 ml of ADP, 1.00 ml of $MgSO_4$, and 0.20 ml of albumin. Pass a slow stream of nitrogen over the liquid (to exclude interference by CO_2 during the titration) and start stirring. Adjust to pH 8.0 with 0.01 M NaOH. Add 10 μl of hexokinase (50–200 units); readjust the pH to 8.00 with 0.01 M NaOH. Add distilled water to a total volume of 8.0 ml. With the "endpoint" of the pH-stat set at 8.00, record the relatively low "blank titration."

After equilibration (3–5 min) and sufficient time to record the blank rate (5–10 min), a 10-μl aliquot (containing 0.1–0.5 μg of myokinase) is added and the reaction rate is recorded.

Calculations: The initial rate is calculated from the linear portion of the curve. The blank rate is subtracted from the overall rate to obtain the enzymic rate, which is expressed in units, 1 U = 1 microequivalent of NaOH consumed per minute.

Impurities: Thus far, significant impurities have not been reported for the enzyme recrystallized 2 to 3 times.[12] Traces of glycolytic enzymes (e.g., triose phosphate isomerase) may be possible contaminants in amorphous preparations. If used as a reagent for ADP determinations at the concentrations employed, the enzyme should be free of ATPase and nucleotidase.

References

1. L. Noda and S. A. Kuby, *J. Biol. Chem.*, **226**, 541 (1957).
2. T. A. Mahowald, E. A. Noltmann, and S. A. Kuby, *J. Biol. Chem.*, **237**, 1138 (1962).
3. L. Noda and S. A. Kuby, *Methods Enzymol.*, **6**, 223 (1963).
4. L. Noda and S. A. Kuby, *J. Biol. Chem.*, **226**, 551 (1957).
5. T. A. Mahowald, E. A. Noltmann, and S. A. Kuby, *J. Biol. Chem.*, **237**, 1535 (1962).
6. L. Noda, *J. Biol. Chem.*, **232**, 237 (1958).
7. A. G. Gornall, C. J. Bardawill, and M. M. David, *J. Biol. Chem.*, **177**, 751 (1949).
8. S. A. Kuby, T. A. Mahowald, and E. A. Noltmann, *Biochemistry*, **1**, 748 (1962).
9. O. E. Olson and S. A. Kuby, *J. Biol. Chem.*, **239**, 460 (1964).
10. R. A. Darrow and S. P. Colowick, *Methods Enzymol.*, **5**, 226 (1962).
11. M. Kunitz and M. R. McDonald, *J. Gen. Physiol.*, **29**, 393 (1946).
12. L. Noda, *Enzymes*, **6**, 139 (1962).

Acknowledgments

Original compilation of Criteria: S. A. Kuby. Contributions by: O. E. Olson, L. Noda, and E. A. Noltmann.

* IUB Classification: 2.7.4.3, ATP:AMP phosphotransferase; commonly used names: adenylate kinase, ATP–AMP transphosphorylase.

† Note that this is equivalent to the IUB unit for the case where two identical molecules react; 1 unit = 2 μmol of ADP reacted per minute.

E-30
Papain*
(Papaya Latex)

Reactions:

Catalyzes the hydrolysis of numerous peptide, amide, and ester bonds. R is preferentially the residue of an L-amino acid bearing a nonpolar side chain;[1] R' can be the side chain of argenine, lysine, glutamine, histidine, glycine, or tyrosine;[2] R'' can be any of numerous amino acid, peptide, or alcohol resdiues.

Method of Isolation: Crystallized from an aqueous solution containing cysteine–NaCl.[3]

CRITERIA (Tentative)

Physical Constant: Molecular weight, based on crystallographic analysis, 23,000 (calculated from Drenth *et al.*[4]).

Procedure for Handling:

Storage: Suspensions of the crystalline enzyme in water or dilute buffers are stable for about 6 months at 4 °C. Solutions of the enzyme (>3 mg/ml) in dilute acetate buffer, pH 5.0, are stable for several days at 5 °C.[5]

Activation: The enzyme must be activated prior to assay. One method is to dilute suspensions of the crystalline enzyme to a concentration of 2–3 mg/ml in 10 mM citrate buffer, pH 5.8, containing 1 mM EDTA and 40 mM cysteine. Assay[6] within 5 h.

Commercially Available Substrates: Numerous proteins and peptides, *N*-benzoyl-L-arginine ethyl ester, *N*-benzoyl-L-lysinamide, *N*-benzoylglycinamide, and benzyloxycarbonylglycine *p*-nitrophenyl ester.

Specific Activity: One unit of activity for this enzyme catalyzes the hydrolysis of 1 μmol of benzyloxycarbonylglycine *p*-nitrophenyl ester per minute at pH 6.8 and 25 °C. Twice crystallized and suitably activated, papain has[7] a specific activity of 6 to 8 U/mg.

Protein Assay:[3] $A_{280} \times 0.416$ = mg enzyme/ml.

Activity Assay:

Method: The rate of hydrolysis of benzyloxycarbonylglycine *p*-nitrophenyl ester is monitored spectrophotometrically at 400 nm, at pH 6.8 and 25 °C.[7,8]

Procedure: Place 0.6 ml of 100 mM phosphate buffer, pH 6.8, 0.3 ml of 10 mM EDTA, and 10 to 20 μl of enzyme solution (~0.2 mg/ml) in a cuvette and dilute to 2.8 ml with distilled water. The reaction is initiated by stirring in, with a plastic spatula, 0.2 ml of 1.5 mM benzyloxycarbonylglycine *p*-nitrophenyl ester in acetonitrile. Read A_{400} at 15-s intervals, or monitor with a recording spectrophotometer to obtain the initial reaction rate. Allow the reaction to proceed to completion (3–5 min) to obtain the final value of A_{400}. Correct the reaction rate by use of a blank without enzyme.

Calculations: The final value of A_{400} is the absorbance of 0.3 μmol of DNP/3 ml. Therefore, number of units (U) in reaction vessel

$$= 0.3 \times \frac{\Delta[A_{400}(\text{sample}) - A_{400}(\text{blank})]/\Delta t(\text{min})}{A_{400}(\text{final})}$$

References

1. I. Schechter and A. Berger, *Biochem. Biophys. Res. Commun.*, **32**, 898 (1968).
2. A. N. Glazer and E. L. Smith, *Enzymes*, **3**, 502 (1971).
3. J. R. Kimmel and E. L. Smith, *Biochem. Prep.*, **6**, 61 (1958).
4. J. Drenth, J. N. Jansonius, R. Koekoek, H. M. Swen, and B. G. Wolthers, *Nature*, **218**, 929 (1968).

5. C. Worthington, personal communication.
6. A. C. Henry and J. F. Kirsch, *Biochemistry*, **6**, 3536 (1967).
7. J. F. Kirsch and M. Ingelström, *Biochemistry*, **5**, 783 (1966).
8. I. B. Klein and J. F. Kirsch, *J. Biol. Chem.*, **244**, 5928 (1969).

Acknowledgments

Original compilation of Criteria: J. F. Kirsch and Leroy Baker. Contributions by: J. B. Neilands.

* IUB Classification: 3.4.17.2.

E-31
Pepsin*
(Pig Gastric Mucosa)

Equation: $R'R''CHCOR''' \rightarrow R'R''CHCOO^- + {}^+H_3N-R$

where R' usually is an acylamido group; R'' is an amino acid side chain, preferably, but not necessarily, that of L-phenylalanine, L-leucine, or L-tyrosine;[1] and R''' is —NHR, R being preferably, but not necessarily, a hydrophobic amino acid side chain, or R''' is —OR (e.g., β-phenyl-L-lactic acid methyl ester[2]).

Reaction: Catalyzes the hydrolysis of certain peptide and ester bonds; pH optimum 1.5 to >4, depending on the substrate. Also catalyzes transpeptidation.

Method of Isolation: The active enzyme is prepared from autolyzed pig gastric mucosa[3] or from purified commercial pepsinogen by rapid activation[4] at pH 2.

CRITERIA (Tentative)

Physical Constants: Reported molecular weights: 32,700 (from sedimentation data[5]) and 34,163 (from amino acid analysis[4]).

Procedure for Handling:

Storage: At −20 °C in 50 mM, pH 4.4, sodium acetate buffer,[4] or as the lyophilized powder at 4 °C.

Dilution: Dilutions are made daily in acetate buffer, 50 mM, pH 5.3, at 0 °C. Autodigestion occurs below pH 4 and denaturation above pH 7.

Dialysis: Between pH 5 and 6, at 4 °C.

Commercially Available Substrates: N-Acetyl-L-phenylalanyl-L-tyrosine, N-acetyl-L-phenylalanyl-L-diiodotyrosine (APADIT), hemoglobin, and casein, as well as many other proteins and synthetic peptide derivatives.

Specific Activity:[6] One unit of activity for this enzyme catalyzes the hydrolysis of 1 μmol of APADIT per minute at 37 °C under the conditions specified in "Activity Assay." Commercial pepsin (twice-crystallized) has a specific activity of 0.13 U/mg. The activity of pepsin freshly prepared from pepsinogen is about 25% greater.[4]

Protein Assay:[7] $A_{278} \times 0.676 = $ mg enzyme/ml.

Activity Assay:

Method: The assay[6,8,9] is based on the estimation, by the use of the ninhydrin reaction, of diiodotyrosine liberated from APADIT.

Procedure: The assay mixture consists of 3.2 ml of buffered substrate solution (10.2 mg of APADIT dissolved in 3 ml of 20 mM NaOH at 37 °C) poured slowly, with stirring, into 80 ml of 5 mM sodium phosphate buffer, pH 2.0, containing 3% of methanol at 37 °C and diluted to 100 ml with the same buffer (this substrate solution should be used within 30 min after preparation) and 0.2 ml of pepsin solution containing about 0.3 mg enzyme/ml of 1 mM acetic acid. The reaction is terminated after 8 min at 37 °C by the addition of 0.1 ml of 0.8 M NaOH. Portions, 0.5 ml each,

of ninhydrin solution (3%, in peroxide-free methylCellosolve) and freshly prepared acetate–cyanide buffer (1 ml of 10 mM NaCN and 49 ml of sodium acetate buffer, pH 5.3; 2.65 M in sodium ion) are added, and the color is developed by heating for 15 min in a boiling-water bath. The A_{570} of the solution is determined after dilution with 2.5 ml of 2-propanol. The diiodotyrosine liberated is estimated after correction for appropriate enzyme and substrate blanks. An A_{570} of 0.268 per 0.1 μmol of diiodotyrosine has been found by using the above procedure. Number of units (U) in reaction vessel = amount of diiodotyrosine (μmol) produced per minute.

Impurities: Commercial pepsin contains active and inactive autodigestion products,[4] as well as other pepsins derived from separate zymogens of the gastric mucosa.[10]

References

1. J. Tang, *Nature*, **199**, 1094 (1963).
2. K. Inouye and J. S. Fruton, *J. Am. Chem. Soc.*, **89**, 187 (1967).
3. J. H. Northrop, *J. Gen. Physiol.*, **30**, 177 (1946).
4. T. G. Rajagopalan, S. Moore, and W. H. Stein, *J. Biol. Chem.*, **241**, 4940 (1966).
5. R. C. Williams, Jr., and T. G. Rajagopalan, *J. Biol. Chem.*, **241**, 4951 (1966).
6. H. M. Lang and B. Kassell, *Biochemistry*, **10**, 2296 (1971).
7. O. O. Blumenfeld, J. Leonis, and G. E. Perlmann, *J. Biol. Chem.*, **235**, 379 (1960).
8. W. T. Jackson, M. Schlamowitz, and A. Shaw, *Biochemistry*, **4**, 1537 (1965).
9. H. Rosen, *Arch. Biochem. Biophys.*, **67**, 10 (1957).
10. A. P. Ryle, *Biochem. J.*, **98**, 485 (1966).

Acknowledgments

Original compilation of Criteria: B. Kassell. Contributions by: J. E. Folk, G. E. Perlman, and A. H. Mehler.

* IUB Classification: 3.4.18.1, New name: Pepsin A.

E-32
Peroxidase*
(Horseradish)

Reaction: Peroxidase, a hemoprotein, catalyzes the oxidation by hydrogen peroxide of a variety of hydrogen donors, including phenols, aromatic primary, secondary, and tertiary amines, indole, and L-ascorbic acid.[1]

Equation: $AH_2 + H_2O_2 \rightarrow A + 2 H_2O$

where AH_2 is the hydrogen donor, and H_2O_2 is the oxidant.

Method of Isolation: Crystallized from ammonium sulfate solution.[2-5] Seven isozymes have been separated and characterized.[6,7]

CRITERIA (Tentative)

Physical Constants: Reported molecular weights: 40,200[4] and 39,800.[8] The one molecule of protohemin IX per molecule of peroxidase[4,6] is readily removed by treatment with acidic acetone.[9] Carbohydrate, composed of sugars and amino sugars, accounts for 18–28% of the enzyme[6,10] by weight.

Procedure for Handling:

Storage and Dilution: Lyophilized powder is stable for more than 1 year when stored at 0 °C. Solutions at room temperature and pH 7.0 are stable for weeks. When warmed for 15 min,[4] the activity is stable up to 63 °C. Fluoride and other halides, as well as CN^- and N_3^-, decrease the stability at pH <5.5.

Commercially Available Substrates: o-Dianisidine† (3,3'-dimethoxybenzidine) recrystallized from ethanol and H_2O; H_2O_2 (30%).

Specific Activity: One unit of activity for this enzyme consumes 1 μmol of H_2O_2 per minute. The major component from a DEAE-Sephadex column, of material passing through a (carboxymethyl)-Sephadex column, has been reported to have a specific activity of 4,000 U/mg.[10] Activities reported for the seven isozymes range from 2,500 to 57,500 U/mg at 30 °C (calculated from Kay et al.).[7]

Protein Assay:[6] $A_{401} \times 0.4 =$ mg protein/ml.‡

Activity Assay: Peroxidase activity is measured by following the change in absorbance at 460 nm due to oxidation of o-dianisidine in the presence of H_2O_2 and the enzyme.

Reaction Medium:

o-Dianisidine	1% in methanol or 50 mM HCl (fresh, in amber bottle)	0.05 ml
H_2O_2	100 mM	0.1 ml
Sodium acetate	50 mM, pH 5.4	2.75 ml

Procedure: To 2.9 ml of the reaction medium at 30 °C in a spectrophotometer cell, add 0.1 ml of enzyme containing 0.02–0.2 μg/ml. The assays are not linear with enzyme concentration above $\Delta A_{460}/\Delta t = 0.7$/min.

Calculations: The activity is calculated from the initial slope of the curve of the absorbance at 460 nm vs. time. Number of units per cuvette volume $= 0.266 \times [\Delta A_{460}/\Delta t(\text{min})]$.

References

1. B. C. Saunders, A. G. Holmes-Siedle, and B. P. Stark, in *Peroxidase*, Butterworth, Washington, D. C. (1964).
2. H. Theorell and A. C. Maehly, *Acta Chem. Scand.*, **4**, 422 (1950).
3. D. Keilin and E. F. Hartree, *Biochem. J.*, **49**, 33 (1951).
4. A. C. Maehly, *Methods Enzymol.*, **2**, 801 (1955).
5. R. H. Kenten and P. J. G. Mann, *Biochem. J.*, **57**, 347 (1954).
6. L. M. Shannon, E. Kay, and J. Y. Lew, *J. Biol. Chem.*, **241**, 2166 (1966).
7. E. Kay, L. M. Shannon, and J. Y. Lew, *J. Biol. Chem.*, **242**, 2470 (1967).
8. R. Cecil and A. G. Ogston, *Biochem. J.*, **49**, 105 (1951).
9. A. C. Maehly, *Biochim. Biophys. Acta*, **8**, 1 (1952).
10. R. Flora, personal communication.

Acknowledgments

Original compilation of Criteria: C. H. Suelter. Contributions by: L. M. Shannon.

* IUB Classification: 1.11.1.7, Peroxidase.

† o-Dianisidine is highly toxic and can cause skin irritation and sensitization. It should be handled with care.

‡ The absorbance ratio A_{401}/A_{275} (sometimes called the RZ number, Reinheitzahl) has been used as criterion of purity; it varies from 3.15 to 4.19 for the 7 isozymes in 0.05M acetate, pH 5.8. The demonstration that this ratio is influenced by buffer and pH detracts from its usefulness as a criterion of purity.[7]

E-33
Phosphorylase a*
(Rabbit Muscle)

Equation: $G_n + x(\alpha\text{-D-glucose 1-phosphate}^{2-}) \rightleftharpoons G_{n+1} + x(\text{HPO}_4^{2-})$ where G_n is an α-D-glucan, such as glycogen, containing n D-glucose residues. AMP is not required.

Method of Isolation: Crystallized from rabbit muscle[1-3] or by conversion of purified b form[4] with phosphorylase b kinase.[5]

CRITERIA (Tentative)

Physical Constants: Reported molecular weight:[6,7] 370,000. One molecule of the crystalline enzyme contains 4 molecules of pyridoxal phosphate[8,9] and 4 molecules of L-serine esterified with phosphate.[10]

Procedures for Handling: Available as a crystalline suspension in L-cysteine–DL-glycerophosphate buffer. May be stored at 4 °C; freezing destroys the activity.

Commercially Available Substrates: Glycogen, α-D-glucose 1-phosphate, soluble starch, amylopectin.

Specific Activity: The units are arbitrary, and correspond to 1 U = 1,000 k, where k is the first-order velocity constant.[2] Preparations having the highest purity have a specific activity of $\sim$1,500–2,000 U/mg. Activities can also be expressed as amount of inorganic phosphate in micromoles released from D-glucose 1-phosphate per minute per milligram of protein. (See "Activity Assay.")

Protein Assay: $A_{279} \times 0.79 =$ mg protein/ml in 0.1 M sodium glycerophosphate or phosphate (pH 7.0).[11]

Activity Assay:

Method: Measurement[12] of the rate of release of orthophosphate from α-D-glucose 1-phosphate in the presence of glycogen and AMP, at pH 6.5 and 30 °C.

Enzyme: About 25 μg of phosphorylase per ml in mercaptoethanol (40 mM) and maleate (100 mM), at pH 6.5, containing 1 mg of crystalline bovine serum albumin/ml.

Substrate: 2% of glycogen,† 150 mM D-glucose 1-phosphate, with or without 2 mM AMP, in 100 mM maleate, pH 6.5.

Procedure: To 0.2 ml of the enzyme, 0.2 ml of substrate is added at 30 °C. After 5 min at 30 °C stop the reaction by adding acidified ammonium molybdate solution, and determine the inorganic phosphate released by the method of Fiske and Subba-Row[13] and the usual procedure of Illingworth and Cori.[2] Under these conditions, one unit of activity (U) is defined as the amount of enzyme causing the release of 1 micromole of inorganic phosphate from D-glucose 1-phosphate per minute. In the presence of AMP, and by use of the protein assay described herein, the specific activity is 88 U/mg. In the absence of AMP, phosphorylase a has a specific activity of 59 U/mg.

Impurities: To remove amylo-1,6-glucosidase, seven to ten recrystallizations are required.[14] The first batch of crystals may contain phosphorylase b and phosphatase.

References

1. A. A. Green and G. T. Cori, *J. Biol. Chem.*, **151**, 21 (1943).
2. B. Illingworth and G. T. Cori, *Biochem. Prep.*, **3**, 1 (1953).
3. G. T. Cori, B. Illingworth, and P. J. Keller, *Methods Enzymol.*, **1**, 200 (1955).
4. E. G. Krebs and E. H. Fischer, *Methods Enzymol.*, **5**, 373 (1962).
5. E. H. Fischer and E. G. Krebs, *Methods Enzymol.*, **5**, 369 (1962).
6. V. L. Seery, E. H. Fischer, and D. C. Teller, *Biochemistry*, **6**, 3315 (1967).
7. D. L. DeVincenzi and J. L. Hedrick, *Biochemistry*, **6**, 3489 (1967).
8. T. Baranowski, B. Illingworth, D. H. Brown, and C. F. Cori, *Biochim. Biophys. Acta*, **25**, 16 (1957).
9. A. B. Kent, E. G. Krebs, and E. H. Fischer, *J. Biol. Chem.*, **232**, 549 (1958).
10. E. G. Krebs and E. H. Fischer, *Advan. Enzymol.*, **24**, 263 (1962).
11. C. Y. Huang and D. J. Graves, *Biochemistry*, **9**, 660 (1970).
12. J. L. Hedrick and E. H. Fischer, *Biochemistry*, **4**, 1337 (1965).
13. C. H. Fiske and Y. SubbaRow, *J. Biol. Chem.*, **66**, 375 (1925).
14. G. T. Cori and J. Larner, *J. Biol. Chem.*, **188**, 17 (1951).

Acknowledgments

Original compilation of Criteria: J. Larner. Contributions by: E. G. Krebs. Revision by: D. J. Graves.

* IUB Classification: 2.4.1.1, 1,4-α-D-Glucan : orthophosphate hydrolase. Committee Note: As given in the IUB report, the configuration is not indicated. The name should be α-D-(1 → 4)-glucan:orthophosphate-D-glucosyltransferase.

† AMP may be removed with a mixed-bed, ion-exchange resin. Traces of AMP, frequently present as a contaminant in commercial glycogen, are sufficient to cause a considerable activation of phosphorylase b.

E-34
Phosphorylase b*
(Rabbit Muscle)

Equation: $G_n + x\,(\alpha\text{-D-glucose 1-phosphate}^{2-}) \rightleftharpoons G_{n+1} + x\,(HPO_4^{2-})$ where G_n is an α-D-glucan, such as glycogen, containing n D-glucose residues. AMP is required.

Method of Isolation: Crystallized from rabbit muscle[1] as the AMP–Mg^{2+} complex, namely, [(phosphorylase b) (Mg^{2+}) (AMP)].[2]

CRITERIA (Tentative)

Physical Constant: Reported molecular weight:[3,4] 185,000. One molecule of the crystalline enzyme contains 2 molecules of pyridoxal phosphate.[5]

Procedures for Handling: Available as a suspension of crystals. AMP may be removed by passing a 1–2% solution of the protein through a 1:1 mixture of charcoal and cellulose powder on a small column. The charcoal that has been used is Norit A (acid-washed), and the cellulose powder is Whatman cellulose powder (coarse grade). The weight of charcoal required is approximately equal to that of the enzyme to be treated. Complete removal of the AMP is indicated when the A_{260}/A_{280} ratio equals 0.53, as measured on solutions of the protein in water.[1]

Commercially Available Substrates: Glycogen, α-D-glucose 1-phosphate, soluble starch, amylopectin.

Specific Activity: The units are arbitrary, and correspond to 1 U = 1,000 k, where k is the first-order velocity constant. Preparations having the highest purity have a specific activity of $\sim$1,600 U/mg.[6] Activities can also be expressed as amount of inorganic phosphate in micromoles released from D-glucose 1-phosphate per minute per milligram of protein. (See "Activity Assay.")[7]

Protein Assay: $A_{279} \times 0.76$ = mg protein/ml in 0.1 M sodium glycerophosphate or phosphate (pH 7.0).[8] This applies to protein solutions free from nucleotide. Nucleotides may be removed by passing enzyme solutions through a charcoal–cellulose column. (See "Procedures for Handling.")

Activity Assay:

Method: Measurement of the rate of release[7] of orthophosphate from α-D-glucose 1-phosphate in the presence of glycogen and AMP, at pH 6.5 and 30 °C.

Enzyme: About 25 μg of phosphorylase per ml in mercaptoethanol (40 mM) and maleate (100 mM), at pH 6.5, containing 1 mg of crystalline bovine serum albumin/ml.

Substrate: 2% of glycogen,† 150 mM D-glucose 1-phosphate, 2 mM AMP in 100 mM maleate, pH 6.5.

Procedure: To 0.2 ml of enzyme, 0.2 ml of substrate is added at 30 °C. After 5 min at 30 °C, the reaction is stopped by addition of acidified ammonium molybdate solution, and the inorganic phosphate released is determined by the method of Fiske and SubbaRow[9] by following the usual procedure of Illingworth and Cori.[6] Under these conditions, one unit of activity is defined as the amount of enzyme causing the release of 1 micromole of inorganic phosphate from D-glucose 1-phosphate per minute. In the presence of AMP, and by using the protein assay described herein, the specific activity for phosphorylase b is 88 U/mg. In the absence of AMP, phosphorylase b has a specific activity of <0.9 U/mg.

Impurities: Amylo-1,6-glucosidase is destroyed[10] in the alkaline incubation step used to isolate and purify the enzyme in the procedure of Fischer and Krebs.[1]

References

1. E. H. Fischer and E. G. Krebs, *Methods Enzymol.*, **5**, 369 (1962).
2. A. B. Kent, E. G. Krebs, and E. H. Fischer, *J. Biol. Chem.*, **232**, 549 (1958).
3. V. L. Seery, E. H. Fischer, and D. C. Teller, *Biochemistry*, **6**, 3315 (1967).
4. D. L. DeVincenzi and J. L. Hedrick, *Biochemistry*, **6**, 3489 (1967).
5. D. H. Brown and C. F. Cori, *Enzymes*, **5**, 207 (1961).
6. B. Illingworth and G. T. Cori, *Biochem. Prep.*, **3**, 1 (1953).
7. J. L. Hedrick and E. H. Fischer, *Biochemistry*, **4**, 1337 (1965).
8. A. M. Gold, *Biochemistry*, **7**, 2106 (1968).
9. C. H. Fiske and Y. SubbaRow, *J. Biol. Chem.*, **66**, 375 (1925).
10. J. Larner and F. Huijing, unpublished data.

Acknowledgments

Original compilation of Criteria: J. Larner. Contributions by: E. G. Krebs. 1970 Revision by: D. J. Graves.

* IUB Classification: 2.4.1.1, 1,4-α-D-Glucan: orthophosphate hydrolase.

† AMP may be removed with a mixed-bed, ion-exchange resin. Traces of AMP, frequently present as a contaminant in commercial glycogen, are sufficient to cause a considerable activation of phosphorylase b.

E-35
Phosphoglucomutase*
(Rabbit Skeletal Muscle)

Equation:† α-D-glucose 1-phosphate $\rightleftharpoons$ α-D-glucose 6-phosphate

Method of Isolation: Ammonium sulfate fractionation, heat precipitation, and O-(carboxymethyl)cellulose chromatography of muscle extracts.[1,2]

CRITERIA (Tentative)

Physical Constants: Reported molecular weight: 62,000,[3] 64,900.[4]

Procedure for Handling:

Storage: Maximal stability is found at pH $\sim$5. The enzyme can be stored[5] for one week or more in 150 mM acetate, pH 5.0, and is relatively stable between pH 4.5 and 8.5. Stable for up to 1 month as a suspension in cold 2.5 M $(NH_4)_2SO_4$ containing 50 mM sodium acetate. Stable indefinitely, if frozen and stored in liquid nitrogen as pellets of buffered solutions.[1]

Commercially Available Substrates: α-D-Glucose 1-phosphate (some commercial preparations contain enough α-D-glucose 1,6-bisphosphate to satisfy most of the assay requirement); D-glucose 6-phosphate; α-D-glucose 1,6-bisphosphate. D-Glucose 1,6-bisphosphate can readily be prepared by the procedure of Hanna and Mendicino.[6]

Specific Activity: One unit of activity (U) of this enzyme catalyzes the conversion of 1 μmol of D-glucose 1-phosphate to D-glucose 6-phosphate per minute under the conditions specified below. The most active enzyme has a specific activity of 740 U/mg at 30 °C and pH 7.5 (calculated from the data of Ray *et al.*[7]).

Protein Assay:[5] $A_{278} \times 1.30$ = mg protein/ml.

Activity Assay: The difference in acid-labile phosphate, that is, D-glucose 1-phosphate, in the presence and absence of enzyme, can be measured by a modified Fiske–SubbaRow procedure.[8] Maximum activity requires[9,10] a sufficiently high concentration of Mg^{2+} to convert all of the enzyme into the Mg^{2+} complex, about 1 mM.

Enzymic Reaction: The enzyme is diluted to about 0.08 U/ml in a solution, at pH 7.5, containing 2.0 mM Mg^{2+}, 1.0 mM EDTA, 100 mM imidazole, and 50 mM tris–hydrochloride, and that contains 0.15 mg/ml of crystalline bovine serum albumin. The solution is prepared daily from crystalline albumin and two stock solutions, one containing magnesium chloride (sulfate is a good, competitive inhibitor) and the second, EDTA, tris, and imidazole. After standing for at least 5 min at room temperature, 0.1 ml of enzyme is added to 0.4 ml of an assay mixture, at 30 °C and pH 7.5, containing 630 μM D-glucose 1-phosphate, 6 μM D-glucose 1,6-bisphosphate, 2.5 mM Mg^{2+}, 1.3 mM EDTA, and 25 mM tris–hydrochloride. After 10 min at 30 °C, the reaction is stopped by addition of 0.5 ml of 1.0 M sulfuric acid, and acid-labile phosphate is measured.

Acid-labile Phosphate Determination:

Reagents: Fiske–SubbaRow Reagent—Dissolve 5.7 g of sodium hydrogen sulfite, 0.2 g of sodium sulfite, and 0.1 g of purified 1-amino-2-naphthol-4-sulfonic acid in 100 ml of H_2O. Filter, and store in the dark in a refrigerator; this solution is stable for 1–2 weeks in a refrigerator (or until crystals appear).

Acid ammonium molybdate solution—Add 28 ml of concentrated H_2SO_4 to ~800 ml of H_2O. Dissolve 3 g of ammonium molybdate in H_2SO_4 and dilute to 1 liter.

Inorganic Phosphate Standard—Dilute 1.4 ml of concentrated H_2SO_4 with 100 ml of water, and dissolve 136 mg of dried KH_2PO_4 therein. Dilute 1 to 10 with 0.25 M H_2SO_4 to give 1 μmol Pi/ml standard.

Procedure: To the acidified reaction mixture (1 ml) from the enzyme reaction, and to the blank (acid added prior to addition of the enzyme), add 0.6 ml of Fiske–SubbaRow reagent and 10.0 ml of acid ammonium molybdate solution. Cap each tube with a marble, and heat in a *boiling*-water bath for 10 min. Cool the tubes to room temperature, and read optical absorbance at 830 nm. The color is stable for hours. Phosphate standards are developed simultaneously with the enzyme assays. The phosphate standard (0.2 ml) is added to 0.8 ml of water and treated as above. The standard, thus, contains 0.2 μmol inorganic phosphate.

Calculation of Activity: Number of units (U) in aliquot = $(A_b - A_{10})/10 A_{Pi}$, where A_{Pi} refers to the absorbance of 1 μmol inorganic phosphate under the assay conditions, A_b is the absorbance of the blank (see above), and A_{10} is the absorbance of the assay mixture after the 10-min reaction period.

References

1. J. A. Yankeelov, H. R. Horton, and D. E. Koshland, Jr., *Biochemistry*, **3**, 349 (1964).
2. J. G. Joshi and P. Handler, *J. Biol. Chem.*, **244**, 3343 (1969).
3. D. L. Filmer and D. E. Koshland, Jr., *Biochim. Biophys. Acta*, **77**, 334 (1963).
4. S. Harshman and H. R. Six, *Biochemistry*, **8**, 3423 (1969).
5. V. S. Najjar, *Methods Enzymol.*, **1**, 294 (1955).
6. R. Hanna and J. Mendicino, *J. Biol. Chem.*, **245**, 4031 (1970).
7. W. J. Ray, Jr., G. A. Roscelli, and D. S. Kirkpatrick, *J. Biol. Chem.*, **241**, 2603 (1966).
8. G. R. Bartlett, *J. Biol. Chem.*, **234**, 466 (1959).
9. W. J. Ray, Jr., *J. Biol. Chem.*, **244**, 3740 (1969).
10. E. J. Peck, Jr., and W. J. Ray, Jr., *J. Biol. Chem.*, **244**, 3748 (1969).

Acknowledgments

Original compilation of Criteria: E. J. Peck, Jr. Contributions by: W. J. Ray, Jr.

* IUB Classification: 2.7.5.1, α-D-Glucose 1,6-bisphosphate:α-D-glucose 1-phosphate phosphotransferase.

† Although free D-glucose 1,6-bisphosphate is not an obligatory reaction intermediate, under assay conditions, it is needed to prevent dead-end inhibition of the enzyme due to the dissociation of the enzyme–D-glucose 1,6-bisphosphate complex. A ratio of D-glucose 1-phosphate:D-glucose 1,6-bisphosphate of 1,000:1 should be used for attaining maximal activity.

E-36
Pyruvate Kinase*
(Rabbit Skeletal Muscle)

Equation:

$$\text{(K}^+\text{, Mg}^{2+}\text{)}$$
$$H^+ + \text{phospho}enol\text{pyruvate}^{3-} + \text{ADP}^{3-} \rightleftharpoons \text{pyruvate}^- + \text{ATP}^{4-}$$

Method of Isolation: Crystallized from ammonium sulfate solution.[1,2]

CRITERIA (Tentative)

Physical Constant: Reported molecular weight:[3] 237,000. Dissociation into 4 subunits has been reported.[4]

Procedures for Handling:

Storage: A suspension of crystals in $(NH_4)_2SO_4$ solution near neutral pH is stable for at least 2 months at 0–4 °C.

Dilution: Solubilization of enzyme, centrifuged to remove most of the $(NH_4)_2SO_4$, and dilution may be made in 50 mM potassium phosphate, imidazole or tris buffer of pH 7–9. Stability at a concentration of 10 μg/ml is enhanced by 100 mM KCl and 1 mM $MgCl_2$. The same buffer may be used for gel filtration and dialysis.

Commercially Available Substrates: The trisodium, tricyclohexylammonium,[5] and barium silver salts of phospho*enol*pyruvic acid are available.

Specific Activity: One unit of activity (U) for this enzyme catalyzes the phosphorylation of 1 μmol of ADP per minute at 30 °C and pH 7.5. Crystallized enzyme has been reported[2,6] to have a specific activity of 250–300 U/mg.

Protein Assay:[1] $A_{280} \times 1.85$ = mg enzyme/ml at pH 7.

Activity:

Method: The activity is determined by reaction of phospho*enol*pyruvate with ADP, by use of lactate dehydrogenase and NADH. Near optimum conditions are 100 mM K^+, 1 mM Mg^{2+}, 1 mM PEP, 4 mM ADP, 160 μM NADH, an excess of pyruvate kinase (400 U/mg) free of lactate dehydrogenase, and 0.05–0.2 μg of pyruvate kinase.

Reaction Medium:

KCl	1 M	0.3 ml
$MgCl_2$	10 mM	0.3 ml
ADP	40 mM, pH 7.5	0.3 ml
PEP	100 mM, pH 7.5	0.3 ml
tris·HCl or imidazole·HCl	500 mM, pH 7.5	0.3 ml
NADH	1.6 mM, pH 7.5	0.3 ml
Lactate dehydrogenase	0.5 mg/ml	0.1 ml
H_2O		1 ml

Procedure: To 2.9 ml of reaction medium at 30 °C, add 0.1 ml of enzyme containing 0.05–0.2 μg of pyruvate kinase per ml.

Calculations: Calculation of the activity is made from the initial slope of curve of the absorbance at 340 nm vs. time. Number of units (U) per reaction vessel (3 ml) = $0.483 \times [\Delta A_{340}/\Delta t(\text{min})]$.

Impurities: Contaminating enzymes that may be present are lactate dehydrogenase, adenylate kinase, enolase, phosphoglycerate mutase, and 3-phosphoglycerate kinase. For use in the determination of ADP in the presence of AMP by measurement of pyruvate produced, the fraction of adenylate kinase should be so low that no detectable pyruvate is formed in the presence of AMP + ATP (ADP-free) during the period required to make the assays.

References

1. T. Bücher and G. Pfleiderer, *Methods Enzymol.*, **1**, 435 (1955).
2. A. Tietz and S. Ochoa, *Arch. Biochem. Biophys.*, **78**, 477 (1958).
3. R. C. Warner, *Arch. Biochem. Biophys.*, **78**, 494 (1958).
4. M. A. Steinmetz and W. C. Deal, Jr., *Biochemistry*, **5**, 1399 (1966).

5. F. Wold and C. E. Ballou, *J. Biol. Chem.*, **227**, 301 (1957).
6. A. Reynard, L. F. Hass, D. D. Jacobsen, and P. D. Boyer, *J. Biol. Chem.*, **236**, 2277 (1961).

Acknowledgments

Original compilation of Criteria: P. D. Boyer. Contributions by: H. U. Bergmeyer, M. Coon, and S. Ochoa. 1970 revision by C. H. Suelter.

* IUB Classification: 2.7.1.40, ATP:Pyruvate phosphotransferase.

E-37
Ribonuclease A*
(Bovine Pancreas)

Reaction: a. Cleavage of ribonucleates at the bonds between pyrimidine nucleotides, by formation of pyrimidine nucleoside 2′:3′-cyclic phosphates.

b. Hydrolysis of 2′:3′-cyclic phosphates of pyrimidine nucleosides to the 3′-phosphates.

Methods of Isolation: Crystalline ribonuclease may be obtained from acid extracts of pancreas.[1] Ribonuclease A may be isolated by chromatography of crystalline ribonuclease.[2]

CRITERIA (Tentative)

Physical Constant: Reported molecular weight:[3] 13,683.

Procedure for Handling:

Storage: Lyophilized preparations may be kept at −10 °C for at least 2 years without detectable alteration.

Handling: The enzyme is stable. Lyophilization, particularly from 50% acetic acid, induces aggregation; this is reversed, without loss of activity,[4] by heating to 65 °C. The enzyme is not retained by certain commercial dialysis membranes.[5]

Commercially Available Substrates: Cytidine 2′:3′-cyclic phosphate can be obtained as the free acid (98% pure) and as the sodium, ammonium, or cyclohexylguanidinium salt (95–99% pure). (Uridine 2′:3′-cyclic phosphate is also available as the sodium or cyclohexylguanidinium salt.) The purity of these substrates should be sufficiently high to make any one of them acceptable as a standard.

Specific Activity: One unit (U) catalyzes the hydrolysis of 1 μmol of substrate per min at 25 °C under the conditions specified in "Activity Assay." The specific activity is approximately 1.8 U/mg.

Protein Assay:[6] $A_{277.5} \times 1.40$ = mg enzyme/ml.

Activity Assay:

Method: The assay[7] is based on that of Crook *et al.*[8]

Substrate Solution: The substrate (either the free acid or the water-soluble salts already listed) is dissolved in tris buffer (see below) to give a concentration of 275 μM; 3 ml of the solution is placed in a 1-cm cuvette and warmed to 25 °C in the thermostated cell compartment of a spectrophotometer. Ribonuclease is added at a level of 0.05–0.2 mg, and the initial rate of change in absorbance at 286 nm is observed by recording the time required to give a change of 0.030 absorbance units. The assay is most reliably performed in a double-beam instrument having a recorder span of 0–0.1 absorbance units. Under these conditions, the change in absorbance at 286 nm per micromole of substrate cleaved is 1.37.[7]

Tris Buffer: 100 mM tris-hydrochloride containing 200 mM NaCl, pH 7.0, at 25 °C.

Impurities: Contaminating activities have not been reported in preparations of ribonuclease A. Preparations are usually not sterile.

References

1. M. Kunitz, *J. Gen. Physiol.*, **24**, 15 (1940).
2. A. M. Crestfield, W. H. Stein, and S. Moore, *J. Biol. Chem.*, **238**, 618 (1963).
3. C. H. W. Hirs, S. Moore, and W. H. Stein, *J. Biol. Chem.*, **219**, 623 (1956).
4. A. M. Crestfield, S. Moore, and W. H. Stein, *Arch. Biochem. Biophys.*, Suppl. 1, 257 (1962).
5. L. C. Craig, T. P. King, and A. Stracher, *J. Am. Chem. Soc.*, **79**, 3729 (1957).
6. C. C. Bigelow, *J. Biol. Chem.*, **236**, 1706 (1961).
7. A. M. Murdock and C. H. W. Hirs, unpublished data (1964).
8. E. M. Crook, A. P. Mathias, and B. R. Rabin, *Biochem. J.*, **74**, 234 (1960).

Acknowledgments

Original compilation of Criteria: C. H. W. Hirs. Contributions by: W. H. Stein, A. M. Crestfield, and R. K. Brown. Revision by: F. Wold.

* IUB Classification: 2.7.7.16, Ribonucleate pyrimidinenucleotido-2′-transferase (cycling).

E-38
Subtilisin*
(*Bacillus subtilis*)

Equation: $R'R''CHCOR''' \rightarrow R'R''CHCOO^- + HOR''' + H^+$

where R′ = H, NH₂, or an acylamino group;

R″ = an amino acid side chain; and

R‴ = − NHR or OR.

Reactions: Catalyzes the hydrolysis of numerous peptide, amino acid ester, and aliphatic ester bonds. The specificity is apparently more dependent on enzyme concentration and reaction conditions than on any inherent structural feature.

Method of Isolation: Crystallization from acetone solution.[3,4]

CRITERIA (Tentative)

Physical Constant: Reported molecular weight:[5,6] 27,600.

Procedures for Handling:

Storage: Crystalline powders and frozen solutions containing calcium ion are stable for 6–12 months. Aqueous, salt-free solutions undergo rapid autolysis and denaturation.

Dilution: The enzyme dissolved in 200 mM acetate buffer, pH 6.0, containing 20 mM CaCl₂, is relatively stable[2] for several days at 4 °C. Dilutions for assay should be made in 20 mM sodium acetate, pH 7.5, containing 10 mM calcium acetate.

Commercially Available Substrates: N-Acetyl-L-tyrosine ethyl ester, N-benzoyl-L-arginine ethyl ester, methyl butyrate, methyl valerate, casein, and numerous other synthetic substrates[7] and natural proteins.

Specific Activity: One unit of activity (U) catalyzes the hydrolysis of 1 μmol of N-acetyl-L-tyrosine ethyl ester per min at a substrate concentration of 20 mM at 37 °C and pH 8.0. For subtilisin BPN′, a specific activity of 375 U/mg has been reported[8] at a substrate concentration of 20 mM by use of the conditions described below.

Protein Assay:[5] $A_{278} \times 0.856$ = mg enzyme/ml; solvent, 50 mM sodium acetate, pH 6.9.

Activity:

Method:[2] The rate of hydrolysis of N-acetyl-L-tyrosine ethyl ester is measured titrimetrically, at pH 8.0 and 37 °C.

Procedure: Standardize an autotitrator at 37 °C. Add to the

reaction vessel 5 ml of 20 mM N-acetyl-L-tyrosine ethyl ester in 100 mM KCl containing 8% p-dioxane (by volume). Initiate the reaction by adding 1–5 μg of subtilisin. Record the volume of standardized 0.020 M NaOH required to maintain the pH at 8.0 until 5% of the substrate has been hydrolyzed.

Calculation: Number of units (U) = [volume of NaOH (ml)/ t(min)] × molar conc. (mol/l) × 10^3.

Impurities: Crystalline subtilisin preparations may contain as much as 30% of low-molecular-weight, dialyzable impurity produced by autolysis.[2,7]

References

1. S. A. Olaitan, R. J. DeLange, and E. L. Smith, J. Biol. Chem., 243, 5296 (1968).
2. A. N. Glazer, J. Biol. Chem., 242, 433 (1967).
3. B. Hagihara, Ann. Rep. Fac. Sci. (Osaka Univ.), 2, 35 (1954).
4. B. Hagihara, H. Matsubara, M. Nakai, and K. Okunuki, J. Biochem. (Tokyo), 45, 185 (1958).
5. H. Matsubara, C. B. Kasper, D. M. Brown, and E. L. Smith, J. Biol. Chem., 240, 1125 (1965).
6. E. L. Smith, F. S. Markland, C. B. Kasper, R. J. DeLange, M. Landon, and W. H. Evans, J. Biol. Chem., 241, 5974 (1966).
7. A. O. Barel and A. N. Glazer, J. Biol. Chem., 243, 1344 (1968).
8. A. N. Glazer, personal communication.

Acknowledgments

Original compilation of Criteria: Leroy Baker. Contributions by: A. N. Glazer.

* IUB Classification: 3.4.16.12, Subtilisin. [Subtilisin BPN' and subtilisin Novo are identical (but subtilisin Carlsberg is different) in amino acid sequence[1] and kinetic properties.[2]]

E-39
Taka-Amylase A*
(*Aspergillus oryzae*)

Reaction: The enzyme randomly hydrolyzes α-D-(1 → 4)-glucosidic bonds in polysaccharides, and hydrolyzes the synthetic substrate phenyl α-maltoside to phenol and maltose.
Method of Isolation: Crystallized from Takadiastase extracts.[1]

CRITERIA (Tentative)

Physical Constants: Reported molecular weight:[2,3] 51,000. Taka-amylase A contains 1 atom of calcium per molecule. The calcium is essential for its enzymic activity.[4]
Procedures for Handling: Crystalline suspensions in 30–40% acetone, or lyophilized powders, are stable at 5 °C for several years.
Commercially Available Substrates: Soluble starch, amylose, and phenyl α-maltoside.
Specific Activity: One unit (U) produces 1 μmol of phenol per minute from phenyl α-maltoside. The enzyme of highest purity has specific activity of 0.097 U/mg, under the conditions described in "Activity Assay."
Protein Assay:[2] A_{280} × 0.452 = mg protein/ml (pH 7.0).
Activity Assay:[5] The amount of phenol liberated from phenyl α-maltoside is measured, by using a phenol reagent.

Enzyme: 10–50 μg/ml H_2O.
Substrate: 0.5% Phenyl α-maltoside in H_2O.
Buffer: 200 mM Acetate buffer (pH 5.3).
Color Reagents: 5% Na_2CO_3; phenol reagent[6]—Dilute stock 2 M solution (may be purchased from most reagent supply firms) to 1 M with water.
Procedure: Add 0.5 ml of substrate solution to a mixture of 0.5

ml of enzyme and 0.5 ml of buffer. Incubate at 37 °C for 30 min. Add 1.5 ml of 5% Na_2CO_3 solution and 0.5 ml of phenol reagent; allow to stand at 37 °C for 20 min. Read absorbance at 660 nm. A calibration curve for phenol standard solution is used to convert absorbance into amount of phenol (μmol).

Impurities: No data are available.

References

1. S. Akabori, B. Hagihara, and T. Ikenaka, J. Biochem. (Tokyo), 41, 577 (1954).
2. T. Isemura and S. Fujita, J. Biochem (Tokyo), 44, 443 (1957).
3. K. Narita, H. Murakami, and T. Ikenaka, J. Biochem. (Tokyo), 59, 170 (1966).
4. A. Oikawa and A. Maeda, J. Biochem. (Tokyo), 44, 745 (1957).
5. S. Matsubara, T. Ikenaka, and S. Akabori, J. Biochem. (Tokyo), 46, 425 (1959).
6. O. Folin and V. Ciocalteu, J. Biol. Chem., 73, 627 (1927).

Acknowledgments

Original compilation of Criteria: H. Toda. Contributions by: J. Folk.

* IUB Classification: 3.2.1.1, α-(1 → 4)-Glucan 4-glucanohydrolase, α-amylase. Also, see footnote under α-amylase. This is a proprietary preparation. The enzyme is widely used, but the properties of preparations may vary somewhat from one lot to another.

E-40
Trypsin*
(Bovine Pancreas)

Equation: $R'R''CHCOR''' \rightarrow R'R''COCOO^- + NH_4^+$, ^+H_3NR, or HOR
where
R' = (usually) an acylamido group,
R'' = side chain of L-lysine or L-arginine, and
R''' = —NH$_2$, —NHR, or —OR.
Reaction: Catalyzes the hydrolysis of peptide, amide, or ester bonds at the carboxyl linkage of L-lysine or L-arginine; pH optimum, 7–9.
Method of Isolation: The active enzyme is routinely isolated after tryptic activation of the zymogen, trypsinogen.[1,2]

CRITERIA (Tentative)

Physical Constants: Reported molecular weights:[3–5] 23,000 to 25,000. Certain values of molecular weight reported are low or high, probably due to autolysis or polymerization, respectively.[5]
Procedures for Handling:
Storage: Lyophilized (salt-free or with $MgSO_4$) preparations are stable when stored at 4 °C.
Dilution: In cold 1 mM HCl, 1 to 10 mg/ml.
Dialysis: Against 1 mM HCl at 4 °C.
Commercially Available Substrates: α-N-p-Tolylsulfonyl-L-arginine methyl ester hydrochloride (I), α-N-benzoyl-L-arginine ethyl ester hydrochloride, L-lysine ethyl ester, α-N-benzoyl-L-argininamide, hemaglobin, casein, and many other protein and synthetic substrates are available.
Specific Activity: One unit (U) catalyzes the hydrolysis of 1 μmol I/min at 25 °C and pH 8.1 in the presence† of 10 mM Ca^{2+}. Preparations of highest purity have a specific activity of about 200 U/mg by the following method:
Protein Assay: A_{280} × 0.64 = mg enzyme/ml; solvent, 1 mM HCl (calculated from data of Smillie and Kay[6]).
Activity Assay:
Method: The rate of hydrolysis of I is measured by the increase in absorbance at 274 nm (based on the method of Hummel[7]).

Procedure: The assay is conducted as follows:

Take I, 10 mM in H_2O, 0.3 ml; tris buffer, 80 mM, pH 8.1, containing 20 mM $CaCl_2$, 1.5 ml; and H_2O, 1.2 ml. Add enzyme, 5 to 20 μl containing 0.1 to 0.5 μg of protein, to the substrate mixture in a 1-cm quartz cuvette, and record A_{247} at 30-s intervals.

The hydrolysis of 1 μmol of I causes an increase in A_{247} of 0.18. Rates must be calculated from ΔA values of not over 0.32. Number of units (U) = $[\Delta A_{247}/\Delta t(\text{min})] \times 5.55$.

Impurities: The purest samples have slight activity toward chymotrypsin substrates. This appears to be an inherent property of trypsin. The presence of ribonuclease activity in commercial crystalline trypsin has been reported.[8]

References

1. M. Kunitz and J. H. Northrop, *J. Gen. Physiol.*, **19**, 991 (1936).
2. M. R. McDonald and M. Kunitz, *J. Gen. Physiol.*, **29**, 155 (1946).
3. S. M. Green and H. Neurath, *Proteins*, **2**, 1057 (1954).
4. M. Laskowski, *Methods Enzymol.*, **2**, 26 (1955).
5. P. Desnuelle, *Enzymes*, **4**, 93, 119 (1960).
6. L. B. Smillie and C. M. Kay, *J. Biol. Chem.*, **236**, 112 (1961).
7. B. C. W. Hummel, *Can. J. Biochem. Physiol.*, **37**, 1393 (1959).
8. M. Tunis, *Science*, **162**, 912 (1968).

Acknowledgments

Original compilation of Criteria: J. E. Folk. Contributions by: F. W. Putnam. C. H. W. Hirs, and R. L. Hill.

* IUB Classification: 3.4.16.3, Trypsin.
† Calcium ions enhance the activity and stability pH > 6.

E-41
Urease*
(Jack Bean)

Equation:[1,2] $H_2NCONH_2 + H_2O \rightarrow NH_4^+ + NH_2COO^-$ (at pH 9)

The enzyme is almost completely specific; hydroxy- and dihydroxy-urea are hydrolyzed at a low rate.[3]

Method of Isolation: The enzyme is crystallized from extract of jack beans.[1,4] Further purification by chromatography has been described.[5,6]

CRITERIA (Tentative)

Physical Constant: Reported molecular weight:[1,7] 480,000; this is an aggregate of at least 6 subunits.[8]

Procedure for Handling: Crystals stored in citrate–acetone mother liquor usually lose <5% of their activity in 1 month. A 1% solution in 20 mM phosphate–EDTA (see "Reagents") loses about 10% of its activity in 1 week; more dilute solutions are less stable.[9]

Commercially Available Substrate: Urea.

Specific Activity: One unit of activity (U) hydrolyzes 1 μmol of urea per minute under the conditions described in "Activity Assay." Three independent reports[1,6,9] give the specific activity of carefully purified preparations as 750–850 U/mg.†

Protein Assay:[6,7,9] $A_{278} \times 1.4$ = mg enzyme/ml. Somewhat discrepant values have been reported. Values may be uncertain by as much as ±15%.

Activity Assay:

Method: The enzyme is allowed to react with the substrate for 2 min. An excess of HCl is then added, to stop the reaction and to hydrolyze carbamate (overall stoichiometry: $H_2NCONH_2 + H_2O + 2H^+ \rightarrow 2NH_4^+ + CO_2$). The excess of acid is titrated with NaOH.

Reagents: (Reagents and water used should be of high purity, because the enzyme is very susceptible to inactivation by heavy metals.[10]) Phosphate, 20 mM, pH 7 (1.66 g Na_2HPO_4, 1.15 g $NaH_2PO_4 \cdot H_2O$, and 372 mg EDTA disodium salt dihydrate, per liter); tris, 100 mM, pH 9 [12.1 g tris(hydroxymethyl)aminomethane, 114 ml 100 mM HCl, and 372 mg EDTA disodium salt hydrate, per liter]. Substrate, urea, 500 mM, in tris buffer.

Procedure: Dilute the enzyme sample to 7–20 U/ml, and allow to stand for 2 h, in which time a 10–20% increase in activity may be observed. Mix 1 ml of enzyme solution with 1 ml of substrate. After exactly 2 min, add 2.00 ml of 100 mM HCl. Titrate with 500 mM NaOH, using methyl orange as the indicator. Also, titrate a blank, containing 1 ml phosphate–EDTA, 1 ml substrate, and 2 ml HCl.

Calculations: V_b = volume of NaOH (ml) consumed by blank; V_s = volume (ml) consumed by sample, and c = molar concentration of NaOH (mol/liter).

Activity (U) = $500 c (V_b - V_s)/2 (\text{min}) = 250 c (V_b - V_s)$.

Impurities: A critical study of contaminating impurities has not yet been made. Many commercial preparations suitable for urea determination are crude extracts of jack bean meal and contain only a small percentage of urease.

References

1. J. B. Sumner, *Enzymes*, **1**, 873 (1951).
2. J. E. Varner, *Enzymes*, **4**, 247 (1960).
3. W. N. Fishbein, *J. Biol. Chem.*, **244**, 1188 (1969).
4. G. Mamiya and G. Gorin, *Biochim. Biophys. Acta*, **105**, 382 (1965).
5. K. R. Lynn, *Biochim. Biophys. Acta*, **146**, 205 (1967).
6. R. L. Blakeley, E. C. Webb, and B. Zerner, *Biochemistry*, **8**, 1984 (1969).
7. F. J. Reithel and J. E. Robbins, *Arch. Biochem. Biophys.*, **120**, 158 (1967).
8. F. J. Reithel, J. E. Robbins, and G. Gorin, *Arch. Biochem. Biophys.*, **108**, 409 (1964).
9. G. Gorin and C. C. Chin, *Anal. Biochem.*, **17**, 49 (1966).
10. W. H. R. Shaw, *J. Am. Chem. Soc.*, **76**, 2160 (1954).

Acknowledgments

Original compilation of Criteria: G. Gorin. Contributions by: F. J. Reithel and W. H. Fishbein.

* IUB Classification: 3.5.1.5, Urea amidohydrolase.
† Although considerably higher values were reported by K. R. Lynn,[5] it has been noted that those values were 10 times too high.[6]

E-42
Xanthine Oxidase*
(Cream)

Equations: hypoxanthine + H_2O + $O_2 \rightarrow$ xanthine + H_2O_2

xanthine + H_2O + $O_2 \rightarrow$ urate + H_2O_2

R—CHO + H_2O + $O_2 \rightarrow$ R—COOH + H_2O_2

R stands for any of a broad variety of aromatic and aliphatic groups.

Xanthine oxidase catalyzes the oxidation of numerous purines, aldehydes, and heterocyclic compounds to the corresponding hydroxylated compounds.[1–7]

Methods of Isolation: The enzyme has been isolated from cream, and crystallized.[8–11]

CRITERIA (Tentative)

Physical Constants: The reported molecular weights[9–13] range from 275,000 to 362,000. Per molecule of enzyme, there are 2 molecules of FAD, 2 atoms of molybdenum, 8 atoms of iron, and 8 atoms of acid-labile sulfide.[10]

Procedures for Handling:

Storage: As a suspension in saturated ammonium sulfate at -20 °C. The enzyme is stable to repeated freezing and thawing in 50 mM potassium phosphate containing 100 μM EDTA (pH 7.8). It is also stable to dilution with, or dialysis against, this buffer. Salicylate stabilizes the enzyme,[14] and should be present at a concentration of 1 mM during storage.

Commercially Available Substrates: Purine, hypoxanthine, xanthine, salicylaldehyde, and numerous other aldehydes and heterocyclic compounds.

Specific Activity: One unit (U) represents the amount of activity that converts 1 μmol of xanthine to urate per minute at 25 °C and pH 7.8. A specific activity of 3.5 U/mg has been reported[10] when the enzyme is assayed as described in "Activity Assay." An optimum of pH of 8.3 has been reported by Bray[15] and of 8.9 by Palmer et al.[16] However, as K_m for xanthine varies drastically with pH, whereas V_m varies hardly at all,[7] it is clear that the optimum pH must be a function of the substrate concentration used.

Protein Assay†: $\Delta A_{450} \times 4.35$ (Ref. 8) = mg enzyme/ml (in 50 mM phosphate at pH 7.8); $A_{280}/A_{450} = 5.0$–5.2 (Ref. 13).

Activity Assay: The rate of production of urate from xanthine is followed at 295 nm. The assay is conducted as follows.[6]

Xanthine (1 mM), 0.1 ml

K or Na pyrophosphate (100 mM, pH 8.3), 0.5 ml

EDTA (1 mM), 0.1 ml; H_2O, 0.29 ml.

The final pH should be 8.3. Equilibrate with air, and add 10 μl of enzyme at a concentration yielding ΔA_{295} of 0.02–0.05 per min. The formation of 1 μmol of urate/ml yields an increase in A_{295} of 9.6 (Ref. 13). Number of units (U) in cuvette ($l = 1$ cm, $V = 1$ ml) = $[\Delta A_{295}/\Delta t(\text{min})] \times 0.104$.

References

1. V. H. Booth, *Biochem. J.*, **32**, 503 (1938).
2. F. Bergmann and S. Dikstein, *J. Biol. Chem.*, **223**, 765 (1956).
3. F. Bergmann and H. Ungar, *J. Am. Chem. Soc.*, **82**, 3957 (1960).
4. H. Ungar-Waron and F. Bergmann, *Israel J. Med. Sci.*, **1**, 138 (1965).
5. F. Bergmann and H. Kwietny, *Biochim. Biophys. Acta*, **33**, 29 (1959).
6. P. Feigelson, J. D. Davidson, and R. K. Robbins, *J. Biol. Chem.*, **226**, 993 (1957).
7. L. Greenlee and P. Handler, *J. Biol. Chem.*, **239**, 1090 (1964).
8. P. G. Avis, F. Bergel, and R. C. Bray, *J. Chem. Soc.*, 1100 (1955).
9. C. A. Nelson and P. Handler, *J. Biol. Chem.*, **243**, 5368 (1968).
10. V. Massey, P. E. Brumby, H. Komai, and G. Palmer, *J. Biol. Chem.*, **244**, 1682 (1969).
11. L. I. Hart, M. A. McGartoll, H. R. Chapman, and R. C. Bray, *Biochem. J.*, **116**, 851 (1970).
12. P. Andrews, R. C. Bray, P. Edwards, and K. V. Shooter, *Biochem. J.*, **93**, 627 (1964).
13. P. G. Avis, F. Bergel, and R. C. Bray, *J. Chem. Soc.*, 1219 (1956).
14. F. Bergel and R. C. Bray, *Nature*, **178**, 88 (1956).
15. R. C. Bray, *Enzymes*, **7**, 533 (1963).
16. G. Palmer, R. C. Bray, and H. Beinert, *J. Biol. Chem.*, **239**, 2657 (1964).

Acknowledgments

Original compilation of Criteria: C. Worthington. Contributions by: I. Fridovich, R. C. Bray, and R. Flora. Revision by: I. Fridovich.

* IUB Classification: 1.2.3.2, Xanthine:oxygen oxidoreductase.

† Salicylate, sometimes added as a preservative, interferes with the spectrophotometric determination of protein, and should be removed by dialysis.

Lipids and Related Compounds

GENERAL REMARKS

The lipids listed herein were prepared by the manufacturers either from natural sources or from commercial products of various degrees of purity. They may have undergone a very thorough isolation procedure, or merely a purification. Chemical reactions are involved in many instances. The methods most commonly used for the preparations include extraction, saponification, esterification, interesterification, crystallization, urea-adduct crystallization, fractional and batch distillation, countercurrent distribution, liquid–solid, liquid–liquid, and gas–liquid chromatography, hydrogenation, elaidinization, and bromination. More complex synthetic procedures may also be involved. Methods are described in many versions, and the books by Hilditch and Williams[1] and Markley[2] serve as guides.

The users of lipids listed should be warned that many commercially available samples of such substances are grossly impure. Directly before use, it is advisable to check the purity, particularly of unsaturated compounds, as these are subject to autoxidation if proper precautions have not been taken in shipping and storing. Such samples should be shipped under an inert gas or sealed *in vacuo*. They should be restored to these conditions after each withdrawal of aliquots and should be stored at low temperature. Even with these precautions, samples should be rechecked for purity after repeated withdrawals have been made from them. Any autoxidizable sample should be regarded with suspicion if these precautions have been neglected.

ANALYTICAL PROCEDURES

Fatty Acids and Methyl Esters

Traditionally, the purity of fatty acids and their derivatives has been established by measurement of the acid (or saponification) value, iodine value, melting point, and refractive index. These may be determined by the Official Methods of the American Oil Chemists' Society.[3] However, these methods are not particularly valuable for establishing purity at the levels in which lipids are available today; these methods are too insensitive, subject to confusion by certain mixtures including those of isomers, and require too much sample. Generally, the purity of fatty acids and their derivatives is best established by gas–liquid chromatography (GLC) and thin-layer chromatography (TLC).

Gas–Liquid Chromatography

GLC is best conducted with an instrument having a hydrogen-flame detector. Fatty acids may be directly chromatographed on "free fatty acid phase (FFAP)," a reaction product of Carbowax 20M with 2-nitroterephthalic acid, but, for the other stationary phases suggested, they must be converted into methyl esters: Secure a reagent consisting of 2% by weight of sulfuric acid *or* 12–15% boron trifluoride, in anhydrous methanol. Weigh ~10 mg of fatty acid into a 15-ml glass-stoppered centrifuge tube. Add 1 ml of reagent to the tube. If the sulfuric acid reagent is used, heat the stoppered tube for 1 h at 55 °C. If the boron trifluoride reagent is used, boil the tube contents for 2 min. Short-chain fatty acids may be selectively lost in such a conversion, and, when they are present, the conversion should be conducted in a tightly sealed tube. Long-chain acids (C_{20}–C_{30}) may be difficult to dissolve in the reagents, and benzene or chloroform may be added to provide higher solubility. After the appropriate heating period, cool the reaction mixture, and add 5 ml of water and 1 ml of reagent-grade carbon disulfide, chloroform, or hexane. Stopper the tube and shake vigorously. Centrifuge to separate the two layers, and inject an aliquot of the organic layer into the gas chromatograph.

The homogeneity of the methyl esters should be checked by GLC on at least two columns, one of which has a polar (polyester) and the other a nonpolar (hydrocarbon or silicone) phase. Polar stationary phases recommended are ethylene glycol succinate (EGS), ethylene glycol succinate and methylsilicone copolymer (EGSS-X), and diethylene glycol succinate (DEGS). The packing should consist of 10–15% by weight of these phases on acid-washed Chromosorb W or an equivalent support. Nonpolar phases recommended are Apiezon L (ApL) and GC-grade SE-30 (polydimethylsiloxane). The packing should consist of 3–15% of the latter phases on Chromosorb W(HP) or an equivalent support. For unesterified fatty acids, 10% FFAP on Chromosorb W(AW), or an equivalent support, may be used; this should be considered to be a polar phase. The columns should be 2–3 meters long and may be of metal or glass, except in the case of FFAP which should be in glass. They should have a minimum of 2,000 theoretical plates. The number of theoretical plates may be measured by drawing straight lines from the linear segment of the sides of a peak to the base line and measuring the width, W, at the base line. The distance, D, from the midpoint of the peak to the injection point is also measured. The number of plates, N, is given by:

$$N = 16(D/W)^2$$

The number of plates should be determined by measurements on a symmetrical peak that has a width of at least 1 cm.

Samples to be injected are dissolved in reagent-grade hexane, carbon disulfide, or chloroform. The solvent giving the shortest peak duration and the steadiest baseline for a particular column and instrument should be used. To make sure that the sample is completely and reproducibly injected, use a 10-μl syringe and charge it with approximately 1 μl of pure solvent before the aliquot of sample is drawn into the syringe.

Compounds may be contaminated with solvent, particularly when they are liquid at room temperature; this is checked by injecting ~1 μl of the neat compound before it has undergone any physical or chemical treatment. This test needs to be done on only one sort of column.

Gas flow rates, amounts of stationary phase, and temperatures should be so chosen as to give a convenient retention time for the main peak and to make sure that impurities are not obscured by the solvent peak or by too long a retention time. These variables can all affect the number of theoretical plates. Many polyester phases lose plate efficiency fairly rapidly at temperatures exceeding 200 °C. A retention time of ~5–10 min for the main component is usually optimal and practical.

Determination of the level of impurities requires either attenuation of the recorder signal or use of an internal standard. For the first of these methods, the detector output should be recorded on a strip chart that has a full-scale deflection of at least 9 in. The attenuator is set, and the amount injected is so chosen that the main peak gives a deflection between 50 and 100% of the full scale. The attenuation is decreased as much as possible up to 50-fold before the main peak leaves the baseline and after it returns, the pen being kept on the chart at all times. The attenuator should be calibrated over the range being used.

When an internal standard is to be used, a compound must be selected that has a retention time different from any of those found in the chromatogram of the compound being tested. About 10% (by weight) of the internal standard is added to the sample. The amount of internal standard added must be known with a relative error not exceeding 10%. The attenuator is so adjusted that the internal standard gives a recorder response of 50–100% of full scale. The main peak is allowed to go off scale.

The area under the peaks may be measured by means of a strip-chart integrator or a planimeter. The procedures have a sensitivity $\geq 0.1\%$, depending on the retention time of the impurity and its resolution from the solvent and main component peaks.

The preceding methods establish the homogeneity of sample by GLC. The *identity* of the sample is best established by retention times relative to those of suitable standards. Either of two reference systems may be used for fatty acids and their derivatives: (*a*) the equivalent chain-length (ECL) system,[4] in which a homologous series of fatty acids or their appropriate derivatives is used as standards, or (*b*) the relative retention index (RI) or Kovats' system,[5] in which normal hydrocarbons are used as standards. The ECL is established by determining the retention times of a series of normal saturated fatty acids or their derivatives. The derivative selected should be of the compound being tested; that is, if the sample is a fatty acid, the standards should be fatty acids, and if the sample is a methyl ester, the standards should be methyl esters. The compound to be tested is chromatographed under the same conditions as are applied to the standards. A plot of the logarithms of the retention times (measured from the beginning of the peak for the solvent or air) of the standards vs. their chain lengths gives a straight line. The ECL of the sample is read from the graph by noting the chain length equivalent to its retention time. For normal saturated fatty acids and their derivatives, the ECL, by definition, is the same as the chain length. Branching, or unsaturation, usually changes the ECL to a fractional value. The value for such compounds is a function of the stationary phase, but is relatively independent of other GLC conditions. Mixtures suitable for establishing the ECL standard curve are available commercially. For short-chain fatty acids and their derivatives, convenient standards may be prepared from milk fat. For polyunsaturated acids and their derivatives, cod liver oil is a convenient source of standards.[6]

The RI is calculated from the equation

$$\mathrm{RI} = 200 \frac{\log V_{\mathrm{sample}} - \log V_{\mathrm{C}_z}}{\log V_{\mathrm{C}_{z+2}} - \log V_{\mathrm{C}_z}} + 100\, z,$$

where V is the retention volume (corrected for that of an unabsorbed substance), C_z is a normal hydrocarbon of chain length z, and C_{z+2} is a normal hydrocarbon 2 carbon atoms longer; $V_{\mathrm{C}_z} \le V_{\mathrm{sample}} \le V_{\mathrm{C}_{z+2}}$.

In this system, only the RI of the n-hydrocarbons is set by definition; that is, 600 for hexane or 1,800 for octadecane. The value for all other compounds is a function of the stationary phase but is relatively independent of other GLC conditions. Pure hydrocarbons suitable for establishing standard curves are available commercially.

Thin-Layer Chromatography

Homogeneity established by GLC should be supplemented by TLC studies, as the compounds may be contaminated with nonvolatile materials not detected by GLC. This is particularly true of unsaturated fatty acids and their derivatives, which may undergo oxidative polymerization. One of the frequent contaminants of a free fatty acid is the methyl ester, from which the acids are usually prepared. Contamination by methyl esters can be detected by GLC of the fatty acids on FFAP or by TLC.

To test homogeneity of fatty acids and derivatives by TLC, use plates of TLC-grade silica gel containing 10% calcium sulfate as a binder. The adsorbent should be about 200 μm thick. The chromatographic chamber should be saturated with the solvent vapors by lining it with filter paper. The plates are developed in 80:20:1

(v/v) hexane–ethyl ether–acetic acid. Spots are detected by spraying the plates with a saturated solution of potassium dichromate in 50% aqueous sulfuric acid and then heating them at 225 °C for 1 h. Approximately 0.1 μg of a lipid may be detected by this method; consequently, if 0.1 mg of the sample is applied to the plate, ~0.1% impurity may be detected. The amount of impurity may be determined by densitometry. A control spot containing 0.1 μg of a lipid should be applied to all plates. Saturated fatty acids (and their derivatives) having chain lengths of less than 12 carbon atoms evaporate partially before charring and, therefore, their detection is less sensitive; however, the method is still satisfactory for detecting nonvolatile organic impurities in these materials.

Isomeric Unsaturation

Unsaturated fatty acids and their derivatives may be homogeneous by both GLC and TLC but actually contain positional and geometric isomers. Positional isomers are best detected by ozonolysis,[7] and geometric isomers, by TLC on silica gel plates impregnated with silver nitrate.[8]

For ozonolysis, fatty acids should be converted into their methyl esters. The ozone may be generated by an apparatus similar to that described by Bonner,[9] or by one of those commercially available. A solution of 20 mg of the methyl ester in 1 ml of carbon disulfide is cooled to −70 °C in dry ice–acetone, and ozone is passed through it. The presence of an excess of ozone may be detected by passing the gas leaving the reaction vessel into a solution of 5% potassium iodide in 5% sulfuric acid containing a starch indicator.* The solution turns blue in the presence of ozone, and when ozone is no longer absorbed by the reaction mixture, the indicator will show the presence of ozone in the exit gas. When ozone absorption is complete, an excess of triphenylphosphine (100 mg) is added, and the reaction mixture is allowed to warm up to room temperature. The triphenylphosphine reduces the ozonides to the corresponding aldehydes. An aliquot of the carbon disulfide solution is gas-chromatographed on a column of 5% Carbowax 20M on Chromosorb W or its equivalent. For elution of both the aldehyde and ω-formyl methyl ester fragments, it is necessary to use temperature programming from 50 to 200 °C at ~5−10 °C/min. When the main peaks are not emerging, it is desirable to diminish the attenuation of the recorder to detect minor components. Monounsaturated esters yield only

---

* Sudan Red dissolved directly in the reaction medium has also been recommended as an indicator [W. Stoffel, *J. Am. Oil Chem. Soc.*, **42**, 586 (1965)]; it gives a faster response than KI/starch necessarily contained in a separate connecting vessel.

two peaks, corresponding to a terminal aldehyde and an ω-formyl methyl ester. Polyunsaturated esters that are methylene-interrupted also yield only two peaks, as the malonaldehyde that should be formed from that portion of the chain between two double bonds is not detected. Positional double-bond isomers are detected as additional peaks.

For the detection of geometrical isomers, chromatography on silica-gel plates impregnated with silver ions is used. Fatty acids should first be converted into methyl esters for this analysis. The coating on the plates should contain about 7.5 g of silver nitrate per 30 g of silica gel. The silver nitrate is dissolved in the water used for dispersing the silica gel. The coating should be about 200 μm thick. The plates are developed in hexane–ethyl ether. The ratio of hexane to ether may be selected empirically to give a convenient R_f for the main component; a ratio of 4:1 (v/v) is usually suitable. Spots are best detected by spraying the plates with 50% sulfuric acid, and charring them at 225 °C for 1 h. (Chromic acid cannot be used with silver-ion plates because of the formation of silver chromate.) The minimum amount of fatty acid detectable by this method is ∼5–10 μg; consequently, to detect 1% of an impurity, a 0.5–1.0-mg sample should be applied. The R_f value of the methyl esters tends to depend on the degree of loading, and, in some instances, there may be insufficient resolution between the main component and the expected impurity at these loads. Should this present a problem, the plate may be sprayed with 2,7-dichlorofluorescein instead of 50% sulfuric acid. With fluorescein, ultraviolet light reveals the spots on the plates. Areas where the main component may have tailed into an area in which the presence of an impurity is suspected may be scraped off, and the methyl ester may be eluted with ethyl ether and rechromatographed on a silver-ion plate. The loading should thereby be lessened sufficiently to afford good resolution; however, it is necessary to work rapidly and to minimize exposure to air during these operations; otherwise, artifacts caused by autoxidation will be detected.

Geometric isomers may also be detected by infrared spectroscopy.[10]

The presence of conjugated isomers in methylene-interrupted polyunsaturated fatty acids and their derivatives is a good test of their exposure to oxidation or other abuse. The presence of conjugated diene is detected by ultraviolet absorption according to[3] the Official Method of the American Oil Chemists' Society Ti la-64. A solution is prepared containing about 0.01 mg of sample per milliliter of isooctane, and its absorbance at 233 nm is measured. The proportion of conjugated

diene present is calculated by the following equation:

$$\% \text{ Diene} = \frac{M}{294} 0.84 \left(\frac{A}{l\rho} - k \right),$$

where M is the molecular weight of the sample, A is the absorbance at 233 nm, l is the length of the light path length in cm, ρ is the mass concentration in mg/ml of solution, and k is a constant introduced to correct for background absorbance. The value of k is 0.07 for esters and 0.03 for acids.

Hydroxy Fatty Acids and Alcohols

Long-chain alcohols and most hydroxy fatty acids can be gas chromatographed directly on FFAP columns (see "Fatty Acids and Methyl Esters"). On nonpolar columns, it is necessary to employ a derivative in which the acid and alcohol functions are protected, to prevent tailing. A hydroxy acid may be converted into its methyl ester by the techniques given in "Fatty Acids and Methyl Esters." The alcohol function may be converted into its acetic ester by reaction with acetic anhydride.[3] To acetylate the hydroxyl group boil 10 mg of the sample and 1 ml of acetic anhydride in a 15-ml standard-taper centrifuge tube under reflux for 2 h. Cool the mixture, add 10 ml of water, and boil for 15 min under reflux to decompose any excess of acetic anhydride. Cool, add 1 ml of hexane, mix thoroughly, centrifuge, and remove the water layer. Wash with two 2-ml portions of water. Analyze the derivative by GLC on SE-30 or Apiezon L (see "Fatty Acids and Methyl Esters"). Make sure that any seeming impurity is not acetic anhydride or acetic acid that had not been removed. Trimethylsilyl and dimethylsilyl groups are also frequently used to modify the hydroxyl groups. Reagents for these conversions are available commercially. Alcohols for GLC standards are available commercially, or may be prepared by reduction, with lithium aluminum hydride, of the appropriate fatty acids, ethyl esters, or natural triglyceride (such as milk fat).

Such fatty acids as ricinoleic acid, which contain a hydroxyl group in the β-position with respect to a double bond, are readily dehydrated during GLC, even if the hydroxyl group has been converted to a derivative. To test the purity of ricinoleic acid or methyl ricinoleate, it is best to use TLC on silica-gel plates. The plates should be developed with 7:3 (v/v) hexane–ethyl ether, with detection by spraying with chromic acid followed by charring. A similar system may be used for detecting nonvolatile impurities in alcohols.

Glycerides

The presence of solvent and the fatty-acid composition of simple and mixed glycerides may be ascertained by the methods already listed for fatty acids and methyl esters, but additional methods are required for verifying the positions occupied by the fatty acid residues on the glycerol residue. Positional isomers of partial glycerides may be checked by TLC on silica gel–boric acid,[10] and triglycerides are best tested by degradation with a Grignard reagent.[11,12]

The 1- and 2-monoglycerides and 1,2- and 1,3-diglycerides may be separated by TLC on plates of 19:1 (w/w) silica gel–boric acid. The boric acid is dissolved in the water that is to be used to disperse the silica gel. For monoglycerides, the plates should be developed with 145:50:1:4 (v/v) chloroform–acetone–acetic acid–methanol; for diglycerides, 24:1 (v/v) chloroform or chloroform–acetone may be used. The presence of nonpolar contaminants in triglycerides may be checked[13] by TLC on silica gel plates developed 47:3 (v/v) with hexane–ethyl ether. Spots are best detected by charring with chromic acid, as described for fatty acids and methyl esters. Standards for the TLC of glycerides may be obtained commercially or be prepared by the glycerolysis of corn oil. To prepare partial glyceride standards, heat 10 g of corn oil, 3 g of glycerol, and 65 mg of sodium methoxide under nitrogen at 225 °C for 2 h with stirring. Cool the mixture, and add dilute acetic acid. The upper layer is a glyceride mixture and normally consists of components migrating in the following order of decreasing R_f: triglyceride, 1,3-diglyceride, 1,2-diglyceride, 2-monoglyceride, and 1-monoglyceride.

To detect the presence of positional isomers in triglycerides,[11,12] dissolve 40 mg of the sample in 2 ml of ether. Add 1 ml of freshly prepared, 0.5 M ethylmagnesium bromide with mixing. After exactly 1 min, add 0.05 ml of acetic acid. Wash the ether layer with 2 ml of water, 2 ml of 2% sodium carbonate, and 2 ml of water. Dry with sodium sulfate, and streak the sample on a 19:1 (w/w) silica gel–boric acid plate. Develop the plate with 1:1 (v/v) hexane–ethyl ether. Locate the bands on the plate with 2,7-dichlorofluorescein. Four bands should be visible; from top to bottom, these are fatty alcohols plus unreacted triglyceride, 1,3-diglyceride, 1,2-diglyceride, and monoglycerides. Scrape the 1,2-diglyceride band from the plate, and place it in a medicine dropper whose tip has been plugged with glass wool. Pour 2 ml of methanol through the medicine-dropper microcolumn to elute the diglyceride. Add one drop of concentrated sulfuric acid to the eluate, and heat in a stoppered tube at 55 °C for 1 h to convert the diglyceride into the methyl esters. Find the percentage of fatty acids by GLC on an appropriate column, and compare the results with the specifications of the preparation.

Sterols

Traditionally, the purity of sterols and their esters has been established by determining the melting point and optical rotation, but, as a rule, these methods are too insensitive to detect anything but gross contamination. Spectroscopic methods of all kinds, although giving more information, are likewise too insensitive to reveal impurities, and they require specialized equipment and techniques not usually available in biochemical laboratories. For suitable systems under study, differential scanning calorimetry is capable of detecting 0.1 mol-% of impurity, but it is not available in most laboratories. The purity of sterols and their esters is best established by GLC[14-18] and TLC.[8,14,19]

Gas–Liquid Chromatography

GLC should be conducted on an instrument fitted with a hydrogen-flame detector. GLC of free sterols may be performed directly on the sterol or after the hydroxyl group has been protected by treatment with a suitable reagent. Acetyl and trimethylsilyl groups are usually used to protect the hydroxyl group. Acetates may be prepared by the method already described for hydroxy fatty acids. Reagents for preparing trimethylsilyl ethers may be obtained commercially. Stationary phases recommended for free sterols are 1% cyclohexanedimethanol succinate (Hi-EFF-8B) at 225 °C, 3% OV-17 (phenyl silicone) at 240 °C, and 5% OV-101 (methyl silicone) at 250 °C. The support should be Chromosorb W(HP) or its equivalent, and, for Hi-EFF-8B, the support should first be baked to remove bound water and coated with 2% of poly(vinylpyrrolidinone) before the cyclohexanedimethanol succinate is applied. For trimethylsilyl ethers, 1% of the silicone polymers SE-52 or OV-3 are recommended.

Steryl esters of long-chain fatty acids may be gas chromatographed directly on 60-cm columns packed with 3% of the silicone polymers QF-1 or OV-210 on Chromosorb W(HP) or its equivalent at 290 °C. However, this procedure separates by molecular weight and will not resolve, for example, a mixture of cholesteryl stearate, cholesteryl oleate, and cholesteryl linoleate. Steryl esters of long-chain fatty acids should also be converted into the free sterols and the methyl esters of the fatty acids by the boron trifluoride–methanol procedure (see "Fatty Acids and Methyl Esters"). The methyl esters can then be checked by the methods already given for methyl esters, and the purity of the sterols can be checked by the procedures given in this section.

To test for homogeneity, GLC conditions are varied

and other procedures are chosen in accordance with the instructions given under fatty acids and methyl esters. The identity of the compounds is established by their retention times relative to those of other sterols and sterol derivatives. To establish relative retention times accurately, all of the GLC conditions must be rigorously standardized.

Thin-Layer Chromatography

The homogeneity of sterols and steryl esters should also be checked by TLC, to detect nonvolatile impurities and oxidation products. For free sterols, TLC is conducted on layers of silica gel that are 200-μm thick. The plates should be activated for 1 h at 110 °C. The plates are developed in chloroform or 9:1 (v/v) hexane–ethyl ether, and the spots are detected by spraying with 20% sulfuric acid in methanol and charring at 225 °C for 1 h. As little as 0.3 μg of a sterol can be detected; therefore, with a 50-μg sample, as little as 0.6% of impurity is revealed.

INSTRUCTIONS FOR USE OF THIS SECTION

Because many of the lipids are homologous series or represent other groupings for which the same considerations of purity apply, general criteria and specifications have been written for classes of lipids. These are followed by lists of compounds commercially available that should meet the various specifications. Information unique to each compound is included in the tables.

The following abbreviations not elsewhere defined are used in the tables:

ECL: Equivalent chain length
FFAP: Free fatty acid phase
RI: Relative retention index
ApL: Apiezon L
DEGS: Diethylene glycol succinate
EGS: Ethylene glycol succinate
EGSSX: Ethylene glycol succinate–methyl silicone copolymer
SE-30: A commercial polydimethylsiloxane

GENERAL CRITERIA AND SPECIFICATIONS 1 FOR FATTY ACIDS C_2 THROUGH C_5

CRITERIA:
Homogeneity:
 GLC: One component on FFAP. No detectable solvent.
 TLC: No nonvolatile components detectable on silica gel plates by charring.
Identity:
 Gives the appropriate RI.
 Gives the theoretical value for the neutralization equivalent.

SPECIFICATIONS: Not more than 1% of impurity by GLC. Theoretical neutralization equivalent (within 0.5% relative error).
Storage: No special precautions, except to prevent loss by evaporation.

TABLE 1   Compounds for Which General Criteria and Specifications 1 Are Applicable

Number	Common name (Systematic name)	Formula	Formula wt. (neutralization eq.)	Structure	RI
L–1	Acetic acid (ethanoic acid)	$C_2H_4O_2$	60.05	CH_3COOH	1470
L–2	Propionic acid (propanoic acid)	$C_3H_6O_2$	74.08	CH_3CH_2COOH	1550
L–3	Butyric acid (butanoic acid)	$C_4H_8O_2$	88.11	$CH_3(CH_2)_2COOH$	1640
L–4	Isobutyric acid (2-methylpropanoic acid)	$C_4H_8O_2$	88.11	$(CH_3)_2CHCOOH$	1560
L–5	Crotonic acid (m.p. 72 °C) (*trans*-2-butenoic acid)	$C_4H_6O_2$	86.09	$CH_3C\overset{H}{\underset{H}{=}}CCOOH$	1780
L–6	3-Hydroxybutyric acid (*rac*-3-hydroxybutanoic acid)	$C_4H_8O_3$	104.11	$CH_3CH(OH)CH_2COOH$	2180
L–7	Valeric acid (pentanoic acid)	$C_5H_{10}O_2$	102.13	$CH_3(CH_2)_3COOH$	1750

GENERAL CRITERIA AND SPECIFICATIONS 2 FOR NORMAL SATURATED SHORT-CHAIN FATTY ACIDS C$_6$ THROUGH C$_{13}$

CRITERIA:
Homogeneity:
    GLC: One component on FFAP and, after conversion into methyl esters, on a nonpolar phase. No detectable solvent.
    TLC: No impurities detectable on silica gel plates by charring.
Identity:
    ECL: Same as chain length on any phase.
    TLC: Migrates with fatty acids on silica gel plates.

SPECIFICATIONS: Not more than 1% of impurity.
Storage: No special precautions.

TABLE 2   Compounds for Which General Criteria and Specifications 2 Are Applicable

Number	Common name (Systematic name)	Formula	Formula wt.	Structure	Likely contaminants
L-8	Caproic acid (hexanoic acid)	$C_6H_{12}O_2$	116.16	$CH_3(CH_2)_4COOH$	caprylic acid
L-9	Enanthic acid (heptanoic acid)	$C_7H_{14}O_2$	130.19	$CH_3(CH_2)_5COOH$	caproic and caprylic acids
L-10	Caprylic acid (octanoic acid)	$C_8H_{16}O_2$	144.22	$CH_3(CH_2)_6COOH$	caproic and capric acids
L-11	Pelargonic acid (nonanoic acid)	$C_9H_{18}O_2$	158.24	$CH_3(CH_2)_7COOH$	caprylic and capric acids
L-12	Capric acid (decanoic acid)	$C_{10}H_{20}O_2$	172.27	$CH_3(CH_2)_8COOH$	caprylic and lauric acids
L-13	Hendecanoic acid (undecanoic acid)	$C_{11}H_{22}O_2$	186.30	$CH_3(CH_2)_9COOH$	capric, lauric, and 10-undecenoic acids
L-14	Lauric acid (dodecanoic acid)	$C_{12}H_{24}O_2$	200.32	$CH_3(CH_2)_{10}COOH$	capric and myristic acids
L-15	— (tridecanoic acid)	$C_{13}H_{26}O_2$	214.35	$CH_3(CH_2)_{11}COOH$	undefined

GENERAL CRITERIA AND SPECIFICATIONS 3 FOR SATURATED LONG-CHAIN FATTY ACIDS $\geq C_{14}$

CRITERIA:
Homogeneity:
   GLC: After conversion into the methyl ester, one component on polar and nonpolar phases. FFAP may be used without esterification and substitute for the polar phase. No detectable solvent.
   TLC: One component on silica gel plates.
Identity:
   ECL: The same as chain length for normal-chain acids on any phase. For branched-chain acids, the ECL is specified under each compound.
   TLC: Migrates with fatty acids on silica gel plates.
SPECIFICATIONS: Not more than 1% of impurity.
Storage: No special precautions.

TABLE 3 Compounds for Which General Criteria and Specifications 3 Are Applicable

Number	Common name (Systematic name)	Formula	Formula wt.	Structure	ECL Methyl esters Ref.	Likely contaminants
L–16	Myristic acid (tetradecanoic acid)	$C_{14}H_{28}O_2$	228.38	$CH_3(CH_2)_{12}COOH$	—	lauric and palmitic acids
L–17	Isomyristic acid (12-methyltridecanoic acid)	$C_{14}H_{28}O_2$	228.38	$(CH_3)_2CH(CH_2)_{10}COOH$	12.40 EGS (20) 13.20 SE-30 (20)	undefined
L–18	— (pentadecanoic acid)	$C_{15}H_{30}O_2$	242.40	$CH_3(CH_2)_{13}COOH$	—	homologous acids
L–19	— (12-methyltetradecanoic acid)[a]	$C_{15}H_{30}O_2$	242.40	$CH_3CH_2CH(CH_3)(CH_2)_{10}COOH$	14.50 EGS (20) 14.30 SE-30 (20)	undefined
L–20	Palmitic acid (hexadecanoic acid)	$C_{16}H_{32}O_2$	256.43	$CH_3(CH_2)_{14}COOH$	—	stearic acid
L–21	Isopalmitic acid (14-methylpentadecanoic acid)	$C_{16}H_{32}O_2$	256.43	$(CH_3)_2CH(CH_2)_{12}COOH$	15.40 EGS (20) 15.20 SE-30 (20)	undefined
L–22	Margaric acid (heptadecanoic acid)	$C_{17}H_{34}O_2$	270.46	$CH_3(CH_2)_{15}COOH$	—	homologous acids
L–23	Anteisomargaric acid (14-methylhexadecanoic acid)[a]	$C_{17}H_{34}O_2$	270.46	$CH_3CH_2CH(CH_3)(CH_2)_{12}COOH$	16.55 EGS (20) 16.30 SE-30 (20)	undefined
L–24	Stearic acid (octadecanoic acid)	$C_{18}H_{36}O_2$	284.49	$CH_3(CH_2)_{16}COOH$	—	palmitic and oleic acids
L–25	Isostearic acid (16-methylheptadecanoic acid)	$C_{18}H_{36}O_2$	284.49	$(CH_3)_2CH(CH_2)_{14}COOH$	17.40 EGS (20) 17.20 SE-30 (20)	undefined
L–26	— (nonadecanoic acid)	$C_{18}H_{38}O_2$	298.51	$CH_3(CH_2)_{17}COOH$	—	homologous acids
L–27	Arachidic acid (eicosanoic acid)	$C_{20}H_{40}O_2$	312.54	$CH_3(CH_2)_{18}COOH$	—	stearic and behenic acids
L–28	Phytanic acid (3,7,11,15-tetramethyl-hexadecanoic acid)[a]	$C_{20}H_{40}O_2$	312.54	$(CH_3)_2CH[(CH_2)_3CH(CH_3)]_3CH_2COOH$	17.03 BDS[b] (21)	undefined
L–29	— (heneicosanoic acid)	$C_{21}H_{42}O_2$	326.57	$CH_3(CH_2)_{19}COOH$	—	undefined
L–30	Behenic acid (docosanoic acid)	$C_{22}H_{44}O_2$	340.60	$CH_3(CH_2)_{20}COOH$	—	arachidic acid
L–31	— (tricosanoic acid)	$C_{23}H_{46}O_2$	354.62	$CH_3(CH_2)_{21}COOH$	—	undefined
L–32	Lignoceric acid (tetracosanoic acid)	$C_{24}H_{48}O_2$	368.65	$CH_3(CH_2)_{22}COOH$	—	undefined
L–33	Cerotic acid (hexacosanoic acid)	$C_{26}H_{52}O_2$	396.70	$CH_3(CH_2)_{24}COOH$	—	undefined

[a] Optical isomerism of commercial samples not specified.

[b] Butanediol succinate.

GENERAL CRITERIA AND SPECIFICATIONS 4 FOR UNSATURATED LONG-CHAIN FATTY ACIDS

CRITERIA:

Homogeneity:

GLC: After conversion into the methyl ester, one component on polar and nonpolar phases. No detectable solvent.

TLC: After conversion into the methyl ester, one component on silica gel and silica gel–silver ion plates.

Ozonolysis: After conversion into the methyl ester, yields only an aldehyde of chain length equal to the number of carbon atoms from the methyl end of the chain to the terminal double bond and a methyl ω-formylalkanoate (methyl ester, semialdehyde) of chain length equal to the number of carbon atoms from the carboxyl group of the chain to the first double bond.

Ultraviolet Spectrum: No detectable conjugation of double bonds.

Identity:

ECL: Specified for each compound.

TLC: Migrates with fatty acids on silica gel plates. After conversion into methyl esters, *trans*-isomers migrate faster than the corresponding *cis*-isomers on silica gel–silver ion plates.

SPECIFICATIONS: Not more than 1% of impurity, unless otherwise noted.

Storage: Below 0 °C under vacuum or an inert gas.

TABLE 4 Compounds for Which General Criteria and Specifications 4 Are Applicable

Number	Common name (Systematic name)	Formula	Formula wt.	Structure	ECL Methyl esters	Ref.	Likely contaminants
L-34	Myristoleic acid (*cis*-9-tetradecenoic acid)	$C_{14}H_{26}O_2$	226.36	H H $CH_3(CH_2)_3C{=}C(CH_2)_7COOH$	14.80 DEGS 14.75 EGS 13.38 ApL	(22) (22) (22)	myristic acid
L-35	Palmitoleic acid (*cis*-9-hexadecenoic acid)	$C_{16}H_{30}O_2$	254.42	H H $CH_3(CH_2)_5C{=}C(CH_2)_7COOH$	16.55 DEGS 16.56 EGS 16.61 EGSSX 15.70 ApL	(22) (22) (6) (22)	palmitic acid
L-36	Palmitelaidic acid (*trans*-9-hexadecenoic acid)	$C_{16}H_{30}O_2$	254.42	H $CH_3(CH_2)_5C{=}C(CH_2)_7COOH$ H	16.6 DEGS 16.6 EGS 16.6 EGSSX 15.7 ApL	(22) (22) (6) (22)	palmitic and palmitoleic acids
L-37	Petroselinic acid (*cis*-6-octadecenoic acid)	$C_{18}H_{34}O_2$	282.47	H H $CH_3(CH_2)_{10}C{=}C(CH_2)_4COOH$	18.57 DEGS 18.54 EGS 17.71 ApL	(22) (22) (22)	oleic acid
L-38	Oleic acid (*cis*-9-octadecenoic acid)	$C_{18}H_{34}O_2$	282.47	H H $CH_3(CH_2)_7C{=}C(CH_2)_7COOH$	18.51 DEGS 18.50 EGS 18.33 EGSSX 17.71 ApL	(22) (22) (6) (22)	stearic, palmitic, and linoleic acids
L-39	Elaidic acid (*trans*-9-octadecenoic acid)	$C_{18}H_{34}O_2$	282.47	H $CH_3(CH_2)_7C{=}C(CH_2)_7COOH$ H	18.47 DEGS 18.47 EGS 18.30 EGSSX 17.7 ApL	(22) (22) (6) (22)	oleic, palmitic, and stearic acids
L-40	*cis*-Vaccenic acid (*cis*-11-octadecenoic acid)	$C_{18}H_{34}O_2$	282.47	H H $CH_3(CH_2)_5C{=}C(CH_2)_9COOH$	18.6 DEGS 18.5 EGS 17.8 ApL	(22) (22) (22)	undefined
L-41	*trans*-Vaccenic acid (*trans*-11-octadecenoic acid)	$C_{18}H_{34}O_2$	282.47	H $CH_3(CH_2)_5C{=}C(CH_2)_9COOH$ H	18.58 DEGS 18.53 EGS 17.80 ApL	(22) (22) (22)	undefined
L-42	Linoleic acid (*cis*-9-*cis*-12-octadecadienoic acid)	$C_{18}H_{32}O_2$	280.46	H H $CH_3(CH_2)_4(C{=}CCH_2)_2(CH_2)_6COOH$	19.30 DEGS 19.22 EGS 19.00 EGSSX 17.53 ApL	(22) (22) (6) (22)	oleic, stearic, and palmitic acids; conjugated isomers
L-43	Linoelaidic acid (*trans*-9-*trans*-12-octadecadienoic acid)	$C_{18}H_{32}O_2$	280.46	H $CH_3(CH_2)_4(C{=}CCH_2)_2(CH_2)_6COOH$ H	19.3 DEGS 19.2 EGS 19.0 EGSSX 17.5 ApL	(22) (22) (6) (22)	*cis*-isomers
L-44	Linolenic acid (all-*cis*-9,12,15-octadecatrienoic acid)	$C_{18}H_{30}O_2$	278.44	H H $CH_3CH_2(C{=}CCH_2)_3(CH_2)_6COOH$	20.40 DEGS 20.13 EGS 19.80 EGSSX 17.51 ApL	(22) (22) (6) (22)	*trans*- and conjugated isomers; linoleic acid

TABLE 4 Compounds for Which General Criteria and Specifications 4 Are Applicable (continued)

Number	Common name (Systematic name)	Formula	Formula wt.	Structure	ECL Methyl ester	Ref.	Likely contaminants
L–45	— (cis-5-eicosenoic acid)	$C_{20}H_{38}O_2$	310.52	H H $CH_3(CH_2)_{13}C{=}C(CH_2)_3COOH$	20.4 DEGS 20.4 EGS 20.2 EGSSX 19.8 ApL	(22) (22) (6) (22)	arachidic acid
L–46	Eicosenoic acid (cis-11-eicosenoic acid)	$C_{20}H_{38}O_2$	310.52	H H $CH_3(CH_2)_7C{=}C(CH_2)_9COOH$	20.44 DEGS 20.38 EGS 20.20 EGSSX 19.78 ApL	(22) (22) (6) (22)	arachidic acid
L–47	Arachidonic acid (all-cis-5,8,11,14-eicosatetraenoic acid)	$C_{20}H_{32}O_2$	304.48	H H $CH_3(CH_2)_4(C{=}CCH_2)_4(CH_2)_2COOH$	22.43 DEGS 22.25 EGS 21.70 EGSSX 19.00 ApL	(22) (22) (6) (22)	positional and trans-isomers; trienoic and pentaenoic acids
L–48	Eicosapentaenoic acid (all-cis-5,8,11,14,17-eicosapentaenoic acid)	$C_{20}H_{30}O_2$	302.46	H H $CH_3CH_2(C{=}CCH_2)_5(CH_2)_2COOH$	23.45 DEGS 22.92 EGS 22.30 EGSSX 19.00 ApL	(22) (22) (6) (22)	polyenoic C_{18} and C_{20} acids
L–49	Erucic acid (cis-13-docosenoic acid)	$C_{22}H_{42}O_2$	338.58	H H $CH_3(CH_2)_7C{=}C(CH_2)_{11}COOH$	22.28 DEGS 22.30 EGS 22.00 EGSSX 21.57 ApL	(22) (22) (6) (22)	arachidic and behenic acids
L–50	Docosahexaenoic acid (all-cis-4,7,10,13,16,19-docosahexaenoic acid)	$C_{22}H_{32}O_2$	328.50	H H $CH_3CH_2(C{=}CCH_2)_6CH_2COOH$	26.03 DEGS 25.40 EGS 24.60 EGSSX 20.73 ApL	(22) (22) (6) (22)	C_{22} polyenoic acids
L–51	Nervonic acid (cis-15-tetracosenoic acid)	$C_{24}H_{46}O_2$	366.63	H H $CH_3(CH_2)_7C{=}C(CH_2)_{13}COOH$	24.27 DEGS 24.40 EGS 23.67 ApL	(22) (22) (22)	undefined

GENERAL CRITERIA AND SPECIFICATIONS 5 FOR NORMAL SATURATED SHORT-CHAIN METHYL ESTERS $<C_{14}$

CRITERIA:

Homogeneity:

GLC: One component on polar and nonpolar phases. No detectable solvent.

TLC: No impurities detectable on silica gel plates by charring.

Identity:

ECL: Same as chain length on any phase.

TLC: Migrates with methyl esters on silica gel plates.

SPECIFICATIONS: Not more than 1% of impurity.

Storage: No special precautions.

TABLE 5   Compounds for Which General Criteria and Specifications 5 Are Applicable

Number	Common name (Systematic name)	Formula	Formula wt.	Structure	Likely contaminants
L–52	Methyl butyrate (methyl butanoate)	$C_5H_{10}O_2$	102.13	$CH_3(CH_2)_2COOCH_3$	homologous esters
L–53	Methyl caproate (methyl hexanoate)	$C_7H_{14}O_2$	130.19	$CH_3(CH_2)_4COOCH_3$	methyl caprylate
L–54	Methyl enanthate (methyl heptanoate)	$C_8H_{16}O_2$	144.22	$CH_3(CH_2)_5COOCH_3$	methyl caproate and caprylate
L–55	Methyl caprylate (methyl octanoate)	$C_9H_{18}O_2$	158.24	$CH_3(CH_2)_6COOCH_3$	methyl caproate and caprate
L–56	Methyl pelargonate (methyl nonanoate)	$C_{10}H_{20}O_2$	172.27	$CH_3(CH_2)_7COOCH_3$	methyl caprylate and caprate
L–57	Methyl caprate (methyl decanoate)	$C_{11}H_{22}O_2$	186.30	$CH_3(CH_2)_8COOCH_3$	methyl caprylate and laurate
L–58	Methyl hendecanoate (methyl undecanoate)	$C_{12}H_{24}O_2$	200.32	$CH_3(CH_2)_9COOCH_3$	methyl caprate, laurate, and 10-undecenoate
L–59	Methyl laurate (methyl dodecanoate)	$C_{13}H_{26}O_2$	214.35	$CH_3(CH_2)_{10}COOCH_3$	methyl caprate and myristate
L–60	— (methyl tridecanoate)	$C_{14}H_{28}O_2$	228.38	$CH_3(CH_2)_{11}COOCH_3$	undefined

GENERAL CRITERIA AND SPECIFICATIONS 6 FOR SATURATED LONG-CHAIN METHYL ESTERS $\geq C_{14}$

CRITERIA:
Homogeneity:
  GLC: One component on polar and nonpolar phases. No detectable solvent.
  TLC: One component on silica gel plates.
Identity:
  ECL: The same as chain length for normal-chain esters on any phase. For branched-chain esters, the ECL is specified under each compound.
  TLC: Migrates with methyl esters on silica gel plates.

SPECIFICATIONS: Not more than 1% of impurity.
Storage: No special precautions.

TABLE 6 Compounds for Which General Criteria and Specifications 6 Are Applicable

Number	Common name (Systematic name)	Formula	Formula wt.	Structure	ECL	Ref.	Likely contaminants
L-61	Methyl myristate (methyl tetradecanoate)	$C_{15}H_{30}O_2$	242.40	$CH_3(CH_2)_{12}COOCH_3$	—		methyl laurate and palmitate
L-62	Methyl isomyristate (methyl 12-methyltridecanoate)	$C_{15}H_{30}O_2$	242.40	$(CH_3)_2CH(CH_2)_{10}COOCH_3$	12.40 EGS 13.20 SE-30	(20) (20)	undefined
L-63	— (methyl pentadecanoate)	$C_{16}H_{32}O_2$	256.43	$CH_3(CH_2)_{13}COOCH_3$	—		homologous methyl esters
L-64	— (methyl 12-methyltetradecanoate)	$C_{16}H_{32}O_2$	256.43	$CH_3CH_2CH(CH_3)(CH_2)_{10}COOCH_3$	14.50 EGS 14.30 SE-30	(20) (20)	undefined
L-65	Methyl palmitate (methyl hexadecanoate)	$C_{17}H_{34}O_2$	270.46	$CH_3(CH_2)_{14}COOCH_3$	—		stearic acid
L-66	Methyl isopalmitate (methyl 14-methylpentadecanoate)	$C_{17}H_{34}O_2$	270.46	$(CH_3)_2CH(CH_2)_{12}COOCH_3$	15.40 EGS 15.20 SE-30	(20) (20)	undefined
L-67	Methyl margarate (methyl heptadecanoate)	$C_{18}H_{36}O_2$	284.49	$CH_3(CH_2)_{15}COOCH_3$	—		homologous methyl esters
L-68	Methyl anteisomargarate (methyl 14-methylhexadecanoate)	$C_{18}H_{36}O_2$	284.49	$CH_3CH_2CH(CH_3)(CH_2)_{12}COOCH_3$	16.55 EGS 16.30 SE-30	(20) (20)	undefined
L-69	Methyl stearate (methyl octanoate)	$C_{19}H_{38}O_2$	298.51	$CH_3(CH_2)_{16}COOCH_3$	—		methyl palmitate and oleate
L-70	Methyl isostearate (methyl 16-methylheptadecanoate)	$C_{19}H_{38}O_2$	298.51	$(CH_3)_2CH(CH_2)_{14}COOCH_3$	17.40 EGS 17.20 SE-30	(20) (20)	undefined
L-71	— (methyl nonadecanoate)	$C_{20}H_{40}O_2$	312.54	$CH_3(CH_2)_{17}COOCH_3$	—		homologous methyl esters
L-72	Methyl arachidate (methyl eicosanoate)	$C_{21}H_{42}O_2$	326.57	$CH_3(CH_2)_{18}COOCH_3$	—		methyl stearate and behenate
L-73	— (methyl heneicosanoate)	$C_{22}H_{44}O_2$	340.60	$CH_3(CH_2)_{19}COOCH_3$	—		undefined
L-74	Methyl behenate (methyl docosanoate)	$C_{23}H_{46}O_2$	354.62	$CH_3(CH_2)_{20}COOCH_3$	—		methyl arachidate
L-75	— (methyl tricosanoate)	$C_{24}H_{48}O_2$	368.65	$CH_3(CH_2)_{21}COOCH_3$	—		undefined
L-76	Methyl lignocerate (methyl tetracosanoate)	$C_{25}H_{50}O_2$	382.68	$CH_3(CH_2)_{22}COOCH_3$	—		undefined
L-77	— (methyl pentacosanoate)	$C_{26}H_{52}O_2$	396.70	$CH_3(CH_2)_{23}COOCH_3$	—		undefined
L-78	Methyl cerotate (methyl hexacosanoate)	$C_{27}H_{54}O_2$	410.73	$CH_3(CH_2)_{24}COOCH_3$	—		undefined

GENERAL CRITERIA AND SPECIFICATIONS 7 FOR UNSATURATED LONG-CHAIN METHYL ESTERS

CRITERIA:

Homogeneity:

GLC: One component on polar and nonpolar phases. No detectable solvent.

TLC: One component on silica gel and silica gel–silver ion plates.

Ozonolysis: Yields only an aldehyde of chain length equal to the number of carbon atoms from the methyl end of the chain to the terminal double bond and a methyl ω-formylalkanoate (methyl ester, semialdehyde) of chain length equal to the number of carbon atoms from the carboxyl end of the chain to the first double bond.

Ultraviolet Spectrum: No detectable conjugated ester.

Identity:

ECL: Specified for each compound.

TLC: Migrates with methyl esters on silica gel plates.

SPECIFICATIONS: Not more than 1% of impurity, unless otherwise noted.

Storage: Below 0 °C under vacuum or an inert gas.

TABLE 7 Compounds for Which General Criteria and Specifications 7 Are Applicable

Number	Common name (Systematic name)	Formula	Formula wt.	Structure	ECL	Ref.	Likely contaminants
L–79	Methyl myristoleate (methyl cis-9-tetradecenoate)	$C_{15}H_{28}O_2$	240.39	H H (above) $CH_3(CH_2)_3C{=}C(CH_2)_7COOCH_3$	14.80 DEGS 14.75 EGS 13.38 ApL	(22) (22) (22)	methyl myristate
L–80	Methyl palmitoleate (methyl cis-9-hexadecenoate)	$C_{17}H_{32}O_2$	268.45	H H (above) $CH_3(CH_2)_5C{=}C(CH_2)_7COOCH_3$	16.55 DEGS 16.55 EGS 16.61 EGSSX 15.70 ApL	(22) (22) (6) (22)	methyl palmitate
L–81	Methyl palmitelaidate (methyl trans-9-hexadecenoate)	$C_{17}H_{32}O_2$	268.45	H (above) $CH_3(CH_2)_5C{=}C(CH_2)_7COOCH_3$ H (below)	16.6 DEGS 16.6 EGS 16.6 EGSSX 15.7 ApL	(22) (22) (6) (22)	methyl palmitate and and palmitoleate
L–82	Methyl petroselinate (methyl cis-6-octadecenoate)	$C_{19}H_{36}O_2$	296.50	H H (above) $CH_3(CH_2)_{10}C{=}C(CH_2)_4COOCH_3$	18.57 DEGS 18.54 EGS 17.17 ApL	(22) (22) (22)	methyl oleate
L–83	Methyl oleate (methyl cis-9-octadecenoate)	$C_{19}H_{36}O_2$	296.50	H H (above) $CH_3(CH_2)_7C{=}C(CH_2)_7COOCH_3$	18.51 DEGS 18.50 EGS 18.33 EGSSX 17.71 ApL	(22) (22) (6) (22)	methyl stearate, palmitate, and linoleate
L–84	Methyl elaidate (methyl trans-9-octadecenoate)	$C_{19}H_{36}O_2$	296.50	H (above) $CH_3(CH_2)_7C{=}C(CH_2)_7COOCH_3$ H (below)	18.47 DEGS 18.47 EGS 18.30 EGSSX 17.7 ApL	(22) (22) (6) (22)	methyl oleate, palmitate, and stearate
L–85	Methyl cis-vaccenate (methyl cis-11-octadecenoate)	$C_{19}H_{36}O_2$	296.50	H H (above) $CH_3(CH_2)_5C{=}C(CH_2)_9COOCH_3$	18.6 DEGS 18.5 EGS 17.8 ApL	(22) (22) (22)	undefined
L–86	Methyl trans-vaccenate (methyl trans-11-octadecenoate)	$C_{19}H_{36}O_2$	296.50	H (above) $CH_3(CH_2)_5C{=}C(CH_2)_9COOCH_3$ H (below)	18.58 DEGS 18.53 EGS 17.80 ApL	(22) (22) (22)	undefined
L–87	Methyl linoleate (methyl cis-9-cis-12-octadecadienoate)	$C_{19}H_{34}O_2$	294.48	H H (above) $CH_3(CH_2)_4(C{=}CCH_2)_2(CH_2)_6COOCH_3$	19.30 DEGS 19.22 EGS 19.00 EGSSX 17.53 ApL	(22) (22) (6) (22)	methyl oleate, stearate, and palmitate, and conjugated isomers
L–88	Methyl linoelaidate (methyl trans-9-trans-12-octadecadienoate)	$C_{19}H_{34}O_2$	294.48	H (above) $CH_3(CH_2)_4(C{=}CCH_2)_2(CH_2)_6COOCH_3$ H (below)	19.3 DEGS 19.2 EGS 19.0 EGSSX 17.5 ApL	(22) (22) (6) (22)	cis-isomers

TABLE 7 Compounds for Which General Criteria and Specifications 7 Are Applicable (continued)

Number	Common name (Systematic name)	Formula	Formula wt.	Structure	ECL		Ref.	Likely contaminants
L-89	Methyl linolenate (methyl all-*cis*-9,12,15-octadecatrienoate)	$C_{19}H_{32}O_2$	292.47	H H $CH_3CH_2(C{=}CCH_2)_3(CH_2)_6COOCH_3$	20.40	DEGS	(22)	*trans*- and conjugated isomers, and methyl linoleate
					20.13	EGS	(22)	
					19.80	EGSSX	(6)	
					17.51	ApL	(22)	
L-90	— (methyl *cis*-5-eicosenoate)	$C_{21}H_{40}O_2$	324.55	H H $CH_3(CH_2)_{13}C{=}C(CH_2)_3COOCH_3$	20.4	DEGS	(22)	methyl arachidate
					20.4	EGS	(22)	
					20.2	EGSSX	(6)	
					19.8	ApL	(22)	
L-91	Methyl eicosenoate (methyl *cis*-11-eicosenoate)	$C_{21}H_{40}O_2$	324.55	H H $CH_3(CH_2)_7C{=}C(CH_2)_9COOCH_3$	20.44	DEGS	(22)	methyl arachidate
					20.38	EGS	(22)	
					20.20	EGSSX	(6)	
					19.78	ApL	(22)	
L-92	— (methyl *cis*-11-*cis*-14-eicosadienoate)	$C_{21}H_{38}O_2$	322.54	H H $CH_3(CH_2)_4(C{=}CCH_2)_2(CH_2)_8COOCH_3$	21.36	DEGS	(22)	undefined
					21.13	EGS	(22)	
					20.83	EGSSX	(6)	
					19.48	ApL	(22)	
L-93	Methyl arachidonate (methyl all-*cis*-5,8,11,14-eicosatetraenoate)	$C_{21}H_{34}O_2$	318.50	H H $CH_3(CH_2)_4(C{=}CCH_2)_4(CH_2)_2COOCH_3$	22.43	DEGS	(22)	positional and *trans*-isomers; methyl trienoate and pentaenoate
					22.25	EGS	(22)	
					21.20	EGSSX	(6)	
					19.00	ApL	(22)	
L-94	Methyl eicosapentaenoate (methyl all-*cis*-5,8,11,14,17-eicosapentaenoate)	$C_{21}H_{32}O_2$	316.48	H H $CH_3CH_2(C{=}CCH_2)_5(CH_2)_2COOCH_3$	23.45	DEGS	(22)	C_{18} and C_{20} polyenoates
					22.92	EGS	(22)	
					22.30	EGSSX	(6)	
					19.00	ApL	(22)	
L-95	Methyl erucate (methyl *cis*-13-docosenoate)	$C_{23}H_{44}O_2$	352.60	H H $CH_3(CH_2)_7C{=}C(CH_2)_{11}COOCH_3$	22.28	DEGS	(22)	methyl arachidate and behenate
					22.30	EGS	(22)	
					22.00	EGSSX	(6)	
					21.57	ApL	(22)	
L-96	Methyl docosahexaenoate (methyl all-*cis*-4,7,10,13,16,19-docosahexaenoate)	$C_{23}H_{34}O_2$	342.56	H H $CH_3(CH_2)(C{=}CCH_2)_6CH_2COOCH_3$	26.03	DEGS	(22)	C_{22} polyenoates
					25.40	EGS	(22)	
					24.60	EGSSX	(6)	
					20.73	ApL	(22)	
L-97	Methyl nervonate (methyl *cis*-15-tetracosenoate)	$C_{25}H_{48}O_2$	380.66	H H $CH_3(CH_2)_7C{=}C(CH_2)_{13}COOCH_3$	24.27	DEGS	(22)	undefined
					24.40	EGS	(22)	
					23.67	ApL	(22)	

GENERAL CRITERIA AND SPECIFICATIONS 8 FOR RICINOLEIC ACID AND RELATED COMPOUNDS

CRITERIA:

Homogeneity:

GLC: No detectable solvent.

TLC: One component on silica gel plates. One component by chromatography of the methyl ester on silica gel–silver ion plates.

Identity:

TLC: The acid migrates more slowly than nonhydroxy fatty acids, and the methyl ester migrates more slowly than nonhydroxy methyl esters on silica gel plates. *trans*-Isomer migrates faster than the *cis* on silica gel–silver ion plates.

Ozonolysis: Methyl ester yields methyl 9-formylnonanoate ("methyl ester of azelaic semialdehyde").

SPECIFICATIONS: Not more than 1% of impurity.

Storage: Under vacuum or an inert gas.

Likely Contaminants: For acids: dihydroxy, oleic, stearic, and palmitic acids; for methyl esters: the corresponding methyl esters.

TABLE 8 Compounds for Which General Criteria and Specifications 8 Are Applicable

Number	Common name (Systematic name)	Formula	Formula wt.	Structure
L–98	Ricinoleic acid [(+)12-hydroxy-*cis*-9-octadecenoic acid]	$C_{18}H_{34}O_3$	298.47	$CH_3(CH_2)_5CHCH_2C{=}C(CH_2)_7COOH$ with H H above, OH below
L–99	Methyl ricinoleate [methyl (+)12-hydroxy-*cis*-9-octadecenoate]	$C_{19}H_{36}O_3$	312.50	$CH_3(CH_2)_5CHCH_2C{=}C(CH_2)_7COOCH_3$ with H H above, OH below
L–100	Methyl ricinelaidate [methyl (+)12-hydroxy-*trans*-9-octadecenoate]	$C_{19}H_{36}O_3$	312.50	$CH_3(CH_2)_5CHCH_2C{=}C(CH_2)_7COOCH_3$ with H above, H below, OH below

GENERAL CRITERIA AND SPECIFICATIONS 9 FOR NORMAL SATURATED SHORT-CHAIN ALCOHOLS $< C_{14}$

CRITERIA:
Homogeneity:
   GLC: One component on FFAP and, after acetylation, on a nonpolar phase. No detectable solvent.
   TLC: No nonvolatile impurities detectable on silica gel plates by charring.
Identity:
   ECL: Same as chain length on any phase.
   TLC: Migrates with alcohols on silica gel plates.

SPECIFICATIONS: Not more than 1% of impurity.
Storage: No special precautions.

TABLE 9   Compounds for Which General Criteria and Specifications 9 Are Applicable

Number	Common name (Systematic name)	Formula	Formula wt.	Structure	Likely contaminants
L–101	Caproyl alcohol (1-hexanol)	$C_6H_{14}O$	102.18	$CH_3(CH_2)_4CH_2OH$	capryl alcohol
L–102	Capryl alcohol (1-octanol)	$C_8H_{18}O$	130.23	$CH_3(CH_2)_6CH_2OH$	caproyl and decyl alcohols
L–103	Decyl alcohol (1-decanol)	$C_{10}H_{22}O$	158.29	$CH_3(CH_2)_8CH_2OH$	capryl and lauryl alcohols
L–104	Lauryl alcohol (1-dodecanol)	$C_{12}H_{26}O$	186.34	$CH_3(CH_2)_{10}CH_2OH$	decyl and myristyl alcohols

GENERAL CRITERIA AND SPECIFICATIONS 10 FOR NORMAL SATURATED LONG-CHAIN ALCOHOLS $\geq C_{14}$

CRITERIA:
Homogeneity:
  GLC: One component on FFAP and, after acetylation, on a nonpolar phase. No detectable solvent.
  TLC: One component on silica gel plates.
Identity:
  ECL: The same as chain length on any phase.
  TLC: Migrates with alcohols on silica gel plates.

SPECIFICATIONS: Not more than 1 % of impurity.
Storage: No special precautions.

TABLE 10   Compounds for Which General Criteria and Specifications 10 Are Applicable

Number	Common name (Systematic name)	Formula	Formula wt.	Structure	RI on FFAP	Likely contaminants
L–105	Myristyl alcohol (1-tetradecanol)	$C_{14}H_{30}O$	214.40	$CH_3(CH_2)_{12}CH_2OH$	2180	lauryl and cetyl alcohols
L–106	Cetyl alcohol (1-hexadecanol)	$C_{16}H_{34}O$	242.45	$CH_3(CH_2)_{14}CH_2OH$	2387	myristyl and stearyl alcohols
L–107	Stearyl alcohol (1-octadecanol)	$C_{18}H_{38}O$	270.50	$CH_3(CH_2)_{16}CH_2OH$	2594	cetyl and oleyl alcohols
L–108	Arachidyl alcohol (1-eicosanol)	$C_{20}H_{42}O$	298.56	$CH_3(CH_2)_{18}CH_2OH$	—	undefined
L–109	Behenyl alcohol (1-docosanol)	$C_{22}H_{46}O$	326.61	$CH_3(CH_2)_{20}CH_2OH$	—	undefined
L–110	Lignoceryl alcohol (1-tetracosanol)	$C_{24}H_{50}O$	354.67	$CH_3(CH_2)_{22}CH_2OH$	—	undefined

GENERAL CRITERIA AND SPECIFICATIONS 11 FOR LONG-CHAIN UNSATURATED ALCOHOLS

CRITERIA:

Homogeneity:

GLC: One component on FFAP and, after acetylation, on a nonpolar phase. No detectable solvent.

TLC: One component on silica gel and silica gel–silver ion plates.

Ozonolysis: After conversion into the acetate ester, yields only an aldehyde of chain length equal to the number of carbon atoms from the methyl end of the chain to the terminal double bond and an acetate ester of an ω-formylalkanol of chain length equal to the number of carbon atoms from the alcohol end of the chain to the first double bond.

Ultraviolet Spectrum: No detectable conjugated double bond.

Identity:

ECL: Specified for each compound.

TLC: Migrates with alcohols on silica gel plates.

SPECIFICATIONS: Not more than 1% of impurity.

Storage: Below 0 °C under vacuum or an inert gas.

TABLE 11   Compounds for Which General Criteria and Specifications 11 Are Applicable

Number	Common name (Systematic name)	Formula	Formula wt.	Structure	RI on FFAP	ECL on FFAP	Likely contaminants
L-111	Oleyl alcohol (*cis*-9-octadecen-1-ol)	$C_{18}H_{36}O$	268.49	$\overset{H\ \ \ H}{CH_3(CH_2)_7C=C(CH_2)_7CH_2OH}$	2624	18.26	cetyl and stearyl alcohols
L-112	Elaidyl alcohol (*trans*-9-octadecen-1-ol)	$C_{18}H_{36}O$	268.49	$\overset{H}{\underset{H}{CH_3(CH_2)_7C=C(CH_2)_7CH_2OH}}$	2620	18.2	cetyl, stearyl, and olelyl alcohols
L-113	Linolyl alcohol (*cis*-9,*cis*-12-octadecadien-1-ol)	$C_{18}H_{34}O$	266.47	$\overset{H\ \ \ H}{CH_3(CH_2)_4(C=CCH_2)_2(CH_2)_6CH_2OH}$	2671	18.68	stearyl, oleyl, cetyl, and conjugated alcohols
L-114	Linolenyl alcohol (all-*cis*-9,12,15-octadecatrien-1-ol)	$C_{18}H_{32}O$	264.46	$\overset{H\ \ \ H}{CH_3CH_2(C=CCH_2)_3(CH_2)_6CH_2OH}$	2732	19.35	conjugated and *trans*-isomers

GENERAL CRITERIA AND SPECIFICATIONS 12 FOR MONOGLYCERIDES

CRITERIA:

Homogeneity:

GLC: No detectable solvent.

TLC: One component on silica gel–boric acid plates.

Identity: Conversion into methyl ester yields a compound meeting the criteria for the appropriate methyl ester.

TLC: Migration with the appropriate monoglyceride on silica gel–boric acid plates.

SPECIFICATIONS: Not more than 1% of other than the appropriate acyl group. At least 99% of the appropriate monoglyceride isomer.

Storage: For unsaturated monoglycerides: below 0 °C, under vacuum or an inert gas; for saturated monoglycerides: no special precautions.

Likely Contaminants: Improper monoglyceride isomer; di- and triglycerides; the contaminants listed under the methylester of the constituent acyl group.

TABLE 12   Compounds for Which General Criteria and Specifications 12 Are Applicable

Number	Common name (Systematic name)	Formula	Formula wt.	Structure
L–115	1-Monomyristin (*rac*-1-tetradecanoylglycerol)	$C_{17}H_{34}O_4$	302.46	$CH_2OOC(CH_2)_{12}CH$ | $CHOH$ | CH_2OH
L–116	1-Monopalmitin (*rac*-1-hexadecanoylglycerol)	$C_{19}H_{38}O_4$	330.51	$CH_2OOC(CH_2)_{14}CH_3$ | $CHOH$ | CH_2OH
L–117	2-Monopalmitin (2-hexadecanoylglycerol)	$C_{19}H_{38}O_4$	330.51	CH_2OH | $CHOOC(CH_2)_{14}CH_3$ | CH_2OH
L–118	1-Monostearin (*rac*-1-octadecanoylglycerol)	$C_{21}H_{42}O_4$	358.56	$CH_2OOC(CH_2)_{16}CH_3$ | $CHOH$ | CH_2OH
L–119	1-Monoolein [*rac*-1(*cis*-9-octadecenoyl)glycerol]	$C_{21}H_{40}O_4$	356.55	$CH_2OOC(CH_2)_7\overset{H}{C}{=}\overset{H}{C}(CH_2)_7CH_3$ | $CHOH$ | CH_2OH
L–120	2-Monoolein [2-(*cis*-9-octadecanoyl)glycerol]	$C_{21}H_{40}O_4$	356.55	CH_2OH | $CHOOC(CH_2)_7\overset{H}{C}{=}\overset{H}{C}(CH_2)_7CH_3$ | CH_2OH
L–121	1-Monolinolein [*rac*-1-(*cis*-9,*cis*-12-octadecadienoyl)glycerol]	$C_{21}H_{38}O_4$	354.53	$CH_2OOC(CH_2)_7(\overset{H}{C}{=}\overset{H}{C}CH_2)_2(CH_2)_3CH_3$ | $CHOH$ | CH_2OH

GENERAL CRITERIA AND SPECIFICATIONS 13 FOR DIGLYCERIDES

CRITERIA:
Homogeneity:
GLC: No detectable solvent.
TLC: One component on silica gel–boric acid plates.
Identity: Conversion into methyl esters yields compound(s) meeting the criteria for the appropriate methyl ester(s). If the sample is a mixed diglyceride, GLC should show the methyl esters in 1:1 molar ratio.
TLC: Migration with the appropriate diglyceride on silica gel–boric acid plates.

SPECIFICATIONS: Not more than 1% of other than the appropriate acyl group(s). At least 99% of the appropriate diglyceride isomer. If the sample is a mixed diglyceride, the molar ratio of the two acyl groups should be 1:1 within 10% relative error.
Storage: For unsaturated diglycerides: below 0 °C under vacuum or an inert gas; for saturated diglycerides: no special precautions.
Likely Contaminants: Improper diglyceride isomers, mono- and triglycerides; the contaminants listed under the methyl esters of the constituent acyl group(s).

TABLE 13   Compounds for Which General Criteria and Specifications 13 Are Applicable

Number	Common name (Systematic name)	Formula	Formula wt.	Structure
L–122	1,2-Dimyristin (rac-1,2-ditetradecanoylglycerol)	$C_{31}H_{60}O_5$	512.72	$CH_2OOC(CH_2)_{12}CH_3$ $\mid$ $CHOOC(CH_2)_{12}CH_3$ $\mid$ CH_2OH
L–123	1,3-Dimyristin (1,3-ditetradecanoylglycerol)	$C_{31}H_{60}O_5$	512.72	$CH_2OOC(CH_2)_{12}CH_3$ $\mid$ $CHOH$ $\mid$ $CH_2OOC(CH_2)_{12}CH_3$
L–124	1,2-Dipalmitin (rac-1,2-dihexadecanoylglycerol)	$C_{35}H_{68}O_5$	568.93	$CH_2OOC(CH_2)_{14}CH_3$ $\mid$ $CHOOC(CH_2)_{14}CH_3$ $\mid$ CH_2OH
L–125	1,3-Dipalmitin (1,3-dihexadecanoylglycerol)	$C_{35}H_{68}O_5$	568.93	$CH_2OOC(CH_2)_{14}CH_3$ $\mid$ $CHOH$ $\mid$ $CH_2OOC(CH_2)_{14}CH_3$
L–126	1,2-Distearin (rac-1,2-dioctadecanoylglycerol)	$C_{39}H_{76}O_5$	625.04	$CH_2OOC(CH_2)_{16}CH_3$ $\mid$ $CHOOC(CH_2)_{16}CH_3$ $\mid$ CH_2OH
L–127	1,3-Distearin (1,3-dioctadecanoylglycerol)	$C_{39}H_{76}O_5$	625.04	$CH_2OOC(CH_2)_{16}CH_3$ $\mid$ $CHOH$ $\mid$ $CH_2OOC(CH_2)_{16}CH_3$
L–128	1,2-Diolein [rac-1,2-di(cis-9-octadecenoyl)glycerol]	$C_{39}H_{72}O_5$	621.00	$\overset{H\ \ \ H}{CH_2OOC(CH_2)_7C=C(CH_2)_7CH_3}$ $\mid\ \ \ \ \ \ \ \ \ \ \ \ \ \ H\ \ \ H$ $CHOOC(CH_2)_7C=C(CH_2)_7CH_3$ $\mid$ CH_2OH

139

TABLE 13 Compounds for Which General Criteria and Specifications 13 Are Applicable (Continued)

Number	Common name (Systematic name)	Formula	Formula wt.	Structure
L–129	1,3-Diolein [1,3-di(*cis*-9-octadecenoyl)glycerol]	$C_{39}H_{72}O_5$	621.00	$CH_2OOC(CH_2)_7C{=}C(CH_2)_7CH_3$ (H H) $\vert$ $CHOH$ $\vert$ $CH_2OOC(CH_2)_7C{=}C(CH_2)_7CH_3$ (H H)
L–130	1,3-Dilinolein [1,3-di(*cis*-9,*cis*-12-octadecadienoyl)glycerol]	$C_{39}H_{68}O_5$	616.98	$CH_2OOC(CH_2)_7(C{=}CCH_2)_2(CH_2)_3CH_3$ (H H) $\vert$ $CHOH$ $\vert$ $CH_2OOC(CH_2)_7(C{=}CCH_2)_2(CH_2)_3CH_3$ (H H)

GENERAL CRITERIA AND SPECIFICATIONS 14 FOR TRIGLYCERIDES

CRITERIA:
Homogeneity:
  GLC: No detectable solvent.
  TLC: One component on silica gel and silica gel–boric acid plates.
Identity: Conversion into methyl esters yields compound(s) meeting the criteria for the appropriate methyl ester(s). If the sample is a mixed triglyceride, GLC should show the methyl esters in the appropriate molar ratio.
  Grignard Degradation: If the sample is a mixed triglyceride, this degradation yields 1,2-diglycerides that give methyl esters; GLC should show these in the appropriate molar ratio.
  TLC: Migration with triglyceride on silica gel–boric acid plates.

SPECIFICATIONS: Not more than 1% of other than the appropriate acyl group(s). At least 99% of triglyceride. If the sample is a mixed triglyceride, the molar ratio of the acyl groups should be the appropriate value within 12% relativeer ror, and the ratio of the methyl esters from the Grignard degradation should be the appropriate value within 15% relative error.
Storage: For unsaturated triglycerides: below 0 °C under vacuum or an inert gas; for saturated triglycerides: no special precautions.
Likely Contaminants: The contaminants listed under the methyl ester(s) of constituent acyl groups; for short-chain simple triglycerides: mono- and diglycerides; for long-chain simple triglycerides: O-acetylglycerides; for mixed triglycerides: improper isomers.

TABLE 14   Compounds for Which General Criteria and Specifications 14 Are Applicable

Number	Common name (Systematic name)	Formula	Formula wt.	Structure
L–131	Triacetin (triethanoylglycerol)	$C_9H_{14}O_6$	218.20	CH_2OOCCH_3 | $CHOOCCH_3$ | CH_2OOCCH_3
L–132	Tributyrin (tributanoylglycerol)	$C_{15}H_{26}O_6$	302.37	$CH_2OOC(CH_2)_2CH_3$ | $CHOOC(CH_2)_2CH_3$ | $CH_2OOC(CH_2)_2CH_3$
L–133	Tricaproin (trihexanoylglycerol)	$C_{21}H_{38}O_6$	386.53	$CH_2OOC(CH_2)_4CH_3$ | $CHOOC(CH_2)_4CH_3$ | $CH_2OOC(CH_2)_4CH_3$
L–134	Tricaprylin (trioctanoylglycerol)	$C_{27}H_{50}O_6$	470.70	$CH_2OOC(CH_2)_6CH_3$ | $CHOOC(CH_2)_6CH_3$ | $CH_2OOC(CH_2)_6CH_3$
L–135	Tricaprin (tridecanoylglycerol)	$C_{33}H_{62}O_6$	554.86	$CH_2OOC(CH_2)_8CH_3$ | $CHOOC(CH_2)_8CH_3$ | $CH_2OOC(CH_2)_8CH_3$
L–136	Trilaurin (tridodecanoylglycerol)	$C_{39}H_{74}O_6$	639.02	$CH_2OOC(CH_2)_{10}CH_3$ | $CHOOC(CH_2)_{10}CH_3$ | $CH_2OOC(CH_2)_{10}CH_3$
L–137	Trimyristin (tritetradecanoylglycerol)	$C_{45}H_{86}O_6$	723.18	$CH_2OOC(CH_2)_{12}CH_3$ | $CHOOC(CH_2)_{12}CH_3$ | $CH_2OOC(CH_2)_{12}CH_3$

TABLE 14   Compounds for Which General Criteria and Specifications 14 Are Applicable (continued)

Number	Common name (Systematic name)	Formula	Formula wt.	Structure
L–138	Tripalmitin (trihexadecanoylglycerol)	$C_{51}H_{98}O_6$	807.35	$CH_2OOC(CH_2)_{14}CH_3$ $CHOOC(CH_2)_{14}CH_3$ $CH_2OOC(CH_2)_{14}CH_3$
L–139	Tripalmitolein (tri-*cis*-9-hexadecenoylglycerol)	$C_{51}H_{92}O_6$	801.30	$\quad\quad\quad\quad\text{H H}$ $CH_2OOC(CH_2)_7C{=}C(CH_2)_5CH_3$ $\quad\quad\quad\quad\text{H H}$ $CHOOC(CH_2)_7C{=}C(CH_2)_5CH_3$ $\quad\quad\quad\quad\text{H H}$ $CH_2OOC(CH_2)_7C{=}C(CH_2)_5CH_3$
L–140	1,2-Dipalmitoylstearin (*rac*-1,2-dihexadecanoyl-3-octadecanoylglycerol)	$C_{53}H_{102}O_6$	835.40	$CH_2OOC(CH_2)_{14}CH_3$ $CHOOC(CH_2)_{14}CH_3$ $CH_2OOC(CH_2)_{16}CH_3$
L–141	1,2-Dipalmitoylolein (*rac*-1,2-dihexadecanoyl-3-*cis*-9-octadecenoylglycerol)	$C_{53}H_{100}O_6$	833.38	$CH_2OOC(CH_2)_{14}CH_3$ $CHOOC(CH_2)_{14}CH_3$ $\quad\quad\quad\quad\text{H H}$ $CH_2OOC(CH_2)_7C{=}C(CH_2)_7CH_3$
L–142	1,3-Dipalmitoylolein (1,3-dihexadecanoyl-2-*cis*-9-octadecenoylglycerol)	$C_{53}H_{100}O_6$	833.38	$CH_2OOC(CH_2)_{14}CH_3$ $\quad\quad\quad\quad\text{H H}$ $CHOOC(CH_2)_7C{=}C(CH_2)_7CH_3$ $CH_2OOC(CH_2)_{14}CH_3$
L–143	1,2-Distearoylpalmitin (*rac*-1,2-octadecanoyl-3-hexadecanoylglycerol)	$C_{55}H_{106}O_6$	863.45	$CH_2OOC(CH_2)_{16}CH_3$ $CHOOC(CH_2)_{16}CH_3$ $CH_2OOC(CH_2)_{14}CH_3$
L–144	Tristearin (trioctadecanoylglycerol)	$C_{57}H_{110}O_6$	891.51	$CH_2OOC(CH_2)_{16}CH_3$ $CHOOC(CH_2)_{16}CH_3$ $CH_2OOC(CH_2)_{16}CH_3$
L–145	1,2-Dioleoylstearin (*rac*-1,2-di-*cis*-9-octadecenoyl-3-octadecanoylglycerol)	$C_{57}H_{106}O_6$	887.47	$\quad\quad\quad\quad\text{H H}$ $CH_2OOC(CH_2)_7C{=}C(CH_2)_7CH_3$ $\quad\quad\quad\quad\text{H H}$ $CHOOC(CH_2)_7\ C{=}C(CH_2)_7CH_3$ $CH_2OOC(CH_2)_{16}CH_3$
L–146	Tripetroselinin (tri-*cis*-6-octadecenoylglycerol)	$C_{57}H_{104}O_6$	885.46	$\quad\quad\quad\quad\text{H H}$ $CH_2OOC(CH_2)_4C{=}C(CH_2)_{10}CH_3$ $\quad\quad\quad\quad\text{H H}$ $CHOOC(CH_2)_4C{=}C(CH_2)_{10}CH_3$ $\quad\quad\quad\quad\text{H H}$ $CH_2OOC(CH_2)_4C{=}C(CH_2)_{10}CH_3$

TABLE 14   Compounds for Which General Criteria and Specifications 14 Are Applicable (continued)

Number	Common name (Systematic name)	Formula	Formula wt.	Structure
L–147	Triolein (tri-*cis*-9-octadecenoylglycerol)	$C_{57}H_{104}O_6$	885.46	CH$_2$OOC(CH$_2$)$_7$C=C(CH$_2$)$_7$CH$_3$ (H H) CHOOC(CH$_2$)$_7$C=C(CH$_2$)$_7$CH$_3$ (H H) CH$_2$OOC(CH$_2$)$_7$C=C(CH$_2$)$_7$CH$_3$ (H H)
L–148	Trielaidin (tri-*trans*-9-octadecenoylglycerol)	$C_{57}H_{104}O_6$	885.46	CH$_2$OOC(CH$_2$)$_7$C=C(CH$_2$)$_7$CH$_3$ (H / H) CHOOC(CH$_2$)$_7$C=C(CH$_2$)$_7$CH$_3$ (H / H) CH$_2$OOC(CH$_2$)$_7$C=C(CH$_2$)$_7$CH$_3$ (H / H)
L–149	Trilinolein (tri-*cis*-9,*cis*-12-octadecadienoylglycerol)	$C_{57}H_{98}O_6$	879.41	CH$_2$OOC(CH$_2$)$_7$(C=CCH$_2$)$_2$(CH$_2$)$_3$CH$_3$ (H H) CHOOC(CH$_2$)$_7$(C=CCH$_2$)$_2$(CH$_2$)$_3$CH$_3$ (H H) CH$_2$OOC(CH$_2$)$_7$(C=CCH$_2$)$_2$(CH$_2$)$_3$CH$_3$ (H H)
L–150	Trilinolenin (tri-all-*cis*-9,12,15-octadecatrienoylglycerol)	$C_{57}H_{92}O_6$	873.37	CH$_2$OOC(CH$_2$)$_7$(C=CCH$_2$)$_3$CH$_3$ (H H) CHOOC(CH$_2$)$_7$(C=CCH$_2$)$_3$CH$_3$ (H H) CH$_2$OOC(CH$_2$)$_7$(C=CCH$_2$)$_3$CH$_3$ (H H)
L–151	Triarachidin (trieicosanoylglycerol)	$C_{63}H_{122}O_6$	975.67	CH$_2$OOC(CH$_2$)$_{18}$CH$_3$ CHOOC(CH$_2$)$_{18}$CH$_3$ CH$_2$OOC(CH$_2$)$_{18}$CH$_3$
L–152	Trieicosenoin (tri-*cis*-11-eicosenoylglycerol)	$C_{63}H_{116}O_6$	969.62	CH$_2$OOC(CH$_2$)$_9$C=C(CH$_2$)$_7$CH$_3$ (H H) CHOOC(CH$_2$)$_9$C=C(CH$_2$)$_7$CH$_3$ (H H) CH$_2$OOC(CH$_2$)$_9$C=C(CH$_2$)$_7$CH$_3$ (H H)
L–153	Tribehenin (tridocosanoylglycerol)	$C_{69}H_{134}O_6$	1059.83	CH$_2$OOC(CH$_2$)$_{20}$CH$_3$ CHOOC(CH$_2$)$_{20}$CH$_3$ CH$_2$OOC(CH$_2$)$_{20}$CH$_3$
L–154	Trierucin (tri-*cis*-13-docosenoylglycerol)	$C_{69}H_{128}O_6$	1053.79	CH$_2$OOC(CH$_2$)$_{11}$C=C(CH$_2$)$_7$CH$_3$ (H H) CHOOC(CH$_2$)$_{11}$C=C(CH$_2$)$_7$CH$_3$ (H H) CH$_2$OOC(CH$_2$)$_{11}$C=C(CH$_2$)$_7$CH$_3$ (H H)

GENERAL CRITERIA AND SPECIFICATIONS 15 FOR STEROLS

CRITERIA:
Homogeneity:
   GLC: One component. No detectable solvent.
   TLC: One component on silica gel plates.
Identity: Appropriate relative retention time.

SPECIFICATIONS: Not more than 1% of impurity.
Storage: Below 0 °C under vacuum or an inert gas.

TABLE 15 Compounds for Which General Criteria and Specifications 15 Are Applicable

Number	Common name	Formula	Formula wt.	Structure	Retention time relative to that of cholestane[a]	
					A	B
L-155	Cholesterol	$C_{27}H_{46}O$	386.67		6.95	1.90
L-156	Campesterol	$C_{28}H_{48}O$	400.69		—	2.46
L-157	Ergosterol	$C_{28}H_{44}O$	396.66		10.7	—
L-158	β-Sitosterol	$C_{29}H_{50}O$	414.72		11.6	3.10
L-159	Stigmasterol	$C_{29}H_{48}O$	412.71		9.9	2.77

[a] A, on HiEFF–8B at 225 °C; B, on OV–101 at 250 °C.

GENERAL CRITERIA AND SPECIFICATIONS 16 FOR STEROL ESTERS

CRITERIA:
Homogeneity:
GLC: One component. No detectable solvent.
TLC: One component on silica gel plates.
Identity: Trans-esterification with methanol yields a sterol having the appropriate relative retention time plus a compound meeting the criteria for the appropriate methyl ester.

SPECIFICATIONS: Not more than 1% of other than the indicated acyl group. At least 99% of sterol ester.
Storage: Below 0 °C, under vacuum or an inert gas.

TABLE 16   Compounds for Which General Criteria and Specifications 16 Are Applicable

Number	Common name (Systematic name)	Formula	Formula wt.	Structure
L-160	Cholesteryl pelargonate (cholesteryl nonanoate)	$C_{36}H_{62}O_2$	526.89	$CH_3(CH_2)_7COO-$
L-161	Cholesteryl hendecanoate (cholesteryl undecanoate)	$C_{38}H_{66}O_2$	554.95	$CH_3(CH_2)_9COO-$
L-162	Cholesteryl laurate (cholesteryl dodecanoate)	$C_{39}H_{68}O_2$	568.98	$CH_3(CH_2)_{10}COO-$
L-163	— (cholesteryl tridecanoate)	$C_{40}H_{70}O_2$	583.01	$CH_3(CH_2)_{11}COO-$
L-164	Cholesteryl myristate (cholesteryl tetradecanoate)	$C_{41}H_{72}O_2$	597.03	$CH_3(CH_2)_{12}COO-$

TABLE 16 Compounds for Which General Criteria and Specifications 16 Are Applicable (continued)

Num-ber	Common name (Systematic name)	Formula	Formula wt.	Structure
L-165	— (cholesteryl pentadecanoate)	$C_{42}H_{74}O_2$	611.06	$CH_3(CH_2)_{13}COO-$
L-166	Cholesteryl palmitate (cholesteryl hexadecanoate)	$C_{43}H_{76}O_2$	625.09	$CH_3(CH_2)_{14}COO-$
L-167	Cholesteryl margarate (cholesteryl heptadecanoate)	$C_{44}H_{78}O_2$	639.11	$CH_3(CH_2)_{15}COO-$
L-168	Cholesteryl stearate (cholesteryl octadecanoate)	$C_{45}H_{80}O_2$	653.14	$CH_3(CH_2)_{16}COO-$
L-169	Cholesteryl oleate (cholesteryl cis-9-octadecenoate)	$C_{45}H_{78}O_2$	651.12	$CH_3(CH_2)_7C\overset{H}{=}\overset{H}{C}(CH_2)_7COO-$
L-170	Cholesteryl linoleate (cholesteryl cis-9,cis-12-octadecadienoate)	$C_{45}H_{76}O_2$	649.11	$CH_3(CH_2)_3(CH_2C\overset{H}{=}\overset{H}{C})_2(CH_2)_7COO-$

TABLE 16 Compounds for Which General Criteria and Specifications 16 Are Applicable (continued)

Number	Common name (Systematic name)	Formula	Formula wt.	Structure
L-171	Cholesteryl linolenate (cholesteryl all-*cis*-9,12,15-octadecatrienoate)	$C_{45}H_{74}O_2$	647.09	H H $CH_3(CH_2C{=}C)_3(CH_2)_7COO$—
L-172	(cholesteryl nonadecanoate)	$C_{46}H_{82}O_2$	667.17	$CH_3(CH_2)_{17}COO$—
L-173	Cholesteryl arachidate (cholesteryl eicosanoate)	$C_{47}H_{84}O_2$	681.20	$CH_3(CH_2)_{18}COO$—
L-174	Cholesteryl arachidonate (cholesteryl all-*cis*-5,8,11,14-eicosatetraenoate)	$C_{47}H_{76}O_2$	673.13	H H $CH_3(CH_2)_3(CH_2C{=}C)_4(CH_2)_3COO$—

REFERENCES

1. T. P. Hilditch and P. N. Williams, *The Chemical Constitution of Natural Fats*, 4th ed., John Wiley & Sons, New York (1964).
2. K. S. Markley, *Fatty Acids, Parts II and III*, Interscience Publishers, New York (1961) and (1964).
3. *Official and Tentative Methods of the American Oil Chemists, Society*, Vol. 1 & 2, 1964 ed., ed. and rev. 1965–1971. American Oil Chemists' Society, Chicago, Illinois.
4. T. Y. Miwa, K. L. Mikolajczak, F. R. Earle, and I. A. Wolff, *Anal. Chem.*, **32**, 1739 (1960).
5. L. S. Ettre, *Anal. Chem.*, **36**, 31A (1964).
6. R. G. Ackman and R. D. Burgher, *J. Am. Oil Chem. Soc.*, **42**, 38 (1965).
7. O. S. Privett, *Progr. Chem. Fats Lipids*, **9**, (Part 1), 91 (1966).
8. D. C. Malins, *Progr. Chem. Fats Lipids*, **8**, (Part 3), 1 (1966).
9. W. A. Bonner, *J. Chem. Educ.*, **30**, 452 (1953).
10. A. E. Thomas, III, J. E. Scharoun, and H. Ralston, *J. Am. Oil Chem. Soc.*, **42** 789 (1965).
11. M. Yurkowski and H. Brockerhoff, *Biochim. Biophys. Acta*, **125**, 55 (1966).
12. W. W. Cristie and J. H. Moore, *Biochim. Biophys. Acta*, **176**, 445 (1969).
13. J. G. Quinn, J. Sampugna, and R. G. Jensen, *J. Am. Oil Chem. Soc.*, **44**, 439 (1967).
14. *Nat. Bur. Stand. Tech. Note* 457, 73 (1968).
15. B. A. Knights, *J. Gas Chromatogr.*, **2**, 338 (1964).
16. C. Grunwald, *J. Chromatogr.*, **44**, 173 (1969).
17. A. Rozanski, *Anal. Chem.*, **38**, 36 (1966).
18. L. Swell, *Proc. Soc. Exp. Biol. Med.*, **121**, 1290 (1966).
19. R. D. Bennett and E. Heftmann, *J. Chromatogr.*, **12**, 245 (1963).
20. N. Pelick and J. W. Shigley, *J. Am. Oil Chem. Soc.*, **44**, 121 (1968).
21. R. G. Ackman, J. C. Sipos, and C. S. Tocher, *J. Fisheries Res. Board Can.*, **24**, 635 (1967).
22. H. H. Hofstetter, N. Sen, and R. T. Holman, *J. Am. Oil Chem. Soc.*, **42**, 537 (1965).

Nucleotides and Related Compounds

GENERAL REMARKS AND ANALYTICAL PROCEDURES

The Subcommittee on Nucleotides and Related Compounds was formed to examine the problem of establishing specifications and criteria for the purity of commercially available purine and pyrimidine derivatives. Although high-purity compounds are not always needed in research, a description of procedures that would enable research workers to evaluate the purity of their compounds when necessary is also useful in itself. Initially, the Subcommittee considered 43 compounds; these were described in terms of their molar absorption coefficients and absorbancy ratios at several different wavelengths. In addition, the more probable impurities were listed. It was recognized that these data were really preliminary descriptions of the compounds, rather than actual specifications of purity.

Subsequently, the Subcommittee addressed itself to the following questions:[1] (1) How shall impurities be detected? (2) How shall the impurities be identified? (3) How shall the amount of an impurity be measured quantitatively? (4) What is a permissible level for a given impurity? Procedures were established for testing commercial samples by (a) paper chromatography in four different solvent systems, (b) high-voltage, paper electrophoresis, and (c) ultraviolet spectrophotometry. Purity specifications were then defined as follows: In order to be described as meeting National Research Council (NRC) specifications, a sample must (1) show, under ultraviolet light, no visible impurities on paper chromatograms obtained under the conditions described, and (2) agree to within $\pm 3\%$ with the values of ϵ_{max} given in the data sheets.

In this section, we consider these questions in some detail, and present the results of an extensive testing program. We have also brought up to date the information given on the data sheets, and have included additional reference data.

Detection of Impurities

The potential impurities in the compounds under consideration can be divided into three general groups: (1) organic compounds absorbing in the ultraviolet region; (2) organic compounds that do *not* appreciably absorb in the ultraviolet region above 230 nm; and (3) inorganic compounds. Of these, the ultraviolet-absorbing contaminants are by far the easiest to detect, and most techniques for observing impurities (after paper, thin-layer, and ion-exchange chromatography and paper electrophoresis, for example) depend on this property. Sensitivity of detection varies widely, depending on the molar absorption coefficient of the impurity, its fluorescence, if any, and the precise way in which ultraviolet absorption is observed. We studied the detection, by paper chromatography, of potential impurities added to adenine, for example, and could readily detect 0.1% by weight of fluorescing compounds, but could not detect 1% of a nonfluorescent compound having a relatively low molar absorption coefficient ($\epsilon_{260} =$

$6,000 \ mol^{-1} \ l \ cm^{-1}$). Detection sensitivity is approximately twice as great in a viewing cabinet equipped with both short- and longwave ultraviolet lamps as with an ultraviolet lamp in a darkened room. To test for ultraviolet-absorbing impurities, new samples of the 57 compounds previously tested that had *not* met NRC specifications were examined by paper chromatography in four different solvent mixtures. In addition, anion- or cation-exchange chromatograms were obtained for those compounds that showed no impurities by paper chromatography.

Detection of impurities that are not ultraviolet-absorbing is difficult, as there are few general methods of analysis, and one rarely knows what specific compounds to test for. Previously, we have depended on quantitative measurement of ultraviolet molar absorption coefficients and spectral ratios, but considering the normal precision of spectrophotometry (2–3%), the lack, in many cases, of reliable reference data, the possibility of ultraviolet-absorbing impurities masking nonultraviolet-absorbers, and the difficulty of removing water from hydrated materials, this is a crude method at best. We have found gas-chromatographic analysis (after formation of trimethylsilyl derivatives) to be a very sensitive and effective test in many instances. Gas chromatography is independent of the optical properties of the compound, is a high-resolution technique, has a wide dynamic range (large samples can be analyzed and small amounts of impurity detected), provides data that can assist in identification, and offers the possibility of at least a semiquantitative estimation of the quantity of impurities. This method is, of course, not without its difficulties, the primary one being that only compounds that are volatile, or form volatile derivatives, can be detected. In addition, there is the possibility of artifacts appearing because of incomplete derivatization, the formation of multiple derivatives, or decomposition on the column. In our experience, there is little or no difficulty with bases, most ribonucleosides (except cytidine derivatives), and ribonucleoside monophosphates. Artifacts have been observed with some 2′-deoxyribonucleosides and 2′-deoxyribonucleoside monophosphates; halogenated compounds usually decompose; and we have not been able to obtain well-shaped peaks with di- and triphosphates. Other methods of analysis applicable to nonultraviolet-absorbing compounds that we evaluated were infrared and nuclear magnetic resonance spectroscopy and mass spectrometry, but we are not prepared to recommend any of these for routine testing, although they have been useful in specific instances.

Analysis for many inorganic contaminants, particularly the heavy elements, is readily accomplished by emission spectroscopy, and some manufacturers provide the value for total heavy metals. In view of the profound effect that many metal ions can have on biological systems, we consider such information to be essential, and that it should preferably be given as analyses for individual elements. Determination of alkali and alkaline earth impurities is of less practical importance, as many nucleotides are supplied as salts, and a few analyses of bases and nucleosides did not reveal any significant contamination. However, one compound was supplied with the empirical formula of the tetrasodium salt and the molecular weight of the trisodium salt; a sodium analysis was necessary so that we could ascertain which salt actually had been supplied. No information whatsoever is available concerning inorganic anion impurities, although many bases are capable of forming hydrochloride salts, for example. We intend to examine this matter in the future. One of the most troublesome nonultraviolet-absorbing impurities is water. Very often, commercial samples, as received, are hydrated, or contain adventitious water, or both. In order that meaningful spectral analyses be obtained, the water must either be removed or measured. It has been our experience that there is no universal drying procedure that will remove moisture from all of the compounds tested without causing decomposition of some of them. Each compound presents a unique problem. Consequently, we have not attempted to dry our samples prior to spectral measurements, but have determined the water content by Karl Fischer titration, instead. We must draw attention to the practice followed by some manufacturers of "adjusting" low molar absorption values to the theoretical values by assuming that the sample is hydrated; this is not justified without actual analysis of the water content, as it is possible that some impurity other than water is present.

Identification of Impurities

Identification of impurities is a problem that has still not been adequately resolved, despite its obvious importance. Often, the amount of a particular impurity cannot be measured until the impurity has been characterized. Furthermore, whether a particular sample of a compound can be used for a specific experiment may depend on the nature of the contaminant as well as the amount. At present, identifications are attempted mainly by comparison of R_f values of the impurity with those of compounds considered to be likely contaminants.

Manufacturers usually report the R_f values of impurities and occasionally identify them on this basis. Extensive tables of R_f values that should be of assistance are included here. It should, however, be kept in mind that contaminants can be completely extraneous materials. We have identified such materials as silicone grease, glue components (from the bottle cap), and plasticizers (probably leached from plastic bottles and tubing) in the samples tested. The separation of impurities by ion-exchange methods, with monitoring of the ultraviolet absorbance of the column effluent at several wavelengths simultaneously, provides other useful items of information—the elution volume, wavelength ratios, and a rough estimate of ϵ_{max}. Impurity peaks can, of course, also be collected, but the total quantity of material that can be isolated by column separation on an analytical scale is usually too small to permit adequate characterization. Gas-chromatographic separation is applicable to nonultraviolet-absorbers, also, and provides data on the retention time of the impurity. In addition, gas chromatography in combination with mass spectrometry has been remarkably successful in elucidating the structure of very small quantities of unknown compounds. We feel that this is probably the most feasible means of unambiguously identifying impurities and are actively working along these lines.

Quantitative Measurement of Impurities

Ultraviolet-absorbing impurities that have been identified can be measured quantitatively by spectrophotometric analysis following fractionation by any of the procedures described. Alternatively, when the amount of impurity is too small to be determined accurately by spectrophotometry, spots on paper or thin-layer chromatograms can be measured with a densitometer, or peaks on ion-exchange column chromatograms can be integrated. These are conventional analytical techniques that are well described and discussed in the literature, and so they will not be considered here.

Nonultraviolet-absorbing impurities are best measured by gas chromatography when possible. Because the response of the flame-ionization detector usually employed depends mainly on the number of carbon atoms in the compound studied, and is relatively insensitive to their chemical state, it is not necessary to identify the impurity prior to performing a semi-quantitative analysis (within a factor of 2, for instance). If the retention time (methylene-unit value) of the impurity is not too different from that of the major component, their relative peak areas may simply be measured.

A discussion of methods for the measurement of inorganic compounds is beyond the scope of this discussion. In general, a semi-quantitative analysis by emission spectroscopy should be adequate for indicating to the user whether the compound is sufficiently pure for his purposes.

Permissible Limits of Impurities

In many ways, the problem of setting permissible limits for impurities is the most difficult. Clearly, the acceptable limit will depend on the nature of the impurity and the use to which the compound under consideration is to be put. For example, a 5% contamination is seldom serious if a compound is to be used only as a chromatographic marker, but a few parts per million of a heavy metal can inhibit certain enzymic reactions. Furthermore, it is impractical to set limits of purity that cannot be met by present manufacturing processes. In an attempt to reach a compromise between the very high purity that research workers would like and the levels that are now commercially feasible, the Subcommittee had previously set the following tentative specifications for compounds to be designated as NRC grade: A sample must show no visible ultraviolet-absorbing impurities on paper chromatograms prepared in the specified solvents under the conditions described and agree to within $\pm 3\%$ with the values of ϵ_{max} given in the data sheets. A compound meeting these specifications was assumed to have less than 1% of ultraviolet-absorbing impurities and less than 3% of nonultraviolet-absorbing impurities. These specifications have proved to be reasonably satisfactory to the research community and well within the capabilities of biochemical manufacturers. In our previous tests, 19 of 75 compounds met the specification for designation as NRC grade, and, after our present retests of 57 compounds plus 10 additional compounds not previously tested, a total of 29 met specifications and 6 tentatively met specifications. Manufacturers have made extensive use of these specifications, and generally offer "NRC-grade" compounds. Consequently, we shall retain the specifications described. However, as we are aware that the procedures given will not detect all possible impurities, we now make the following recommendations for additional tests: (1) A gas-chromatographic analysis shall be made when possible; (2) a cation-exchange separation shall be performed on bases and nucleosides and an anion-exchange separation on nucleotides that show no impurities on paper chromatograms; and (3) the compounds shall be analyzed for inorganic contaminants by emission spectroscopy.

It should be emphasized that the specifications of purity described in this section, as well as the procedures to be followed in estimating the purity of a sample, are tentative. The Subcommittee intends to continue its efforts and will welcome recommendations from persons interested in any aspect of this work. It is also possible that this report contains some errors. Recommendations and corrections should be sent to the Subcommittee chairman.

SPECIFIC PROCEDURES

The following procedures were used for testing the commercial samples. The techniques are, in general, conventional analytical procedures and can be carried out in any laboratory that is reasonably well equipped. These procedures, or procedures demonstrated to be equivalent, must be used in establishing whether a specific compound meets the specifications for designation as a NRC-grade sample. We trust that manufacturers will use all of these tests exactly as described and will continue to use such additional methods of quality control as elemental analysis or optical rotation.

Paper Chromatography

All chromatograms were run on strips (7 × 22.5 in.) or sheets (46 × 57 cm) of Whatman No. 40 chromatography paper, by use of the descending technique, in a sealed cabinet whose atmosphere had been presaturated with solvent vapor. The compounds were chromatographed in four solvent systems, designated A, B, C, and D and, occasionally, in a fifth system indicated in the individual data sheets. The R_f values of the compounds examined are listed alphabetically in Table 1, and the compositions of the solvent systems are given in the footnotes to the table.

A minimum of ten A_{260} units (A_{260} = ultraviolet absorbance of the sample measured in a 1-cm quartz cuvette at 260 nm) were applied to the paper in a 5-μl aliquot as a 1.5-cm streak at a distance of 2.5 in. from one end. In a few cases (see individual data sheets), it was not possible to apply the required ten A_{260} units in a single aliquot. In these cases, multiple streaking, with air drying after each application, was used.

The sample size required to give ten A_{260} units per 5-μl aliquot can be calculated from the molar absorption coefficient at 260 nm. Values of ϵ_{260} are usually available from manufacturers' literature or other compilations.[2] Molar absorption coefficients may change drastically with pH, and this should be taken into account when calculating the sample sizes for the different solvents.

The solubilities of different compounds differ markedly, and the actual solvent used for preparing chromatographic samples is given in the data sheets. Whenever possible, the samples were dissolved in water. Insoluble acids were dissolved in 1 M NH_4OH, and insoluble bases in 0.5 M HCl.

Chromatograms in solvents A, B, and C were developed overnight (about 16 h), and in solvent D, for about 8 h, and then air-dried. The ultraviolet-absorbing components were detected in a chromatographic cabinet equipped with both "shortwave" (254 nm) and "long-

wave" (366 nm) ultraviolet lamps. Any fluorescence observed is indicated on the data sheets. It should be noted that the sensitivity of detection depends on the ultraviolet lamp used and the method of viewing. The viewing technique described is considerably more sensitive than an ultraviolet lamp in a darkened room.

The R_f values of all the components in the compounds tested are given in the data sheets, and the R_f values of the major spots are listed in Table 1. The mobility of any impurity detected is expressed as R_M, which is defined as:

$$R_M = \frac{\text{distance traveled by impurity}}{\text{distance traveled by major compound}}.$$

All R_f and R_M values are taken from two chromatograms where they agreed to within ±0.2 unit. The R_f value of any impurity can be calculated by multiplying its R_M value by the R_f value of the major compound. In general, resolution of two components was obtained when their R_f values differed by more than 0.1.

It should be mentioned that we were able to obtain comparable results on thin layers (250 μm) of microcrystalline cellulose, but had considerable difficulty in adjusting the quantity of sample applied to avoid overloading and streaking.

Paper Electrophoresis

Results obtained by paper electrophoresis are not reported, because in no case did we detect impurities that were not detected by paper chromatography. The mobilities of various compounds relative to adenosine 5′-phosphate ($R_{A5'P}$) are included in Table 1. (Those interested in the procedure are referred to Ref. 1.)

Gas Chromatography

Gas-chromatographic analysis has been successfully applied to the trimethylsilyl derivatives of purines and pyrimidines, ribonucleosides and deoxyribonucleosides, and nucleoside monophosphates.[3] The derivatives were formed by dissolving 1–2 mg of sample in 0.2 ml of pyridine or N,N-dimethylformamide and 0.3 ml of bis(trimethylsilyl)trifluoroacetamide containing 1% of chlorotrimethylsilane The reaction mixture was heated at 75 °C for 3 h for bases and ribonucleosides, and overnight for deoxyribonucleosides and nucleotides. Two microliters of the mixture were injected onto the column.

The gas chromatograph was a Tracor Model MT-220 instrument equipped with an all-glass injection system, a glass column (6 ft × 0.25 in.) packed with 5% SE-30 on 80–100 mesh Chromosorb W(HP), and a flame-ionization detector. The inlet and detector tem-

TABLE 1 Paper Chromatography and Electrophoresis

| Compound | Chromatographic solvent systems | | | | Electro-phoresis $R_{A5'P}$ Values[e] |
	A[a] R_f Values	B[b]	C[c]	D[d]	
1. Adenine	0.88	0.70	0.67	0.39	
2. Adenosine	0.83	0.51	0.61	0.56	
3. Adenosine 3':5'-cyclic phosphate	0.65	0.45	0.45	0.77	0.7
4. Adenosine 5'-diphosphate	0.37	0.03	0.14	0.91	1.6
5. Adenosine 2'-phosphate	0.66	0.11	0.25	0.88	0.9
6. Adenosine 2'(3')-phosphate	0.65	0.12	0.24	0.87	0.7
7. Adenosine 3'-phosphate	0.65	0.10	0.22	0.86	0.9
8. Adenosine 5'-phosphate	0.49	0.14	0.18	0.89	1.0
9. Adenosine 5'-triphosphate	0.23	0.05	0.10	0.93	1.9
10. 6-Azauridine	0.50	0.56	0.63	0.89	
11. 5-Bromo-2'-deoxycytidine	0.79	0.69	0.78	0.71	
12. 5-Bromouridine	0.57	0.50	0.72		
13. Cytidine	0.74	0.54	0.70	0.81	
14. Cytidine 2':3'-cyclic phosphate	0.50	0.42	0.53	0.91	0.8
15. Cytidine 5'-diphosphate	0.19	0.04	0.18	0.93	1.6
16. Cytidine 2'-phosphate	0.50	0.24	0.31	0.95	
17. Cytidine 2'(3')-phosphate	0.47	0.18	0.25	0.91	
18. Cytidine 3'-phosphate	0.48	0.21	0.26	0.95	
19. Cytidine 5'-phosphate	0.39	0.14	0.16	0.87	0.6
20. Cytidine 5'-triphosphate	0.20	0.05	0.12	0.94	1.9
21. Cytosine	0.76	0.50	0.70	0.71	
22. 2'-Deoxyadenosine	0.93	0.57	0.66	0.40	
23. 2'-Deoxyadenosine 5'-diphosphate	0.30	0.07	0.26	0.88	1.7
24. 2'-Deoxyadenosine 5'-phosphate	0.56	0.19	0.34	0.87	0.7
25. 2'-Deoxyadenosine 5'-triphosphate	0.39	0.05	0.18	0.93	2.0
26. 2'-Deoxycytidine	0.80	0.62	0.77	0.76	
27. 2'-Deoxycytidine 5'-diphosphate	0.25	0.05	0.20	0.89	1.6
28. 2'-Deoxycytidine 5'-phosphate	0.53	0.16	0.21	0.85	0.7
29. 2'-Deoxycytidine 5'-triphosphate	0.22	0.05	0.18	0.93	2.1
30. 2'-Deoxyguanosine	0.64	0.46	0.68	0.63	
31. 2'-Deoxyguanosine 5'-diphosphate	0.17	0.04	0.13	0.88	2.0
32. 2'-Deoxyguanosine 5'-phosphate	0.39	0.03	0.16	0.85	1.1
33. 2'-Deoxyinosine	0.62	0.44	0.58	0.60	
34. 2'-Deoxyuridine	0.67	0.56	0.76		
35. 2'-Deoxyuridine 5'-phosphate	0.39	0.12	0.24		1.3
36. N^6,N^6-Dimethyladenine	0.87	0.83	0.82	0.51	
37. N^2,N^2-Dimethylguanine	0.56	0.37	0.40	0.20	
38. Guanine	0.66	0.27	0.47	0.40	
39. Guanosine	0.50	0.36	0.53	0.62	
40. Guanosine 2':3'-cyclic phosphate	0.34	0.55	0.50	0.85	1.5
41. Guanosine 5'-diphosphate	0.12	0.02	0.08	0.89	1.7
42. Guanosine 2'-phosphate	0.31	0.10	0.15	0.94	
43. Guanosine 2'(3')-phosphate	0.28	0.05	0.07	0.88	1.3
44. Guanosine 3'-phosphate	0.27	0.09	0.02	0.91	
45. Guanosine 5'-phosphate	0.15	0.06	0.05	0.87	1.4
46. Guanosine 5'-triphosphate	0.20	0.05	0.10	0.94	2.2
47. Hypoxanthine	0.62	0.51	0.66	0.58	
48. Inosine	0.48	0.47	0.59	0.74	
49. Inosine 5'-diphosphate	0.13	0.02	0.11	0.88	2.2
50. Inosine 5'-phosphate	0.28	0.03	0.15	0.90	1.5
51. Inosine 5'-triphosphate	0.20	0.05	0.16	0.95	2.2
52. 5-Iodo-2'-deoxycytidine	0.80	0.63	0.70	0.64	
53. 5-Iodo-2'-deoxyuridine	0.68	0.59	0.84		
54. 5-Iodouridine	0.61	0.53	0.75	0.73	
55. N^6-(Isopent-2-enyl)adenine	0.96	0.84	0.88	0.53	
56. N^6-(Isopent-2-enyl)adenosine	0.93	0.84	0.84	0.64	
57. Kinetin	0.93	0.85	0.85		
58. N^6-Methyladenine	0.89	0.77	0.78	0.46	
59. 5-Methylcytosine	0.80	0.65	0.70	0.71	

TABLE 1   Paper Chromatography and Electrophoresis (continued)

| Compound | Chromatographic solvent systems | | | | Electro-phoresis |
	A[a] R_f Values	B[b]	C[c]	D[d]	$R_{A5'P}$ Values[e]
60. 5-Methyl-2'-deoxycytidine	0.81	0.65	0.77	0.80	
61. 7-Methylguanine	0.82	0.35	0.55	0.45	
62. 1-Methylinosine	0.65	0.51	0.75	0.84	
63. 5-Methyluridine	0.62	0.52	0.76	0.85	
64. Orotic acid	0.25	0.45	0.43	0.84	
65. Pseudouridine, mixed anomers	0.38	0.45	0.69	0.79	
66. Pseudouridine, β anomer	0.44	0.41	0.65	0.82	
67. 9-β-D-Ribosylkinetin	0.88	0.89	0.80	0.68	
68. Thymidine	0.72	0.73	0.81	0.78	
69. Thymidine 3',5'-bisphosphate	0.21	0.04	0.10	0.97	3.1
70. Thymidine 5'-diphosphate	0.22	0.07	0.31	0.91	2.2
71. Thymidine 5'-phosphate	0.39	0.22	0.48	0.89	1.4
72. Thymidine 5'-triphosphate	0.23	0.05	0.20	0.94	2.4
73. Thymine	0.78	0.63	0.72		
74. Uracil	0.63	0.59	0.70	0.70	
75. Uridine	0.51	0.57	0.68	0.80	
76. Uridine 2':3'-cyclic phosphate	0.32	0.38	0.65	0.94	1.7
77. Uridine 5'-diphosphate	0.14	0.02	0.16	0.89	2.2
78. Uridine 2'-phosphate	0.32	0.22	0.33	0.94	
79. Uridine 2'(3')-phosphate	0.28	0.18	0.30	0.87	1.5
80. Uridine 3'-phosphate	0.31	0.23	0.33	0.96	
81. Uridine 5'-phosphate	0.22	0.11	0.28	0.90	1.3
82. Uridine 5'-triphosphate	0.20	0.05	0.19	0.94	2.9
83. Xanthine	0.54	0.32	0.35		
84. Xanthosine	0.37	0.35	0.41	0.71	
85. Xanthosine 5'-phosphate	0.16	0.03	0.15	0.91	1.5

[a] Solvent A: isobutyric acid–0.5 M NH_4OH (5:3, v/v).
[b] Solvent B: isopropyl alcohol–conc. NH_4OH–H_2O (7:1:2, v/v).
[c] Solvent C: 95% ethanol–1 M sodium acetate (7:3, v/v).
[d] Solvent D: H_2O adjusted to pH 10 with NH_4OH.
[e] $R_{A5'P}$ = mobility relative to adenosine 5'-phosphate on paper electrophoresis in 0.02 M citrate buffer at pH 3.5.

peratures were ~300 °C, the carrier gas was helium at a flow rate of 80–90 ml/min, and the temperature was programmed from 100 to 300 °C at 10 °C/min. Retention data, as methylene unit (MU) values, were obtained by chromatographing a mixture of n-alkanes along with the sample and calculating, by linear interpolation, the retention time of the compound in n-alkane carbon equivalents.[4] A compound having the same retention time as the C_{16} alkane, for example, has an MU value of 16.00; one having a retention time halfway between the C_{16} and C_{17} alkanes has an MU value of 16.50. Methylene unit values for the compounds tested, and the MU values and relative peak areas for any impurities found, are given on the individual data sheets.

Anion-Exchange Chromatography

The anion-exchange system used is an adaptation[5] of the nucleotide analyzer developed by Anderson and co-workers.[6] This system consists of a heated stainless steel column of 0.62 cm internal diameter and 160 cm length,

packed with Dowex 1 × 8 resin of particle diameter 5–10 μm or its equivalent. The detector is an ultraviolet spectrophotometer that operates alternately at four preselected wavelengths (250, 260, 280, and 310 nm). The sample of about 2 mg in 2 ml of solution is introduced onto the column by means of an injection valve and eluted first with 200 ml of 0.015 M ammonium acetate–acetic acid buffer, pH 4.4, and then with 1,300 ml of a linear, 0.015 to 6 M gradient of acetate buffer at 60 °C at a flow rate of 30 ml/h. Virtually all of the purine and pyrimidine bases, nucleosides, and nucleoside mono-, di,- and triphosphates are resolved from one another in a 40-h run. Any impurities detected, and tentative identification based on elution position and wavelength ratios, are noted in the individual data sheets.

Cation-Exchange Chromatography

This cation-exchange separation is based on the method of Uziel, Koh, and Cohn.[7] The apparatus consisted of a high-pressure, glass column (0.9 × 40 cm) having an adjustable column head, packed with Dowex 50 × 8 resin of particle diameter 10–15 μm or its equivalent, operating at 50 °C, and a very stable, high-sensitivity, ultraviolet monitor (LDC Corp., Model 1205). The sample of about 250 μg was injected onto the column, and eluted with 0.4 M ammonium formate–formic acid buffer, pH 4.5, at 50 °C, at a flow rate of 30 ml/h. This system provides excellent resolution of most bases and nucleosides in 3–4 h. Any impurities detected, and tentative identification based on elution position, are noted in the individual data sheets.

Spectral Analyses

The ultraviolet spectra of compounds that showed no detectable impurities by paper chromatography were obtained with a Cary Model 14 recording spectrophotometer. No attempt was made to dry the compounds. Instead, the water content was determined by Karl Fischer titration,[8] and the concentration of the sample was corrected.

A quantity of each compound, calculated from values of ϵ_{max} in the literature, was dissolved in 100 ml of water at a concentration such that a dilution of one tenth would give a solution having an absorbance of ~1 for the highest peak. The pH values of the solution were chosen to be at least 1 or, preferably, 2 pH units from the pK_a of the base. For solutions of pH 1 or 2, sufficient 1 M HCl was added to the diluent to bring the concentration of HCl to 0.1M or 0.01 M, respectively; 1 M NaOH was used to adjust the pH to 11 or 12. Solutions at pH 7 were prepared by dilution with 0.5 M potassium phosphate buffer, pH 7. All spectra were obtained within 15 min after dilution. Molar absorption coefficients and wavelength ratios were determined from these curves and compared with spectral data taken from the literature; any discrepancies are noted in the individual data sheets.

RESULTS

Data Sheets

The data sheets are arranged alphabetically, according to the common name of the compound. In parentheses is the systematic name or the name under which the compound is listed in *Chemical Abstracts* if this is different from the common name.

The abbreviations given are those of the IUPAC–IUB Commission on Biochemical Nomenclature.[9]

Formulas, molecular weights, and elemental analyses are for the nonhydrated form of the compounds and for the free acid form of nucleotides.

Melting points of bases and nucleosides are given when they are true melting points and not decomposition points. The literature is not always clear on this distinction, so there may be exceptions.

Specific rotations are given in their conventional form, followed by the mass concentration of the sample (g/100 ml of solution) and the medium in parentheses.

As pK_a values depend critically on temperature, ionic strength, and ionic medium, we report rounded-off values for them. References to more precise data are given. A compilation containing pK_a data is also available.[2]

The R_f values in paper chromatography reported for the compounds retested are new values and are not identical with those previously obtained, although they usually agree to within 0.1 unit. The R_f values for the compounds not retested (those that had previously met specifications), and all of the paper-electrophoresis data are earlier measurements.[1] A single asterisk (*) indicates that the spot fluoresces on excitation with 254-nm light and a double asterisk (**) that it fluoresces on excitation with 366-nm light.

Gas-chromatographic MU values, although relatively independent of such factors as column length, carrier-gas flow rate, and temperature-program rate, depend on the polarity of the liquid phase. The values on the data sheets were obtained by use of a nonpolar liquid phase, and different values would be expected with polar liquid phases. Where the relative peak-area of an impurity is given as "trace," the area is <0.2% of that of the primary peak.

The spectral constants given are taken from the literature references mentioned, λ_{max} values are in nanometers (nm), and the ϵ_{max} values have been multiplied by 10^{-3}, that is, are expressed in units of $(mmol/l)^{-1}cm^{-1}$, to conserve space. These data have been reviewed and have been brought up to date where necessary; in the opinion of the Subcommittee, they are the best data available. In some cases, spectral data are still unavailable for nucleoside di- and triphosphates, and so the data for the monophosphate derivative are used. Spectra obtained at pH values near to a pK_a of a base represent mixed ionic forms and are difficult to reproduce; we have tried to avoid reporting such data. Phosphate ionizations do not usually have much effect on the spectrum. If the spectral reference data available for a compound are reasonably complete, and if the sample met NRC specifications, the actual measurements are not included, as they agreed to within ±3% with the reference data. If the reference data are *not* complete, however, we have included measurements made on the NRC sample to provide additional information.

In the "Remarks," we have indicated whether a sample met specifications and any other information considered of value.

TABLE 2   Compounds Meeting NRC Specifications

Met Specifications

> Adenine
> Adenosine 3′:5′-cyclic phosphate
> Adenosine 5′-phosphate
> 6-Azauridine
> 5-Bromouridine
> Cytidine
> Cytidine 2′:3′-cyclic phosphate
> Cytidine 2′-phosphate
> Cytidine 5′-phosphate
> Cytosine
> 2′-Deoxycytidine
> 2′-Deoxycytidine 5′-phosphate
> 2′-Deoxyuridine
> 2′-Deoxyuridine 5′-phosphate
> Guanine
> Guanosine 5′-phosphate
> Hypoxanthine
> 5-Iodo-2′-deoxycytidine
> 5-Iodo-2′-deoxyuridine
> 5-Iodouridine
> 5-Methylcytosine
> 7-Methylguanine
> Thymidine
> Thymidine 3′,5′-bisphosphate
> Thymine
> Uracil
> Uridine 3′-phosphate
> Uridine 5′-phosphate
> Xanthine

Tentative

> Cytidine 3′-phosphate
> 2′-Deoxyadenosine 5′-phosphate
> 2′-Deoxyguanosine 5′-phosphate
> Guanosine 2′(3′)-phosphate
> N^6-(Isopent-2-enyl)adenosine
> 5-Methyluridine

Compounds Meeting Specifications

Those compounds for which at least one sample met the specifications of purity previously described are listed in Table 2. The appearance of a compound in this list does not ensure that all commercial samples will meet the purity specifications, nor does the failure of a compound to appear on this list imply that no commercial sample meeting these specifications is available. It should be understood that the fact that a sample meets specifications does not guarantee a specific degree of purity, but only that impurities were not detected by the particular tests applied. We consider that samples meeting specifications are at least 95% pure, and that some are almost certainly better than 99% pure, exclusive of water.

In order to be specified as "NRC grade," each lot of a compound must have been tested by paper chromatography in all four of the solvent systems described, and molar absorption coefficients shall have been measured at the pH values prescribed. Designation as "NRC grade" is not justified when all four of the solvent systems have not been employed, or other solvent systems have been substituted, or when molar absorption coefficients have been "adjusted" for hydration without an actual analysis for water having been made and reported.

ACKNOWLEDGMENTS

The Subcommittee acknowledges the valuable contributions of Dr. J. C. Wolford, Dr. W. C. Butts, and Mr. W. B. Cottrell, who collected most of the experimental data reported in the data sheets.

N-1
Adenine

Abbreviation: Ade
Formula: $C_5H_5N_5$
Mol. Wt.: 135.13
Calc. %: C, 44.45
H, 3.73
N, 51.82
pK_a:[10,11] 4.2, 9.8

Paper Chromatography: Dissolve sample in 0.5 M HCl.

Solvent	R_f
A	0.88
B	0.70
C	0.67
D	0.39

Gas Chromatography: Trimethylsilyl derivative.

MU value	Rel. Peak Area
18.52	100

Spectral Constants:

pH	λ_{max}	ϵ_{max}	$\dfrac{A_{250}}{A_{260}}$	$\dfrac{A_{280}}{A_{260}}$	Ref.
1	263	13.2	0.76	0.38	12
7	261	13.4	0.76	0.13	
12	269	12.3	0.57	0.60	

Water Analysis: 0.1%.

Remarks: Three of four samples tested met NRC specifications. Cation-exchange chromatography showed, in only one sample, an impurity tentatively identified as 2-oxyadenine.

N-2
Adenosine
(9-β-D-Ribofuranosyladenine)

Abbreviation: Ado
Formula: $C_{10}H_{13}N_5O_4$
Mol. Wt.: 267.25
Calc. %: C, 44.94
H, 4.90
N, 26.21
O, 23.95
Melting Pt.:[13] 234–235 °C
Specific Rotation: $[\alpha]_D^{11}$ −61.7° (0.7 g/100 ml, H_2O)[13]
pK_a:[14] 3.6, 12.4

Paper Chromatography: Dissolve the sample in 1 M NH_4OH.

Solvent	R_f	R_M	R_M
A	0.83		
B	0.51	0.10	0.18
C	0.61	0.67	
D	0.56		

Gas Chromatography: Trimethylsilyl derivative.

MU value	Rel. Peak Area
26.54	100
25.82	2

Spectral Constants:

pH	λ_{max}	ϵ_{max}	$\dfrac{A_{250}}{A_{260}}$	$\dfrac{A_{280}}{A_{260}}$	Ref.
2	257	15.1	0.86	0.22	15
7	259	15.4	0.79	0.15	
11	259	15.4	0.79	0.15	

N-3
Adenosine 3′:5′-Cyclic Phosphate

Abbreviations: Ado-3′:5′-P, 3′:5′-cyclic AMP
Formula: $C_{10}H_{12}N_5O_6P$
Mol. Wt.: 329.21
Calc. %: C, 33.29
H, 4.33
N, 19.42
O, 34.37
P, 8.59

Paper Chromatography: Dissolve the free acid or salts in H_2O.

Solvent	R_f
A	0.65
B	0.45
C	0.45
D	0.77

Spectral Constants:

pH	λ_{max}	ϵ_{max}	λ_{min}	ϵ_{min}	ϵ_{260}	Ref.
2	256	14.5				16
7	258	14.7				
1.2	256	14.3	225	2.28	13.7	NRC sample
6.9	259	14.6	227	2.60	14.5	

Remarks: The free acid and the barium salt met NRC specifications.

N-4
Adenosine 5′-Diphosphate
(Adenosine 5′-Pyrophosphate)

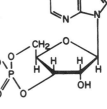

Abbreviations: Ado-5′-P_2, 5′-ADP
Formula: $C_{10}H_{15}N_5O_{10}P_2$
Mol. Wt.: 427.21
Calc. %: C, 28.12
H, 3.54
N, 16.40
O, 37.45
P, 14.50
pK_a:[14] 4.0, 6.4

Paper Chromatography: Dissolve the Li⁺, Na⁺, K⁺, or NH₄⁺ salt in H₂O.

Solvent	R_f	R_M	R_M
A	0.37	0.29	2.2
B	0.03		
C	0.14	1.5	
D	0.91	0.88	

Spectral Constants:

pH	λ_{max}	ϵ_{max}	$\dfrac{A_{250}}{A_{260}}$	$\dfrac{A_{280}}{A_{260}}$	Ref.
2	257	15.0	0.85	0.21	15
7	259	15.4	0.78	0.16	
11	259	15.4	0.78	0.16	

N-5
Adenosine 2'-Phosphate (2'-Adenylic Acid)

Abbreviations: Ado-2'-P; 2'-AMP
Formula: $C_{10}H_{14}N_5O_7P$
Molecular Wt.: 347.22
Calc. %: C, 34.58
H, 4.06
N, 20.17
O, 32.26
P, 8.92
Specific Rotation: $[\alpha]_D^{22}$ −65.4° (0.5 g/100 ml, 0.5 M Na₂HPO₄)[17]
pK_a:[18] 3.8, 6.2

Paper Chromatography: Dissolve the Li⁺, Na⁺, or K⁺ salt in H₂O. This compound should be supplied as a neutral salt, as isomerization of the phosphate group occurs at low pH.

Solvent	R_f	R_M
A	0.66	0.47
B	0.11	4.6
C	0.25	2.7
D	0.88	0.58
F[a]	0.29	0.52 (3'-isomer)

Gas Chromatography: Trimethylsilyl derivative.

MU value	Rel. Peak Area
30.06	100

Spectral Constants:

pH	ϵ_{260}	$\dfrac{A_{250}}{A_{260}}$	$\dfrac{A_{280}}{A_{260}}$	Ref.
2	14.2	0.85	0.23	19
7	15.0	0.80	0.15	
12	15.0	0.80	0.15	

[a] To detect the 3'-isomer, a special solvent is necessary, namely, 79:19:2 (v/v) 90% saturated (NH₄)₂SO₄–0.1 M phosphate buffer (pH 6)–isopropyl alcohol.

N-6
Adenosine 2'(3')-Phosphate, Mixed Isomers

Abbreviations: Ado-2'(3')-P, 2'(3')-AMP
Formula: $C_{10}H_{14}N_5O_7P$
Mol. Wt.: 347.22
Calc. %: C, 34.58
H, 4.06
N, 20.17
O, 32.26
P, 8.92

Paper Chromatography: Dissolve the free acid in 1 M NH₄OH; the Li⁺, Na⁺, or K⁺ salt, in H₂O.

Solvent	R_f	R_M	R_M
A	0.65	0.45	
B	0.12	3.6	4.2
C	0.24	2.7	
D	0.87	0.60	

Spectral Constants:

pH	ϵ_{260}	$\dfrac{A_{250}}{A_{260}}$	$\dfrac{A_{280}}{A_{260}}$	Ref.
2	14.2	0.85	0.23	19
7	15.0	0.80	0.15	
12	15.0	0.80	0.15	

N-7
Adenosine 3'-Phosphate (3'-Adenylic Acid)

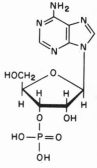

Abbreviations: Ado-3'-P, 3'-AMP
Formula: $C_{10}H_{14}N_5O_7P$
Mol. Wt.: 347.22
Calc. %: C, 34.58
H, 4.06
N, 20.17
O, 32.26
P, 8.92
Specific Rotation: $[\alpha]_D^{22}$ −45.4° (0.5 g/100 ml, 0.5 M Na₂HPO₄)[17]
pK_a:[18] 3.7, 5.9

Paper Chromatography: Dissolve the Li^+, Na^+, or K^+ salt in H_2O. This compound should be supplied as a neutral salt, as isomerization of the phosphate group occurs at low pH.

Solvent	R_f	R_M	R_M
A	0.65	0.45	
B	0.10	4.7	
C	0.22	1.4	2.9
D	0.86	0.60	
F[a]	0.19		

Gas Chromatography: Trimethylsilyl derivative.

MU value	Rel. Peak Area
30.28	100

Spectral Constants:

pH	ϵ_{260}	$\dfrac{A_{250}}{A_{260}}$	$\dfrac{A_{280}}{A_{260}}$	Ref.
2	14.2	0.85	0.22	19
7	15.0	0.80	0.15	
12	15.0	0.80	0.15	

[a] See 2'-AMP (N-5).

N-8
Adenosine 5'-Phosphate
(5'-Adenylic Acid)

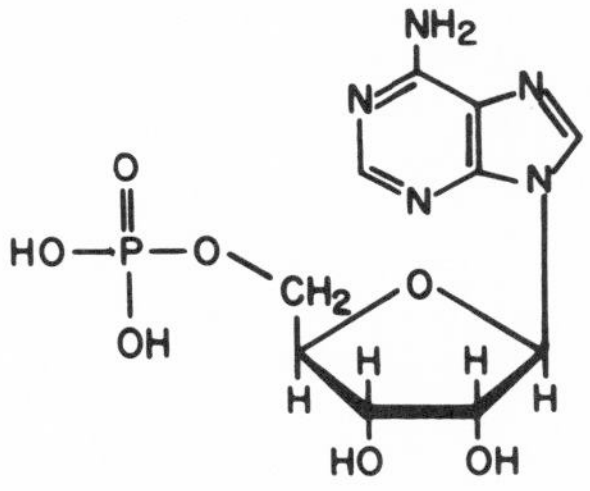

Abbreviations: Ado-5'-P, 5'-AMP
Formula: $C_{10}H_{14}N_5O_7P$
Mol. Wt.: 347.22
Calc. %: C, 34.58
       H, 4.06
       N, 20.17
       O, 32.26
       P, 8.92
Specific Rotation: $[\alpha]_D^{20}$ −26.0° (1.0 g/100 ml, 10% HCl)[20]
pK_a:[14,21] 0.9, 3.8, 6.2, 13.1

Paper Chromatography: Dissolve the free acid in 1 M NH_4OH; the Li^+, Na^+, or K^+ salt, in H_2O.

Solvent	R_f
A	0.49
B	0.14
C	0.18
D	0.89

Gas Chromatography: Trimethylsilyl derivative.

MU value	Rel. Peak Area
31.14	100

Spectral Constants:

pH	λ_{max}	ϵ_{max}	$\dfrac{A_{250}}{A_{260}}$	$\dfrac{A_{280}}{A_{260}}$	Ref.
2	257	15.0	0.84	0.22	15
7	259	15.4	0.79	0.16	
11	259	15.4	0.79	0.16	

Water Analysis: 4.2% (calculated for the monohydrate, 4.9%).

Remarks: Met NRC specifications. Anion-exchange chromatography showed approximately 1% of an unidentified impurity, possibly a purine nucleoside.

N-9
Adenosine 5'-Triphosphate

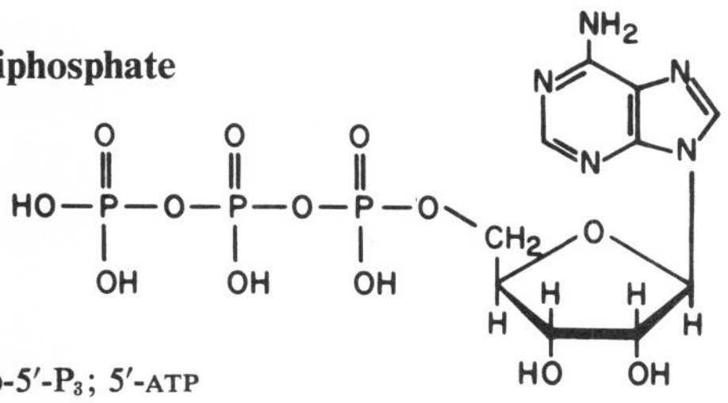

Abbreviations: Ado-5'-P_3; 5'-ATP
Formula: $C_{10}H_{16}N_5O_{13}P_3$
Mol. Wt.: 507.19
Calc. %: C, 23.67
       H, 3.18
       N, 13.80
       O, 41.01
       P, 18.32
pK_a:[14] 4.0, 6.5

Paper Chromatography: Dissolve the Li^+, Na^+, K^+, or NH_4^+ salt in H_2O.

Solvent	R_f	R_M
A	0.23	1.2
B	0.05	
C	0.10	3.4
D	0.93	

Spectral Constants:

pH	λ_{max}	ϵ_{max}	$\dfrac{A_{250}}{A_{260}}$	$\dfrac{A_{280}}{A_{260}}$	Ref.
2	257	14.7	0.85	0.22	15
7	259	15.4	0.80	0.15	
11	259	15.4	0.80	0.15	

N-10
6-Azauridine
[2-β-D-Ribofuranosyl-as-triazine-3,5(2H,4H)-dione]

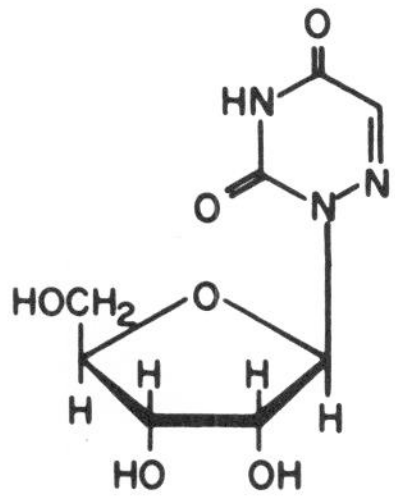

Abbreviation: azaUrd
Formula: $C_8H_{11}N_3O_6$
Mol. Wt.: 245.19
Calc. %: C, 39.19
       H, 4.52
       N, 17.14
       O, 39.15
Melting Pt.:[22] 160 °C
Specific Rotation: $[\alpha]_D^{24}$ −132° (pyridine)[22]
pK_a:[23] 6.7

Paper Chromatography: Dissolve the sample in H_2O.

Solvent	R_f
A	0.50
B	0.56
C	0.63
D	0.89

Spectral Constants:

pH	λ_{max}	ϵ_{max}	λ_{min}	ϵ_{min}	ϵ_{260}	Ref.
13	257	7.0				24
1.2	262	6.05	230	3.0	6.0	NRC sample
11.2	254	6.9	221	2.2	6.6	

Remarks: Met NRC specifications.

N-11
5-Bromo-2′-deoxycytidine

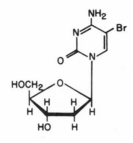

Abbreviation: 5BrdCyd
Formula: $C_9H_{12}BrN_3O_4$
Mol. Wt.: 306.12
Calc. %: C, 35.31
        H, 3.95
        Br, 26.10
        N, 13.73
        O, 20.91
Melting Pt.:[25] 175–179 °C

Paper Chromatography: Dissolve the sample in H_2O.

Solvent	R_f	R_M
A	0.79	
B	0.69	0.0
C	0.78	0.0
D	0.71	0.0

Spectral Constants:

pH	λ_{max}	ϵ_{max}	Ref.
2	300	9.6	25

N-12
5-Bromouridine

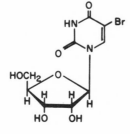

Abbreviation: 5BrUrd
Formula: $C_9H_{11}BrN_2O_6$
Mol. Wt.: 323.10
Calc. %: C, 33.46
        H, 3.43
        Br, 24.73
        N, 8.67
        O, 29.71
Melting Pt.:[26] 213–214 °C

Specific Rotation: $[\alpha]_D^{25}$ −61.8° (2 g/100 ml, H_2O)[27]
pK_a:[28] 8.2

Paper Chromatography: Dissolve the sample in H_2O.

Solvent	R_f
A	0.57
B	0.50
C	0.72

Spectral Constants:

pH	λ_{max}	ϵ_{max}	$\dfrac{A_{250}}{A_{260}}$	$\dfrac{A_{280}}{A_{260}}$	Ref.
1–5	279	9.30	0.54	1.79	28
8.0	279	8.15			
10–12	278	6.40			

Remarks: Met NRC specifications.

N-13
Cytidine
(1-β-D-Ribofuranosylcytosine)

Abbreviation: Cyd
Formula: $C_9H_{13}N_3O_5$
Mol. Wt.: 243.22
Calc. %: C, 44.44
        H, 5.39
        N, 32.89
        O, 17.28
Specific Rotation: $[\alpha]_D^{16}$ +34.2° (2.0 g/100 ml, H_2O)[29]
pK_a:[30] 4.1, 12.2

Paper Chromatography: Dissolve the sample in H_2O.

Solvent	R_f
A	0.74
B	0.54
C	0.70
D	0.81

Gas Chromatography: Methoxime–trimethylsilyl derivative.[3]

MU value	Rel. Peak Area
24.91	100

Spectral Constants:

pH	λ_{max}	ϵ_{max}	$\dfrac{A_{250}}{A_{260}}$	$\dfrac{A_{280}}{A_{260}}$	Ref.
1	280	13.4	0.45	2.10	12
7	271	9.1	0.86	0.93	
13	273	9.2	0.87	1.17	

Remarks: Met NRC specifications.

N-14
Cytidine 2′:3′-Cyclic Phosphate

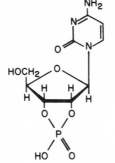

Abbreviations: Cyd-2′:3′-P; 2′:3′-cyclic CMP
Formula: $C_9H_{12}N_3O_7P$
Mol. Wt.: 305.20
Calc. %: C, 35.42
　　　　　H, 3.96
　　　　　N, 13.77
　　　　　O, 36.70
　　　　　P, 10.15
pK_a:[31] 4.2

Paper Chromatography: Dissolve the sample in H_2O.

Solvent	R_f
A	0.50
B	0.42
C	0.53
D	0.91

Spectral Constants:

pH	λ_{max}	ϵ_{max}	λ_{min}	ϵ_{min}	ϵ_{260}	Ref.
7	268	8.4				32
1.5	276	12.2	240	1.1	8.0	NRC sample
7.0	232	7.9	251	7.25	7.6	
	267	8.2				

Remarks: Both the barium and the sodium salt met NRC specifications.

N-15
Cytidine 5′-Diphosphate
(Cytidine 5′-Pyrophosphate)

Abbreviations: Cyd-5′-P$_2$, 5′-CDP
Formula: $C_9H_{15}N_3O_{11}P_2$
Mol. Wt.: 403.18
Calc. %: C, 26.81
　　　　　H, 3.75
　　　　　N, 10.42
　　　　　O, 43.65
　　　　　P, 15.36
pK_a:[15] 4.6, 6.4

Paper Chromatography: Dissolve the Li$^+$, Na$^+$, K$^+$, or NH$_4^+$ salt in H_2O.

Solvent	R_f	R_M	R_M
A	0.19	0.22	2.7
B	0.04		
C	0.18	1.7	
D	0.93		

Spectral Constants:

pH	λ_{max}	ϵ_{max}	$\dfrac{A_{250}}{A_{260}}$	$\dfrac{A_{280}}{A_{260}}$	Ref.
2	280	12.8	0.46	2.07	15
7	271	9.1	0.83	0.98	
11	271	9.1	0.83	0.98	

N-16
Cytidine 2′-Phosphate
(2′-Cytidylic Acid)

Abbreviations: Cyd-2′-P, 2′-CMP
Formula: $C_9H_{14}N_3O_8P$
Mol. Wt.: 323.20
Calc. %: C, 33.44
　　　　　H, 4.37
　　　　　N, 13.00
　　　　　O, 39.60
　　　　　P, 9.58
Specific Rotation: $[\alpha]_D^{20}$ +20.7° (1.0 g/100 ml, H_2O)[33]
pK_a:[34] 4.4, 6.2

Paper Chromatography: Dissolve the Li$^+$ salt in H_2O.

Solvent	R_f
A	0.50
B	0.24
C	0.31
D	0.95

Spectral Constants:

pH	λ_{max}	ϵ_{max}	$\dfrac{A_{250}}{A_{260}}$	$\dfrac{A_{280}}{A_{260}}$	Ref.
2	278	12.7	0.48	1.80	12, 35
7	270	8.9	0.90	0.85	
12	270	8.9	0.90	0.85	

Water Analysis: 11.7% (calculated for the dihydrate, 9.7%).

Remarks: Met NRC specifications. Anion-exchange chromatography showed no 3′-isomer, and only traces of 3 nucleosides and 1 nucleotide.

N-17
Cytidine 2'(3')-Phosphate, Mixed Isomers

Abbreviations: Cyd-2'(3')-P, 2'(3')-CMP
Formula: $C_9H_{14}N_3O_8P$
Mol. Wt.: 323.20
Calc. %: C, 33.44
H, 4.37
N, 13.00
O, 39.60
P, 9.58

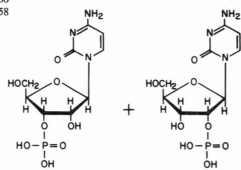

Paper Chromatography: Dissolve the free acid in 1 M NH₄OH; the Li⁺, Na⁺, or K⁺ salt, in H_2O.

Solvent	R_f	R_M	R_M
A	0.47	0.62	1.5
B	0.18	5.4	
C	0.25	2.6	
D	0.91		

Spectral Constants:

pH	λ_{max}	ϵ_{max}	Ref.
2	278	12.7	36
7	270	9.0	

N-18
Cytidine 3'-Phosphate
(3'-Cytidylic Acid)

Abbreviations: Cyd-3'-P, 3'-CMP
Formula: $C_9H_{14}N_3O_8P$
Mol. Wt.: 323.20
Calc. %: C, 33.44
H, 4.37
N, 13.00
O, 39.60
P, 9.58
Specific Rotation: $[\alpha]_D^{20}$ +49.4° (1.0 g/100 ml, H_2O)[33]
pK_a:[34] 4.3, 6.0

Paper Chromatography: Dissolve the Li⁺ salt in H_2O.

Solvent	R_f
A	0.48
B	0.21
C	0.26
D	0.95

Spectral Constants:

pH	λ_{max}	ϵ_{max}	$\dfrac{A_{250}}{A_{260}}$	$\dfrac{A_{280}}{A_{260}}$	Ref.
2	279	(12.4)	0.45	2.00	12 (ratios)
7	270	(8.4)	0.86	0.93	NRC sample (ε values)
12	270	(8.4)	0.86	0.93	

Water Analysis: 5.12% (calculated for the monohydrate, 5.10%).

Remarks: Anion-exchange chromatography showed no 2'-isomer, and only traces of 3 nucleosides. However, literature data on the molar absorptivity are inconsistent, and the purity of this compound is regarded as tentative.

N-19
Cytidine 5'-Phosphate
(5'-Cytidylic Acid)

Abbreviations: Cyd-5'-P, 5'-CMP
Formula: $C_9H_{14}N_3O_8P$
Mol. Wt.: 323.20
Calc. %: C, 33.44
H, 4.37
N, 13.00
O, 39.60
P, 9.58
Specific Rotation: $[\alpha]_D^{14}$ +27.1° (0.54 g/100 ml, H_2O)[37]
pK_a:[15] 4.5, 6.3

Paper Chromatography: Dissolve the free acid in 1 M NH₄OH; the Li⁺, Na⁺, or K⁺ salt, in H_2O.

Solvent	R_f
A	0.39
B	0.14
C	0.16
D	0.87

Spectral Constants:

pH	λ_{max}	ϵ_{max}	$\dfrac{A_{250}}{A_{260}}$	$\dfrac{A_{280}}{A_{260}}$	Ref.
2	280	13.2	0.44	2.09	15
7	271	9.1	0.84	0.98	
11	271	9.1	0.84	0.98	

Remarks: The free acid met NRC specifications.

N-20
Cytidine 5′-Triphosphate

Abbreviations: Cyd-5′-P₃, 5′-CTP
Formula: $C_9H_{16}N_3O_{14}P_3$
Mol Wt.: 483.16
Calc. %: C, 22.37
   H, 3.34
   N, 8.70
   O, 48.36
   P, 19.23
pK$_a$:[16] 4.8, 6.6

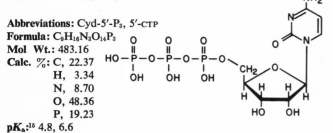

Paper Chromatography: Dissolve the Li⁺, Na⁺, K⁺, or NH₄⁺ salt in H₂O.

Solvent	R_f	R_M	R_M
A	0.20	1.15	1.55
B	0.05		
C	0.12	2.8	
D	0.94		

Spectral Constants:

pH	λ_{max}	ϵ_{max}	$\dfrac{A_{250}}{A_{260}}$	$\dfrac{A_{280}}{A_{260}}$	Ref.
2	280	12.8	0.45	2.12	15
7	271	9.0	0.84	0.97	
11	271	9.0	0.84	0.97	

N-21
Cytosine

Abbreviation: Cyt
Formula: $C_4H_5N_3O$
Mol. Wt.: 111.10
Calc. %: C, 43.24
   H, 4.54
   N, 37.82
   O, 14.40
pK$_a$:[30] 4.6, 12.2

Paper Chromatography: Dissolve the sample in 1 M NH₄OH.

Solvent	R_f
A	0.76
B	0.50
C	0.70
D	0.71

Gas Chromatography: Trimethylsilyl derivative.

MU value	Rel. Peak Area
15.18	100

Spectral Constants:

pH	λ_{max}	ϵ_{max}	$\dfrac{A_{250}}{A_{260}}$	$\dfrac{A_{280}}{A_{260}}$	Ref.
1	276	10.0	0.48	1.53	38
7	267	6.1	0.78	0.58	
14	282	7.9	0.60	3.28	

Remarks: A sample was dried for 8 h at 65 °C for spectral analysis. Met NRC specifications.

N-22
2′-Deoxyadenosine
[9-(2-Deoxy-β-D-ribofuranosyl)adenine]

Abbreviation: dAdo
Formula: $C_{10}H_{13}N_5O_3$
Mol. Wt.: 251.25
Calc. %: C, 47.80
   H, 5.21
   N, 27.88
   O, 19.11
Melting Pt.:[39] 188–190 °C
Specific Rotation: $[\alpha]_D^{24}$ −24.0° (0.37 g/100 ml, H₂O)[39]
pK$_a$:[31] 3.8

Paper Chromatography: Dissolve the sample in H₂O.

Solvent	R_f	R_M
A	0.93	
B	0.57	0.10
C	0.66	
D	0.40	2.1**

Gas Chromatography: Trimethylsilyl derivative.

MU value	Rel. Peak Area
25.91	100
24.95	2

Spectral Constants:

pH	λ_{max}	ϵ_{max}	$\dfrac{A_{250}}{A_{260}}$	$\dfrac{A_{280}}{A_{260}}$	Ref.
1	258	14.6			39, 40
7	260	14.9	0.80	0.16	
13	260	14.9			

N-23
2'-Deoxyadenosine 5'-Diphosphate
(2'-Deoxyadenosine 5'-Pyrophosphate)

Abbreviations: dAdo-5'-P$_2$, 5'-dADP
Formula: $C_{10}H_{15}N_5O_9P_2$
Mol. Wt.: 411.21
Calc. %: C, 29.21
H, 3.68
N, 17.03
O, 35.02
P, 15.06

Paper Chromatography: Dissolve the Li$^+$, Na$^+$, K$^+$, or NH$_4^+$ salt in H$_2$O.

Solvent	R_f	R_M	R_M
A	0.30	0.40	1.8
B	0.07	7.3	
C	0.26	1.6	
D	0.88	0.86	

Spectral Constants: No spectral data in the literature; use the data for 2'-deoxyadenosine 5'-phosphate (N-24).

N-24
2'-Deoxyadenosine 5'-Phosphate

Abbreviations: dAdo-5'-P, 5'-dAMP
Formula: $C_{10}H_{14}N_5O_6P$
Mol. Wt.: 331.23
Calc. %: C, 36.25
H, 4.26
N, 21.14
O, 28.98
P, 9.35
Specific Rotation: $[\alpha]_D^{19}$ −38.0° (0.23 g/100 ml, H$_2$O)[41]
pK_a:[31,42] 3.8, 6.4

Paper Chromatography: Dissolve the Li$^+$, Na$^+$, K$^+$, or NH$_4^+$ salt in H$_2$O.

Solvent	R_f
A	0.56
B	0.19
C	0.34
D	0.87

Gas Chromatography: Trimethylsilyl derivative.

MU value	Rel. Peak Area
30.92	100

Water Analysis: 8.90% (calculated for the dihydrate, 9.00%).

Spectral Constants: Literature data are inconsistent. Our data are given.

pH	λ_{max}	ϵ_{max}	$\frac{A_{250}}{A_{260}}$	$\frac{A_{280}}{A_{260}}$	Ref.
2	258	14.8	0.83	0.23	NRC sample
7	260	15.1	0.77	0.12	
11	260	15.2	0.77	0.12	

Remarks: Anion-exchange chromatography showed traces of 2 nucleosides and 1 nucleotide. In the absence of reliable, spectral reference data, the purity of this compound is regarded as tentative.

N-25
2'-Deoxyadenosine 5'-Triphosphate

Abbreviations: dAdo-5'-P$_3$, 5'-dATP
Formula: $C_{10}H_{16}N_5O_{12}P_3$
Mol. Wt.: 491.19
Calc. %: C, 24.45
H, 3.28
N, 14.26
O, 39.09
P, 18.92

Paper Chromatography: Dissolve the Li$^+$, Na$^+$, K$^+$, or NH$_4^+$ salt in H$_2$O.

Solvent	R_f	R_M	R_M
A	0.39	0.10	1.3
B	0.05		
C	0.18	4.9	
D	0.93	0.40	

Spectral Constants: No spectral data in the literature; use the data for 2'-deoxyadenosine 5'-phosphate (N-24).

N-26
2'-Deoxycytidine
[1-(2-Deoxy-β-D-ribofuranosyl)cytosine]

Abbreviation: dCyd
Formula: $C_9H_{13}N_3O_4$
Mol. Wt.: 227.22
Calc. %: C, 47.57
H, 5.77
N, 18.49
O, 28.16
Melting Pt.:[43] 207–209 °C
Specific Rotation: $[\alpha]_D^{23}$ +57.6° (2 g/100 ml, H$_2$O)[43]
pK_a:[44] 4.3

Paper Chromatography: Dissolve the sample in H_2O.

Solvent	R_f
A	0.80
B	0.62
C	0.77
D	0.76

Gas Chromatography: Trimethylsilyl derivative.

MU value	Rel. Peak Area
23.79	100

Water Analysis: 0.06%.

Spectral Constants:

pH	λ_{max}	ϵ_{max}	$\dfrac{A_{250}}{A_{260}}$	$\dfrac{A_{280}}{A_{260}}$	Ref.
1	280	13.2	0.42	2.15	44
7	271	9.0	0.83	0.97	
11	271	9.0	0.83	0.97	

Remarks: Met NRC specifications. Cation-exchange chromatography showed 1 impurity, probably less than 1%.

N-27

2'-Deoxycytidine 5'-Diphosphate
(2'-Deoxycytidine 5'-Pyrophosphate)

Abbreviations: dCyd-5'-P$_2$, 5'-dCDP
Formula: $C_9H_{15}N_3O_{10}P_2$
Mol. Wt.: 387.18
Calc. %: C, 27.92
         H, 3.90
         N, 10.85
         O, 41.32
         P, 16.00

Paper Chromatography: Dissolve the Li$^+$, Na$^+$, K$^+$, or NH$_4^+$ salt in H_2O.

Solvent	R_f	R_M	R_M
A	0.25	0.27	2.5
B	0.05		
C	0.20	0.57	1.5
D	0.89		

Spectral Constants: No spectral data in the literature; use the data for 2'-deoxycytidine 5'-phosphate (N-28).

N-28

2'-Deoxycytidine 5'-Phosphate

Abbreviations: dCyd-5'-P, 5'-dCMP

Formula: $C_9H_{14}N_3O_7P$
Mol. Wt.: 307.20
Calc. %: C, 35.18
         H, 4.59
         N, 13.67
         O, 36.46
         P, 10.08

Specific Rotation: $[\alpha]_D^{17}$ +38.5° (1.2 g/100 ml, H_2O)[45]
pK_a:[42] 4.6, 6.6

Paper Chromatography: Dissolve the free acid in 1 M NH_4OH; the Li$^+$, Na$^+$, or K$^+$ salt, in H_2O.

Solvent	R_f
A	0.53
B	0.16
C	0.21
D	0.85

Spectral Constants: Literature data on molar absorption coefficients are inconsistent. Our data are given.

pH	λ_{max}	ϵ_{max}	$\dfrac{A_{250}}{A_{260}}$	$\dfrac{A_{280}}{A_{260}}$	Ref.
1.5	279	13.0	0.43	2.12	12 (ratios)
7.0	270	9.1	0.82	0.99	NRC sample (ϵ values)

Remarks: The free acid met NRC specifications.

N-29

2'-Deoxycytidine 5'-Triphosphate

Abbreviations: dCyd-5'-P$_3$, 5'-dCTP
Formula: $C_9H_{16}N_3O_{13}P_3$
Mol. Wt.: 467.16
Calc. %: C, 23.14
         H, 3.45
         N, 8.99
         O, 44.52
         P, 19.89

Paper Chromatography: Dissolve the Li$^+$, Na$^+$, K$^+$, or NH$_4^+$ salt in H_2O.

Solvent	R_f	R_M	R_M
A	0.22	1.3	2.0
B	0.05		
C	0.18	0.50	2.5
D	0.93		

Spectral Constants: No spectral data in the literature; use the data for 2'-deoxycytidine 5'-phosphate (N-28).

N-30
2′-Deoxyguanosine
[9-(2-Deoxy-β-D-ribofuranosyl)guanine]

Abbreviation: dGuo
Formula: $C_{10}H_{13}N_5O_4$
Mol. Wt.: 267.25
Calc. %: C, 44.94
H, 4.90
N, 26.21
O, 23.95
Specific Rotation: $[\alpha]_D^{24} -30.2°$ (0.20 g/100 ml H_2O)[39]
pK_a:[31,46] 2.8, 9.5

Paper Chromatography: Dissolve the sample in 1 M NH_4OH.

Solvent	R_f	R_M
A	0.64	0.78
B	0.46	0.65
C	0.68	0.78
D	0.63	

Gas Chromatography: Trimethylsilyl derivative.

MU value	Rel. Peak Area
28.00	100
19.80	2
21.12	1 (guanine)
27.60	1
29.60	3

Spectral Constants:

pH	λ_{max}	ϵ_{max}	$\dfrac{A_{250}}{A_{260}}$	$\dfrac{A_{280}}{A_{260}}$	Ref.
7	254	13.0	1.16	0.65	39

N-31
2′-Deoxyguanosine 5′-Diphosphate
(2′-Deoxyguanosine 5′-Pyrophosphate)

Abbreviations: dGuo-5′-P_2, 5′-dGDP
Formula: $C_{10}H_{15}N_5O_{10}P_2$
Mol. Wt.: 427.21
Calc. %: C, 28.12
H, 3.54
N, 16.40
O, 37.45
P, 14.50

Paper Chromatography: Dissolve the sample in H_2O.

Solvent	R_f	R_M	R_M
A	0.17	2.4	
B	0.04	7.7	
C	0.13	1.6	
D	0.88	0.32	0.90

Spectral Constants: No spectral data in the literature; use the data for 2′-deoxyguanosine 5′-phosphate (N-32).

N-32
2′-Deoxyguanosine 5′-Phosphate

Abbreviations: dGuo-5′-P, 5′-dGMP
Formula: $C_{10}H_{14}N_5O_7P$
Mol. Wt.: 347.23
Calc. %: C, 34.58
H, 4.06
N, 20.17
O, 32.26
P, 8.92
Specific Rotation: $[\alpha]_D^{19} -31°$ (0.43 g/100 ml, H_2O)[47]
pK_a:[42] 2.9, 6.4, 9.7

Paper Chromatography: Dissolve the free acid in 1 M NH_4OH; the Li^+, Na^+, or K^+ salt, in H_2O.

Solvent	R_f
A	0.39
B	0.03
C	0.16
D	0.85

Gas Chromatography: Trimethylsilyl derivative.

MU value	Rel. Peak Area
32.24	100

Water Analysis: 18.2% (calculated for 4.5 H_2O, 17.1%).

Spectral Constants: Literature data are inconsistent. Our data are given.

pH	λ_{max}	ϵ_{max}	$\dfrac{A_{250}}{A_{260}}$	$\dfrac{A_{280}}{A_{260}}$	Ref.
2	255	12.5	1.05	0.73	NRC sample
7	253	13.8	1.18	0.67	
11	258	11.7	0.93	0.65	

Remarks: Anion-exchange chromatography showed 1 nucleotide impurity (approximately 2%). In the absence of reliable, spectral reference data, the purity of this sample is regarded as tentative.

N-33
2′-Deoxyinosine

Abbreviation: dIno
Formula: $C_{10}H_{12}N_4O_4$
Mol. Wt.: 252.23
Calc. %: C, 47.62
H, 4.80
N, 22.21
O, 25.37
Specific Rotation: $[\alpha]_D^{26} -21.5°$ (1 g/100 ml, H_2O)[43]

Paper Chromatography: Dissolve the sample in H_2O.

Solvent	R_f	R_M
A	0.62	0.30*
B	0.44	
C	0.58	
D	0.60	1.4**

Gas Chromatography: Trimethylsilyl derivative.

MU value	Rel. Peak Area
25.44	100
17.95	2 (hypoxanthine)
28.10	0.4

Spectral Constants:

pH	λ_{max}	ϵ_{max}	Ref.
7	249	12.8	48

N-34
2′-Deoxyuridine

Abbreviation: dUrd
Formula: $C_9H_{12}N_2O_5$
Mol. Wt.: 228.21
Calc. %: C, 47.37
　　　　H, 5.30
　　　　N, 12.28
　　　　O, 35.06
Melting Pt.:[43] 163 °C
Specific Rotation: $[\alpha]_D^{26}$ +30.0° (2 g/100 ml, H_2O)[43]
pK_a:[44] 9.3

Paper Chromatography: Dissolve the sample in H_2O.

Solvent	R_f
A	0.67
B	0.56
C	0.76

Spectral Constants:

pH	λ_{max}	ϵ_{max}	$\dfrac{A_{250}}{A_{260}}$	$\dfrac{A_{280}}{A_{260}}$	Ref.
1	262	10.2	0.72	0.38	12
7	262	10.2	0.72	0.38	
12	262	7.6	0.81	0.31	

Remarks: Met NRC specifications.

N-35
2′-Deoxyuridine 5′-Phosphate

Abbreviations: dUrd-5′-P, 5′-dUMP
Formula: $C_9H_{13}N_2O_8P$
Mol. Wt.: 308.19
Calc. %: C, 35.08
　　　　H, 4.25
　　　　N, 9.09
　　　　O, 41.53
　　　　P, 10.05
pK_a:[31] 9.3

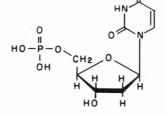

Paper Chromatography: Dissolve the Li⁺, Na⁺, K⁺, or NH₄⁺ salt in H_2O.

Solvent	R_f
A	0.39
B	0.12
C	0.24

Spectral Constants: Values in parentheses are NRC data for this compound.

pH	λ_{max}	ϵ_{max}	$\dfrac{A_{250}}{A_{260}}$	$\dfrac{A_{280}}{A_{260}}$	Ref.
2	260	9.8	0.78	0.41	49
7	260(262)	(9.7)	0.80	0.43	
12	260(261)	(7.5)	0.83	0.32	

Remarks: Met NRC specifications.

N-36
N^6,N^6-Dimethyladenine
(6-Dimethylaminopurine)

Abbreviation: 6Me₂Ade
Formula: $C_7H_9N_5$
Mol. Wt.: 163.18
Calc. %: C, 51.52
　　　　H, 5.56
　　　　N, 42.92
Melting Pt.:[52] 257 °C
pK_a:[52] 3.9, 10.5

Paper Chromatography: Dissolve the sample in 0.5 M HCl.

Solvent	R_f	R_M
A	0.87	0.85*
B	0.83	0.84*,**
C	0.82	0.65*,**
D	0.51	0.18*,**

Gas Chromatography: Trimethylsilyl derivative.

MU value	Rel. Peak Area
17.95	100

Spectral Constants:

pH	λ_{max}	ϵ_{max}	Ref.
2	276	15.5	53
7	275	17.8	
13	281	17.8	

N-37
N²,N²-Dimethylguanine
(2-Dimethylamino-6-hydroxypurine)

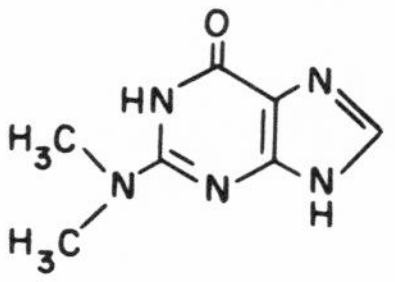

Abbreviation: 2Me₂Gua
Formula: $C_7H_9N_5O$
Mol. Wt.: 179.18
Calc. %: C, 46.92
    H, 5.06
    N, 39.09
    O, 8.93

Paper Chromatography: A solution containing 10 A_{260} units/5 μl could not be made. The data reported here were obtained by dissolving 1.6 mg in 0.3 ml of 0.5 M HCl (2 A_{260} units/5 μl) with heating, and applying 5 separate aliquots to the paper, drying after each application.

Solvent	R_f	R_M	R_M
A	0.56	1.4	
B	0.37	0.32	1.22
C	0.40		
D	0.20	0.09	1.1

Gas Chromatography: Trimethylsilyl derivative.

MU value
20.79

Spectral Constants:

pH	λ_{max}	ϵ_{max}	$\dfrac{A_{250}}{A_{260}}$	$\dfrac{A_{280}}{A_{260}}$	Ref.
1	256	19.0	0.92	0.36	50, 51
11	282	7.5	1.6	1.5	

Remarks: The sample tested was grossly impure, containing among other constituents di-isobutyl phthalate, a plasticizer. Paper-chromatographic data are for the sample tested previously.[1]

N-38
Guanine

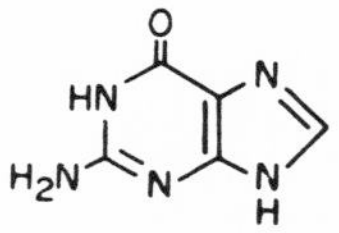

Abbreviation: Gua
Formula: $C_5H_5N_5O$
Mol. Wt.: 151.13
Calc. %: C, 39.73
    H, 3.33
    N, 46.34
    O, 10.59
pK_a:[54] 3.0, 9.3, 12.6

Paper Chromatography: A solution containing 10 A_{260} units/5 μl could not be made. The data reported here were obtained by dissolving 3.0 mg in 0.3 ml of 0.5 M HCl (3.3 A_{260} units/5 μl) with

heating and applying 3 separate aliquots to the paper, drying after each application.

Solvent	R_f
A	0.66
B	0.27
C	0.47
D	0.40

Gas Chromatography: Trimethylsilyl derivative.

MU value	Rel. Peak Area
21.15	100

Spectral Constants:

pH	λ_{max}	ϵ_{max}	$\dfrac{A_{250}}{A_{260}}$	$\dfrac{A_{280}}{A_{260}}$	Ref.
1	248	11.4	1.37	0.84	12
7	246	10.7	1.42	1.04	
11	274	8.0	0.99	1.14	

Remarks: Met NRC specifications.

N-39
Guanosine
(9-*β*-D-Ribofuranosylguanine)

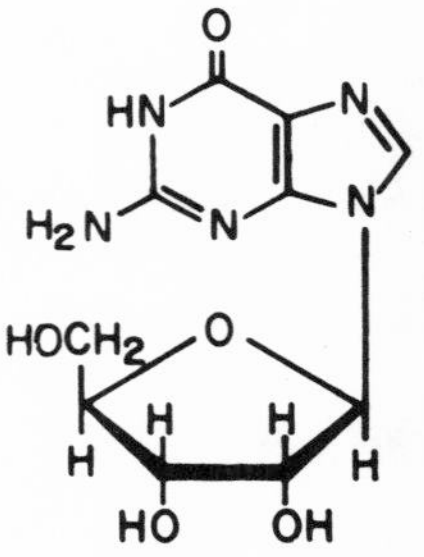

Abbreviation: Guo
Formula: $C_{10}H_{13}N_5O_5$
Mol. Wt.: 283.25
Calc. %: C, 42.40
    H, 4.63
    N, 24.73
    O, 28.24
Specific Rotation: $[\alpha]_D^{26}$ −72° (1.4 g/100 ml, 0.1 M NaOH)[55]
pK_a:[56,57] 2.2, 9.2, 12.3

Paper Chromatography: A solution containing 10 A_{260} units/5 μl could not be made. Data reported here were obtained by dissolving 2.5 mg in 0.1 ml of 1 M NH₄OH (5 A_{260} units/5 μl) with difficulty (i.e., the solid dissolves slowly) and applying 2 separate aliquots to the paper, drying after each application.

Solvent	R_f	R_M
A	0.50	
B	0.36	1.2
C	0.53	
D	0.62	

Gas Chromatography: Trimethylsilyl derivative.

MU value	Rel. Peak Area
28.02	100
21.15	0.5 (guanine)
27.61	1.5

Spectral Constants:

pH	λ_{max}	ϵ_{max}	$\dfrac{A_{250}}{A_{260}}$	$\dfrac{A_{280}}{A_{260}}$	Ref.
1	257	12.2	0.94	0.70	12
7	253	13.7	1.15	0.67	
11	258–266	11.3	0.89	0.61	

N-40
Guanosine 2′:3′-Cyclic Phosphate

Abbreviations: Guo-2′:3′-P; 2′:3′-cyclic GMP
Formula: $C_{10}H_{12}N_5O_7P$
Mol. Wt.: 345.21
Calc. %: C, 34.79
      H, 3.51
      N, 20.29
      O, 32.44
      P, 8.97

Paper Chromatography: Dissolve the Li^+, Na^+, K^+, NH_4^+, or Ba^{2+} salt in H_2O.

Solvent	R_f	R_M	R_M
A	0.34	0.39	0.54
B	0.55	0.33	
C	0.50	0.60	
D	0.85	1.1	

Spectral Constants: There are no reliable data available; use the data for guanosine 5′-phosphate (N-45).

N-41
Guanosine 5′-Diphosphate
(Guanosine 5′-Pyrophosphate)

Abbreviations: Guo-5′-P₂, 5′-GDP
Formula: $C_{10}H_{15}N_5O_{11}P_2$
Mol. Wt.: 443.21
Calc. %: C, 27.10
      H, 3.41
      N, 15.80
      O, 39.71
      P, 13.98
pK_a:[15] 2.9, 6.3, 9.6

Paper Chromatography: Dissolve the Li^+, Na^+, K^+, or NH_4^+ salt in H_2O.

Solvent	R_f	R_M
A	0.12	4.0
B	0.02	
C	0.08	
D	0.89	0.89

Spectral Constants:

pH	λ_{max}	ϵ_{max}	$\dfrac{A_{250}}{A_{260}}$	$\dfrac{A_{280}}{A_{260}}$	Ref.
1	256	12.3	0.95	0.67	15
7	253	13.7	1.15	0.66	
11	257	11.7	0.91	0.61	

N-42
Guanosine 2′-Phosphate
(2′-Guanylic Acid)

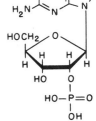

Abbreviations: Guo-2′-P, 2′-GMP
Formula: $C_{10}H_{14}N_5O_8P$
Mol. Wt.: 363.23
Calc. %: C, 33.06
      H, 3.88
      N, 19.28
      O, 35.24
      P, 8.52
pK_a:[31] 9.2

Paper Chromatography: Dissolve the Na^+ salt in H_2O.

Solvent	R_f	R_M	R_M
A	0.31	1.6	3.0
B	0.10	2.9	
C	0.15	3.6	
D	0.94	0.70	
F[a]	0.53	0.77 (3′-isomer)	

Gas Chromatography: Trimethylsilyl derivative.

MU value	Rel. Peak Area
31.15	100

Spectral Constants:

pH	$\dfrac{A_{250}}{A_{260}}$	$\dfrac{A_{280}}{A_{260}}$	Ref.
1	0.90	0.68	12
7	1.15	0.68	
12	0.89	0.60	

[a] See 2′-AMP (N-5).

N-43
Guanosine 2′(3′)-Phosphate, Mixed Anomers

Abbreviations: Guo-2′(3′)-P, 2′(3′)-GMP
Formula: $C_{10}H_{14}N_5O_8P$
Mol. Wt.: 363.23
Calc. %: C, 33.06
      H, 3.88
      N, 19.28
      O, 35.24
      P, 8.52

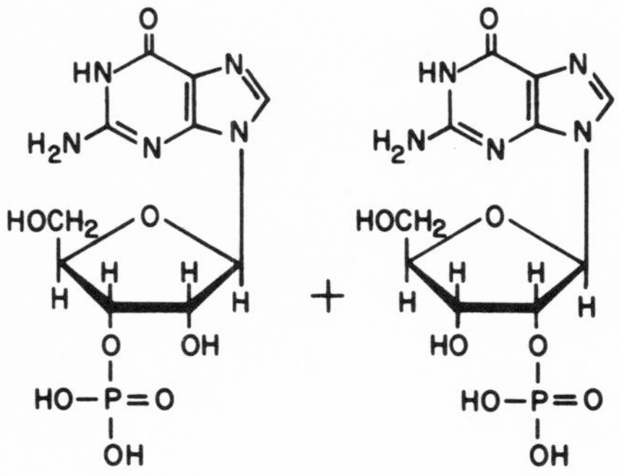

Paper Chromatography: Dissolve the sample in 1 M NH_4OH.

Solvent	R_f
A	0.28
B	0.05
C	0.07
D	0.88

Water Analysis: 5.6% (calculated for the monohydrate, 4.2%).

Spectral Constants: Literature data are inconsistent. Our data are given.

pH	λ_{max}	ϵ_{max}	$\dfrac{A_{250}}{A_{260}}$	$\dfrac{A_{280}}{A_{260}}$	Ref.
1	257	12.3	0.92	0.65	NRC sample
7	253	13.3	1.15	0.67	
12	263	11.2	0.88	0.58	

Remarks: Guanosine (about 1%), traces of 3 nucleosides, and 2'- and 3'-UMP detected by anion-exchange chromatography. In the absence of reliable, spectral reference data, the purity of this compound is regarded as tentative.

N-44
Guanosine 3'-Phosphate
(3'-Guanylic Acid)

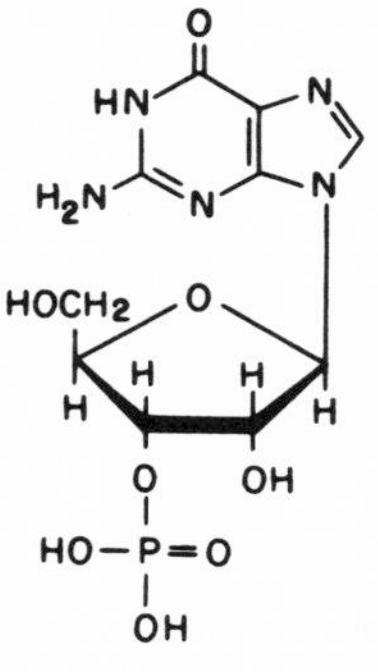

Abbreviations: Guo-3'-P, 3'-GMP
Formula: $C_{10}H_{14}N_5O_8P$
Mol. Wt.: 363.23
Calc. %: C, 33.06
H, 3.88
N, 19.28
O, 35.24
P, 8.52
pK_a:[31] 9.2

Paper Chromatography: Dissolve the Na^+ salt in H_2O.

Solvent	R_f	R_M
A	0.27	1.7
B	0.09	3.5
C	0.02	
D	0.91	0.75
F[a]	0.44	1.25 (2'-isomer)

Gas Chromatography: Trimethylsilyl derivative.

MU value	Rel. Peak Area
31.18	100

Spectral Constants:

pH	λ_{max}	$\dfrac{A_{250}}{A_{260}}$	$\dfrac{A_{280}}{A_{260}}$	Ref.
1	256	0.93	0.69	58
7	264	1.17	0.68	
12	260	0.88	0.58	

[a] See 2'-AMP (N-5).

N-45
Guanosine 5'-Phosphate
(5'-Guanylic Acid)

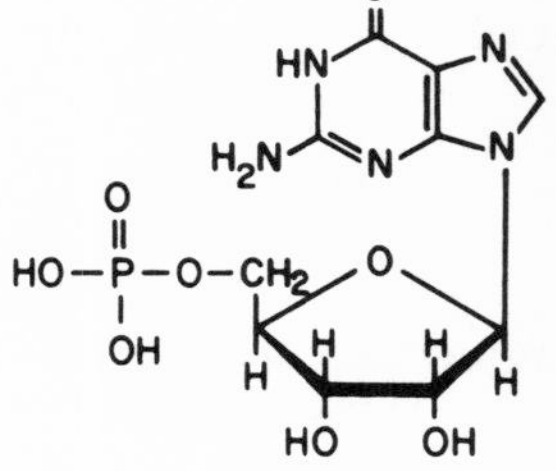

Abbreviations: Guo-5'-P, 5'-GMP
Formula: $C_{10}H_{14}N_5O_8P$
Mol. Wt.: 363.23
Calc. %: C, 33.06
H, 3.88
N, 19.28
O, 35.24
P, 8.52
pK_a:[15] 2.4, 6.1, 9.4

Paper Chromatography: Dissolve the free acid in 1 M NH_4OH; the Li^+, Na^+, or K^+ salt, in H_2O.

Solvent	R_f
A	0.15
B	0.06
C	0.05
D	0.87

Gas Chromatography: Trimethylsilyl derivative.

MU value	Rel. Peak Area
31.79	100

Water Analysis: 6.3% (calculated for the monohydrate, 4.3%).

Spectral Constants:

pH	λ_{max}	ϵ_{max}	$\dfrac{A_{250}}{A_{260}}$	$\dfrac{A_{280}}{A_{260}}$	Ref.
1	256	12.2	0.96	0.67	15
7	252	13.7	1.16	0.66	
11	258	11.6	0.90	0.61	

Remarks: Met NRC specifications.

N-46
Guanosine 5′-Triphosphate

Abbreviations: Guo-5′-P₃, 5′-GTP
Formula: $C_{10}H_{16}N_5O_{14}P_3$
Mol. Wt.: 523.19
Calc. %: C, 22.96
     H, 3.08
     N, 13.39
     O, 42.81
     P, 17.76
pKₐ:[15] 3.3, 6.5, 9.3

Paper Chromatography: Dissolve the Li⁺, Na⁺, K⁺, or NH₄⁺ salt in H₂O.

Solvent	R_f	R_M	R_M
A	0.20	1.2	
B	0.05		
C	0.10	2.6	0.50
D	0.94		

Spectral Constants:

pH	λ_{max}	ϵ_{max}	$\dfrac{A_{250}}{A_{260}}$	$\dfrac{A_{280}}{A_{260}}$	Ref.
1	256	12.4	0.96	0.67	15
7	253	13.7	1.17	0.66	
11	257	11.9	0.92	0.59	

N-47
Hypoxanthine

Abbreviation: Hyp
Formula: $C_5H_4N_4O$
Mol. Wt.: 136.11
Calc. %: C, 44.12
     H, 2.96
     N, 41.17
     O, 11.75
pKₐ:[52] 2.0, 8.9, 12.1

Paper Chromatography: Dissolve the sample in 0.1 M HCl.

Solvent	R_f
A	0.62
B	0.51
C	0.66
D	0.58

Gas Chromatography: Trimethylsilyl derivative.

MU value	Rel. Peak Area
17.92	100

Water Analysis: 0.04%.

Spectral Constants:

pH	λ_{max}	ϵ_{max}	$\dfrac{A_{250}}{A_{260}}$	$\dfrac{A_{280}}{A_{260}}$	Ref.
1	248	10.8	1.45	0.04	12
7	250	10.7	1.32	0.09	
11	259	11.1	0.84	0.12	

Remarks: Met NRC specifications. Cation-exchange chromatography showed a trace of 1 impurity, possibly xanthine.

N-48
Inosine
(9-β-D-Ribofuranosylhypoxanthine)

Abbreviation: Ino
Formula: $C_{10}H_{12}N_4O_5$
Mol. Wt.: 268.23
Calc. %: C, 44.78
     H, 4.51
     N, 20.89
     O, 29.82
Specific Rotation: $[\alpha]_D^{27}$ −58.8° (2.5 g/100 ml, H₂O)[59]
pKₐ:[59,60] 1.0, 8.8, 12.3

Paper Chromatography: Dissolve the sample in H₂O (warming may be necessary to effect dissolution).

Solvent	R_f	R_M
A	0.48	1.3
B	0.47	
C	0.59	0.83**
D	0.74	0.81

Gas Chromatography: Trimethylsilyl derivative.

MU value	Rel. Peak Area
25.96	100

Spectral Constants:

pH	λ_{max}	ϵ_{max}	$\dfrac{A_{250}}{A_{260}}$	$\dfrac{A_{280}}{A_{260}}$	Ref.
0	251	10.9	1.21	0.11	12
6	249	12.3	1.68	0.25	
11	253	13.1	1.05	0.18	

N-49
Inosine 5′-Diphosphate
(Inosine 5′-Pyrophosphate)

Abbreviations: Ino-5′-P₂, 5′-IDP
Formula: $C_{10}H_{14}N_4O_{11}P_2$
Mol. Wt.: 428.19
Calc. %: C, 28.05
     H, 3.30
     N, 13.09
     O, 41.10
     P, 14.47

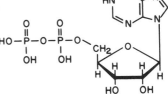

Paper Chromatography: Dissolve the Li⁺, Na⁺, K⁺, or NH₄⁺ salt in H₂O.

Solvent	R_f	R_M
A	0.13	3.0
B	0.02	
C	0.11	1.6
D	0.88	

Spectral Constants: No spectral data in the literature; use the data for inosine 5′-phosphate (N-50).

N-50
Inosine 5′-Phosphate
(5′-Inosinic Acid)

Abbreviations: Ino-5′-P, 5′-IMP
Formula: $C_{10}H_{13}N_4O_8P$
Mol. Wt.: 348.21
Calc. %: C, 34.48
  H, 3.76
  N, 16.09
  O, 36.76
  P, 8.89

Specific Rotation: $[\alpha]_D^{27}$ −36.8° (0.87 g/100 ml, 0.1 M HCl)[61]
pK_a:[59] 1.5, 6.0, 8.9

Paper Chromatography: Dissolve the Li⁺, Na⁺, K⁺, or NH₄⁺ salt in H₂O.

Solvent	R_f
A	0.28
B	0.03
C	0.15
D	0.90

Gas Chromatography: Trimethylsilyl derivative.

MU value	Rel. Peak Area
30.40	100

Water Analysis: 8.9% (calculated for the dihydrate, 8.4%).

Spectral Constants:

pH	λ_{max}	ϵ_{max}	$\dfrac{A_{250}}{A_{260}}$	$\dfrac{A_{280}}{A_{260}}$	Ref.
7	249	12.7	1.66	0.26	62

Remarks: The molar absorptivity at pH 7 was low, even after correction for water content (11.9). Anion-exchange chromatography showed one nucleotide impurity (about 5%).

N-51
Inosine 5′-Triphosphate

Abbreviations: Ino-5′-P₃, 5′-ITP
Formula: $C_{10}H_{15}N_4O_{14}P_3$
Mol. Wt.: 508.17
Calc. %: C, 23.64
  H, 2.98
  N, 11.03
  O, 44.08
  P, 18.28
pK_a:[63] 9.2

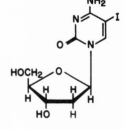

Paper Chromatography: Dissolve the Li⁺, Na⁺, K⁺, or NH₄⁺ salt in H₂O.

Solvent	R_f	R_M	R_M
A	0.20	1.2	
B	0.05		
C	0.16	0.50**	2.1
D	0.95		

Spectral Constants: No spectral data in the literature; use the data for inosine 5′-phosphate (N-50).

N-52
5-Iodo-2′-deoxycytidine
(2′-Deoxy-5-iodocytidine)

Abbreviation: 5IdCyd
Formula: $C_9H_{12}IN_3O_4$
Mol. Wt.: 353.12
Calc. %: C, 30.61
  H, 3.43
  I, 35.94
  N, 11.90
  O, 18.12

Paper Chromatography: Dissolve the sample in H₂O.

Solvent	R_f
A	0.80
B	0.63
C	0.70
D	0.64

Spectral Constants:

pH	λ_{max}	ϵ_{max}	Ref.
1	309	8.15	64
7	293	6.02	
13	294	5.93	

Remarks: Met NRC specifications.

N-53
5-Iodo-2′-deoxyuridine
(2′-Deoxy-5-iodouridine)

Abbreviation: 5IdUrd
Formula: $C_9H_{11}IN_2O_5$
Mol. Wt.: 354.10
Calc. %: C, 30.53
　　　　　H, 3.13
　　　　　I, 35.84
　　　　　N, 7.91
　　　　　O, 22.59
pK_a:[28] 8.2

Paper Chromatography: Dissolve the sample in H_2O.

Solvent	R_f
A	0.68
B	0.59
C	0.84

Spectral Constants:

pH	λ_{max}	ϵ_{max}	Ref.
1	289	7.83	64
7	289	7.69	
13	280	5.79	

Remarks: Met NRC specifications.

N-54
5-Iodouridine

Abbreviation: 5IUrd
Formula: $C_9H_{11}IN_2O_6$
Mol. Wt.: 370.10
Calc. %: C, 29.21
　　　　　H, 3.00
　　　　　I, 34.29
　　　　　N, 7.57
　　　　　O, 25.94
pK_a:[28] 8.5

Paper Chromatography: Dissolve the sample in H_2O.

Solvent	R_f
A	0.61
B	0.53
C	0.75
D	0.73

Spectral Constants:

pH	λ_{max}	ϵ_{max}	Ref.
1	289	8.02	64
7	288	7.96	
13	280	6.17	

Remarks: Met NRC specifications.

N-55
N^6-Isopentenyladenine
[N^6-(3-Methyl-2-butenyl)adenine]

Abbreviation: 6Pe^{i}Ade
Formula: $C_{10}H_{13}N_5$
Mol. Wt.: 203.24
Calc. %: C, 59.09
　　　　　H, 6.45
　　　　　N, 34.46
Melting Pt.:[65] 213–215 °C

Paper Chromatography: Dissolve the sample in 0.5 M HCl.

Solvent	R_f	R_M
A	0.96	
B	0.84	0.89*
C	0.88	
D	0.53	

Gas Chromatography: Trimethylsilyl derivative.

MU value	Rel. Peak Area
21.15	100
23.76	0.5

Spectral Constants:

pH	λ_{max}	ϵ_{max}	Ref.
1	273	18.6	66
7	269	19.4	
12	275	18.1	

N-56
N^6-Isopentenyladenosine
[N^6-(3-Methyl-2-butenyl)adenosine]

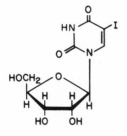

Abbreviation: 6Pe^{i}Ado
Formula: $C_{15}H_{21}N_5O_4$
Mol. Wt.: 335.36
Calc. %: C, 53.72
　　　　　H, 6.31
　　　　　N, 20.88
　　　　　O, 19.08
Melting Pt.:[67] 145–147 °C
Specific Rotation: $[\alpha]_D^{28}$ −103° (0.14 g/100 ml, EtOH)[68]
pK_a:[69] 3.8

Paper Chromatography: Dissolve the sample in H_2O.

Solvent	R_f
A	0.93
B	0.84
C	0.84
D	0.64

Gas Chromatography: Trimethylsilyl derivative.

MU value	Rel. Peak Area
29.50	100
26.09	0.5

Water Analysis: 2.2%

Spectral Constants:

pH	λ_{max}	ϵ_{max}	Ref.
1	265	20.4	66
7	269	20.0	
12	269	20.0	

Remarks: Measured molar absorption coefficients were 5% higher than those listed. The purity of this compound is regarded as tentative until additional spectral reference data are available, including ratios. This compound decomposes on storage, even at $-70\ °C$.

N-57
Kinetin
(N^6-Furfuryladenine)

Formula: $C_{10}H_9N_5O$
Mol. Wt.: 215.22
Calc. %: C, 55.81
　　　　 H, 4.21
　　　　 N, 32.54
　　　　 O, 7.43
Melting Pt.:[70] 265–266 °C
pK_a:[71] 3.8, 10.0

Paper Chromatography: Dissolve the sample in 0.5 M HCl.

Solvent	R_f	R_M
A	0.93	0.45
B	0.85	0.61
C	0.85	0.68**

Gas Chromatography: Trimethylsilyl derivative.

MU value	Rel. Peak Area
22.04	100
18.52	1

Spectral Constants:

pH	λ_{max}	ϵ_{max}	Ref.
0	274	17.0	71
6	267	18.8	
14	273.5	17.6	

N-58
N^6-Methyladenine

Abbreviation: 6MeAde
Formula: $C_6H_7N_5$
Mol. Wt.: 149.16
Calc. %: C, 48.31
　　　　 H, 4.73
　　　　 N, 46.95
Melting Pt.:[72] 319–320 °C
pK_a:[52] 4.2, 10.0

Paper Chromatography: Dissolve the sample in hot H_2O.

Solvent	R_f	R_M
A	0.89	
B	0.77	1.1*
C	0.78	0.10
D	0.46	1.2*
F[a]	0.62	0.10

Gas Chromatography: Trimethylsilyl derivative.

MU value	Rel. Peak Area
17.59	100
20.00	1.3

Spectral Constants:

pH	λ_{max}	ϵ_{max}	Ref.
2	267	15.1	53
7	266	16.2	
12	273	15.8	

[a]Butanol saturated with H_2O.

N-59
5-Methylcytosine

Abbreviation: 5MeCyt
Formula: $C_5H_7N_3O$
Mol. Wt.: 125.13
Calc. %: C, 47.99
　　　　 H, 5.64
　　　　 N, 33.58
　　　　 O, 12.79
pK_a:[73] 4.6, 12.4

Paper Chromatography: A solution containing 10 A_{260} units/μl could not be made. The data reported here were obtained by dissolving 3.6 mg in 0.1 ml of H_2O (3.3 A_{260} units/5 μl) with warming and applying 3 separate aliquots to the paper, drying after each application.

Solvent	R_f
A	0.80
B	0.65
C	0.70
D	0.71

Gas Chromatography: Trimethylsilyl derivative.

MU value	Rel. Peak Area
15.44	100

Water Analysis: 6.0%.

Spectral Constants:

pH	λ_{max}	ϵ_{max}	$\dfrac{A_{250}}{A_{260}}$	$\dfrac{A_{280}}{A_{260}}$	Ref.
1	283	9.8	0.41	2.66	12
7	273	6.2	0.81	1.20	
14	289	8.1	0.85	3.75	

Remarks: Met NRC specifications.

N-60
5-Methyl-2'-deoxycytidine
(2'-Deoxy-5-methylcytidine)

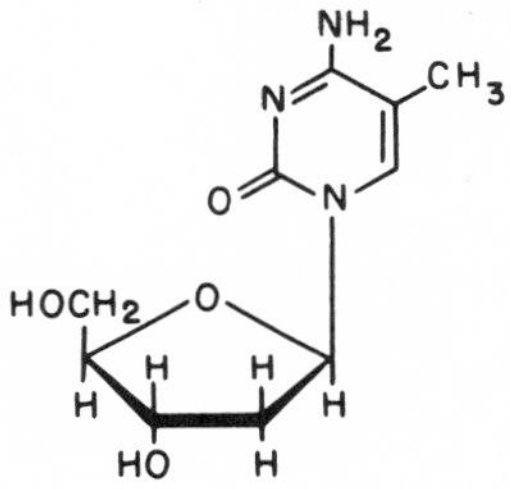

Abbreviation: 5MedCyd
Formula: $C_{10}H_{15}N_3O_4$
Mol. Wt.: 241.25
Calc. %: C, 49.79
             H, 6.27
             N, 17.42
             O, 26.53
Specific Rotation: $[\alpha]_D^{23}$ +62° (1.0 g/100 ml, 1 M NaOH)[73]
pK_a:[73] 4.4

Paper Chromatography: Dissolve the sample in H$_2$O.

Solvent	R_f
A	0.81
B	0.65
C	0.77
D	0.80

Water Analysis: 0.7%.

Spectral Constants:

pH	λ_{max}	ϵ_{max}	$\dfrac{A_{250}}{A_{260}}$	$\dfrac{A_{280}}{A_{260}}$	Ref.
1	287	12.4	0.43	3.00	73
7	277	8.5	0.99	1.54	
14	279	8.8			

Remarks: Molar absorption coefficients were 5–10% too low, even after correction for water content. Supplier claimed 1.5 molecules of water of hydration (10.1%) per molecule. Cation-exchange chromatography showed 1 impurity (about 3%).

N-61
7-Methylguanine

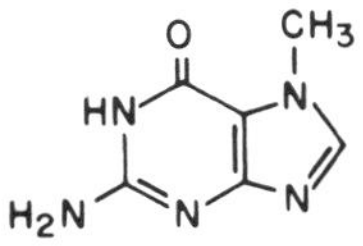

Abbreviation: 7MeGua
Formula: $C_6H_7N_5O$
Mol. Wt.: 165.16
Calc. %: C, 43.63
             H, 4.27
             N, 42.40
             O, 9.69
pK_a:[74] 3.5, 10.0

Paper Chromatography: Dissolve the sample in 0.5 M HCl.

Solvent	R_f
A	0.82
B	0.35
C	0.55
D	0.45

Spectral Constants:

pH	λ_{max}	ϵ_{max}	$\dfrac{A_{280}}{A_{260}}$	Ref.
1	250	11.0	0.79	74, 75
7	283	7.8	1.8	
13	280	7.8	1.9	

Remarks: Met NRC specifications.

N-62
1-Methylinosine
(1-Methyl-9-β-D-ribofuranosylhypoxanthine)

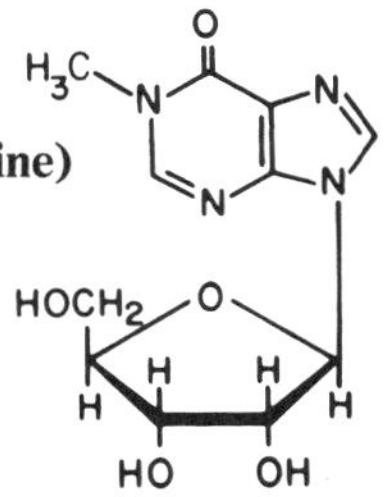

Abbreviation: 1MeIno
Formula: $C_{11}H_{14}N_4O_5$
Mol. Wt.: 282.26
Calc. %: C, 46.81
             H, 5.00
             N, 19.85
             O, 28.34
Melting Pt.:[76] 210–212 °C
Specific Rotation: $[\alpha]_D^{28}$ −49.2° (0.50 g/100 ml, H$_2$O)[76]
pK_a:[60] 1.2

Paper Chromatography: Dissolve the sample in H$_2$O.

Solvent	R_f	R_M	R_M
A	0.65	0.77	1.2
B	0.51	0.59	
C	0.75	0.87	
D	0.84		

Gas Chromatography: Trimethylsilyl derivative.

MU value	Rel. Peak Area
29.80	100
25.88	5 (inosine)

Spectral Constants:

pH	λ_{max}	ϵ_{max}	Ref.
7	251	10.4	77

N-63
5-Methyluridine
(5-Methyl-1-β-D-ribofuranosyluracil)

Abbreviation: 5MeUrd
Formula: $C_{10}H_{14}N_2O_6$
Mol. Wt.: 258.23
Calc. %: C, 46.51
       H, 5.46
       N, 10.85
       O, 37.18
Melting Pt.:[78] 183–185 °C
Specific Rotation: $[\alpha]_D^{31}$ −10° (2 g/100 ml, H_2O)[78]
pK$_a$:[78] 9.7

Paper Chromatography: Dissolve the sample in H_2O.

Solvent	R_f
A	0.62
B	0.52
C	0.76
D	0.85

Gas Chromatography: Trimethylsilyl derivative.

MU value	Rel. Peak Area
25.02	100
25.90	2

Water Analysis: 3.8%.

Spectral Constants: Literature data are inconsistent. Our data are given.

pH	λ_{max}	ϵ_{max}	$\dfrac{A_{250}}{A_{260}}$	$\dfrac{A_{280}}{A_{260}}$	Ref.
1	266	9.7	0.66	0.67	NRC sample
7	267	9.5	0.61	0.79	
13	267	7.3	0.72	0.74	

Remarks: In the absence of reliable, spectral reference data, the purity of this sample is regarded as tentative.

N-64
Orotic Acid

Abbreviation: Oro
Formula: $C_5H_4N_2O_4$
Mol. Wt.: 156.10
Calc. %: C, 38.48
       H, 2.58
       N, 17.94
       O, 41.00
pK$_a$:[38] 2.4, 9.5

Paper Chromatography: A solution containing 10 A_{260} units/5 μl could not be made. Data reported here were obtained by dissolving 4.1 mg in 0.1 ml of 1 M NH_4OH (5 A_{260} units/5 μl) and applying two separate aliquots to the paper, drying after each application.

Solvent	R_f	R_M	R_M
A	0.25	0.20	
B	0.45		
C	0.43	0.40*	0.10
D	0.84		

Gas Chromatography: Trimethylsilyl derivative.

MU value	Rel. Peak Area
17.50	100
18.43	0.2
20.13	0.3

Spectral Constants:

pH	λ_{max}	ϵ_{max}	Ref.
1	280	7.5	38
7	279	7.7	
12	286	6.0	

N-65
Pseudouridine, Mixed Anomers

Abbreviation: Ψrd
Formula: $C_9H_{12}N_2O_6$
Mol. Wt.: 244.21
Calc. %: C, 44.27
       H, 4.95
       N, 11.47
       O, 39.31

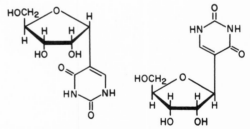

Paper Chromatography: Dissolve the sample in H_2O.

Solvent	R_f	R_M	R_M	R_M
A	0.38	1.9	1.3 (α-anomer)	
B	0.45	0.76	1.5	2.0**
C	0.59	1.2	1.3	
D	0.79	0.24**		

Gas Chromatography: Trimethylsilyl derivative.

MU value	Rel. Peak Area
23.32	37.3 (α anomer)
23.75	61.4 (β anomer)
24.10	0.6
28.86	0.7

Spectral Constants: Use the data for pseudouridine, β-anomer (N-66).

N-66
Pseudouridine, β-Anomer
(5-β-D-Ribofuranosyluracil)

Abbreviation: Ψrd
Formula: $C_9H_{12}N_2O_6$
Mol. Wt.: 244.21
Calc. %: C, 44.27
          H, 4.95
          N, 11.47
          O, 39.31
Melting Pt.:[79] 233–234 °C
Specific Rotation: $[a]_D$ −3° (1.0 g/100 ml, H_2O)[80]
pK_a:[81] 9.1

Paper Chromatography: Dissolve the sample in H_2O.

Solvent	R_f	R_M
A	0.44	1.5
B	0.41	
C	0.65	
D	0.82	

Gas Chromatography: Trimethylsilyl derivative.

MU value	Rel. Peak Area
23.75	100
17.30, 24.13 }	traces
24.89, 26.24 }	

Spectral Constants:

pH	λ_{max}	ϵ_{max}	$\dfrac{A_{240}}{A_{260}}$	$\dfrac{A_{280}}{A_{260}}$	Ref.
0–7	263	8.1	0.36	0.40	81, 82
12	287	7.8			

N-67
9-β-D-Ribosylkinetin
(N^6-Furfuryladenosine)

Formula: $C_{15}H_{17}N_5O_5$
Mol. Wt.: 347.35
Calc. %: C, 51.87
          H, 4.93
          N, 20.16
          O, 23.03
Melting Pt.:[83] 151–152 °C
Specific Rotation: $[\alpha]_D^{25}$ −63.5° (1.13 g/100 ml, EtOH)[84]

Paper Chromatography: Dissolve the sample in 0.1 M HCl.

Solvent	R_f	R_M	R_M	R_M
A	0.88	0.39	0.93*,**	
B	0.89	0.63	0.73	
C	0.80	0.81**	0.65*,**	
D	0.68	0.76**	0.06*,**	1.1

Gas Chromatography: Trimethylsilyl derivative.

MU value	Rel. Peak Area
30.47	100
25.95, 26.43	0.6
29.96, 31.32	0.4

Spectral Constants:

Solvent	λ_{max}	ϵ_{max}	Ref.
Ethanol	268	19.0	84

N-68
Thymidine
[1-(2-Deoxy-β-D-ribofuranosyl)thymine]

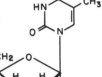

Abbreviation: dThd
Formula: $C_{10}H_{14}N_2O_5$
Mol. Wt.: 242.23
Calc. %: C, 49.58
          H, 5.83
          N, 11.57
          O, 33.03
Melting Pt.:[43] 186–187 °C
Specific Rotation: $[\alpha]_D^{23}$ +18.5° (2 g/100 ml, H_2O)[43]
pK_a:[30] 9.8, 12.9

Paper Chromatography: Dissolve the sample in H_2O.

Solvent	R_f
A	0.72
B	0.73
C	0.81
D	0.78

Water Analysis: 0.1%.

Spectral Constants:

pH	λ_{max}	ϵ_{max}	$\dfrac{A_{250}}{A_{260}}$	$\dfrac{A_{280}}{A_{260}}$	Ref.
1	267	9.65	0.65	0.72	12
7	267	9.65	0.65	0.72	
13	267	7.4	0.75	0.67	

Remarks: Met NRC specifications. Cation-exchange chromatography showed 1 impurity (less than 1%).

N-69
Thymidine 3′,5′-Bisphosphate
(Thymidine 3′,5′-Diphosphate)

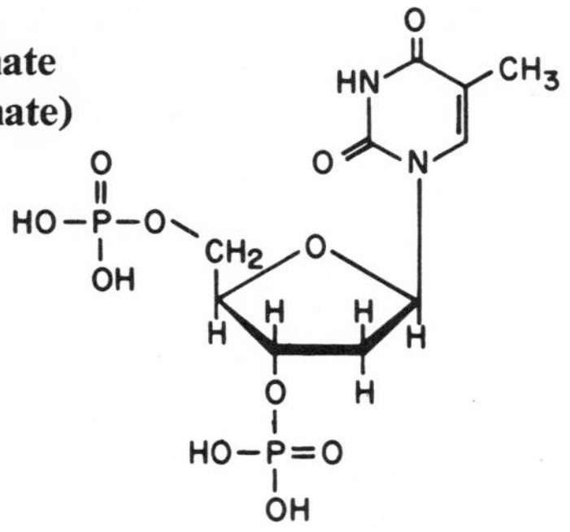

Abbreviations: dThd-3′,5′-P$_2$;
3′,5′-dTDP
Formula: C$_{10}$H$_{16}$N$_2$O$_{11}$P$_2$
Mol. Wt.: 402.19
Calc. %: C, 29.86
H, 4.01
N, 6.97
O, 43.76
P, 15.40

Paper Chromatography: Dissolve the Li$^+$, Na$^+$, K$^+$, or NH$_4^+$ salt in H$_2$O.

Solvent	R_f
A	0.21
B	0.04
C	0.10
D	0.97

Water Analysis: 27.2% (calculated for 6.5 H$_2$O, as on label, 19.3%).

Spectral Constants:

pH	λ_{max}	ϵ_{max}	$\dfrac{A_{250}}{A_{260}}$	$\dfrac{A_{280}}{A_{260}}$	Ref.
1–2	267	9.9	0.65	0.70	85
2	267	10.2	0.66	0.72	NRC sample
7	268	10.4	0.65	0.72	

Remarks: Met NRC specifications. Anion-exchange chromatography showed one nucleoside and two nucleotide impurities, totaling 2–3%. Molecular weight given by supplier for the Na$_4 \cdot$6.5 H$_2$O salt is incorrect.

N-70
Thymidine 5′-Diphosphate
(Thymidine 5′-Pyrophosphate)

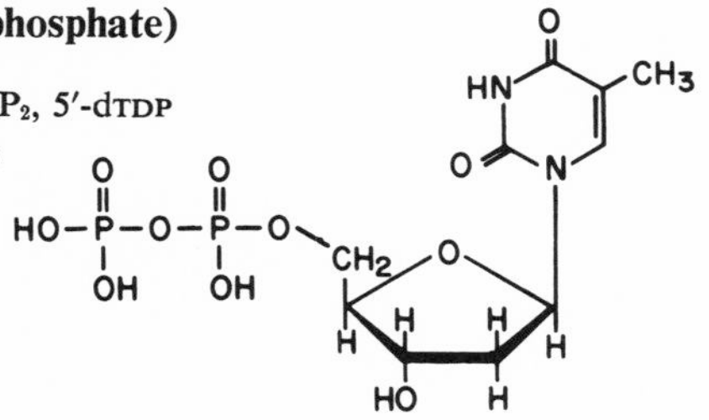

Abbreviations: dThd-5′-P$_2$, 5′-dTDP
Formula: C$_{10}$H$_{16}$N$_2$O$_{11}$P$_2$
Mol. Wt.: 402.19
Calc. %: C, 29.86
H, 4.01
N, 6.97
O, 43.76
P, 15.40

Paper Chromatography: Dissolve the Li$^+$, Na$^+$, K$^+$, or NH$_+^4$ salt in H$_2$O.

Solvent	R_f	R_M
A	0.22	
B	0.07	
C	0.31	1.3
D	0.91	

Spectral Constants: No spectral data in the literature; use the data for thymidine 5′-phosphate (N-71).

N-71
Thymidine 5′-Phosphate
(5′-Thymidylic Acid)

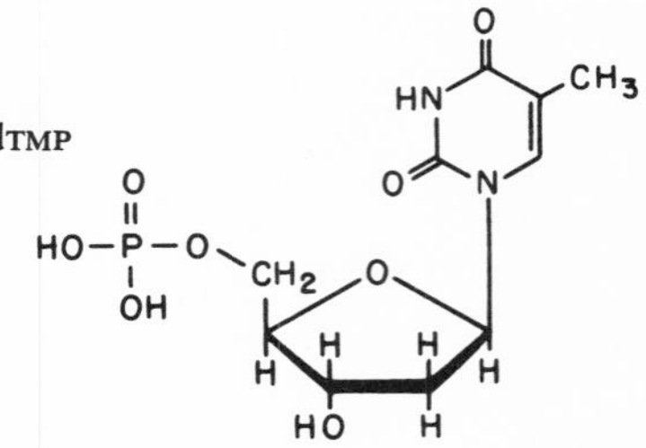

Abbreviations: dThd-5′-P, 5′-dTMP
Formula: C$_{10}$H$_{15}$N$_2$O$_8$P
Mol. Wt.: 322.21
Calc. %: C, 37.27
H, 4.69
N, 8.69
O, 39.73
P, 9.61
Specific Rotation: $[\alpha]_D^{21}$ −4.4° (0.4 g/100 ml, H$_2$O)[86]
pK_a:[42] 6.5, 10.0

Paper Chromatography: Dissolve the Li$^+$, Na$^+$, K$^+$, or NH$_4^+$ salt in H$_2$O.

Solvent	R_f	R_M	R_M
A	0.39		
B	0.22		
C	0.48		
D	0.89	0.36	0.10

Spectral Constants:

pH	λ_{max}	ϵ_{max}	$\dfrac{A_{250}}{A_{260}}$	$\dfrac{A_{280}}{A_{260}}$	Ref.
7	267	9.6	0.65	0.72	62

N-72
Thymidine 5′-Triphosphate

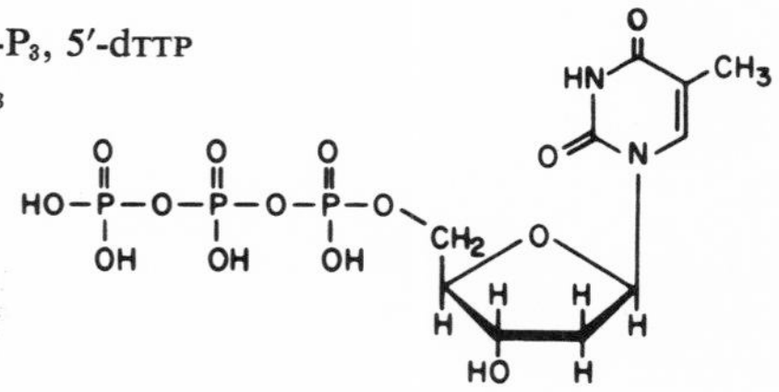

Abbreviations: dThd-5′-P$_3$, 5′-dTTP
Formula: C$_{10}$H$_{17}$N$_2$O$_{14}$P$_3$
Mol. Wt.: 482.17
Calc. %: C, 24.91
H, 3.55
N, 5.81
O, 46.46
P, 19.27
pK_a:[63] 9.8

Paper Chromatography: Dissolve the Li$^+$, Na$^+$, K$^+$, or NH$_4^+$ salt in H$_2$O.

Solvent	R_f	R_M	R_M
A	0.23	1.2	1.7
B	0.05		
C	0.20	2.0	
D	0.94		

Spectral Constants: No spectral data in the literature; use the data for thymidine 5′-phosphate (N-71).

N-73
Thymine
(5-Methyluracil)

Abbreviation: Thy
Formula: $C_5H_6N_2O_2$
Mol. Wt.: 126.12
Calc. %: C, 47.62
       H, 4.80
       N, 22.22
       O, 25.37
pK_a:[30] 9.9

Paper Chromatography: A solution containing 10 A_{260} units/5 μl could not be made. The data reported here were obtained by dissolving 1.4 mg in 0.1 ml of 2 M NH_4OH (5 A_{260} units/5 μl) and applying two separate aliquots to the paper, drying after each application.

Solvent	R_f
A	0.78
B	0.63
C	0.72

Spectral Constants:

pH	λ_{max}	ϵ_{max}	$\dfrac{A_{250}}{A_{260}}$	$\dfrac{A_{280}}{A_{260}}$	Ref.
4	264.5	7.9	0.67	0.53	12
7	264.5	7.9	0.67	0.53	
12	291	5.4	0.65	1.31	

Remarks: Met NRC specifications.

N-74
Uracil

Abbreviation: Ura
Formula: $C_4H_4N_2O_2$
Mol. Wt.: 112.09
Calc. %: C, 42.86
       H, 3.60
       N, 24.99
       O, 28.55
pK_a:[30] 9.5

Paper Chromatography: A solution containing 10 A_{260} units/5 μl could not be made. The data reported here were obtained by dissolving 2.9 mg in 0.2 ml of 1 M HCl (5 A_{260} units/5 μl) with heating and applying two separate aliquots to the paper, drying after each application.

Solvent	R_f
A	0.63
B	0.59
C	0.70
D	0.70

Gas Chromatography: Trimethylsilyl derivative.

MU value	Rel. Peak Area
13.30	100

Spectral Constants:

pH	λ_{max}	ϵ_{max}	$\dfrac{A_{250}}{A_{260}}$	$\dfrac{A_{280}}{A_{260}}$	Ref.
0	260	7.8	0.80	0.30	12
7	260	8.2	0.84	0.18	
12	284	6.2	0.71	1.40	

Remarks: Met NRC specifications.

N-75
Uridine
(1-β-D-Ribofuranosyluracil)

Abbreviation: Urd
Formula: $C_9H_{12}N_2O_6$
Mol. Wt.: 244.21
Calc. %: C, 44.26
       H, 4.95
       N, 11.47
       O, 39.31
Melting Pt.:[29] 163–165 °C
Specific Rotation: $[\alpha]_D^{16}$ +9.6° (2.0 g/100 ml, H_2O)[29]
pK_a:[30] 9.3, 12.6

Paper Chromatography: Dissolve the sample in H_2O.

Solvent	R_f	R_M	R_M
A	0.51	0.75	
B	0.57	0.66	0.74
C	0.68	0.87	
D	0.80		

Gas Chromatography: Trimethylsilyl derivative.

MU value	Rel. Peak Area
24.66	100
13.34	1 (uracil)
23.64	2.4
32	1.6

Spectral Constants:

pH	λ_{max}	ϵ_{max}	$\dfrac{A_{250}}{A_{260}}$	$\dfrac{A_{280}}{A_{260}}$	Ref.
1	262	10.1	0.74	0.35	12, 44
7	262	10.1	0.74	0.35	
12	262	7.45	0.83	0.29	

N-76
Uridine 2′:3′-Cyclic Phosphate

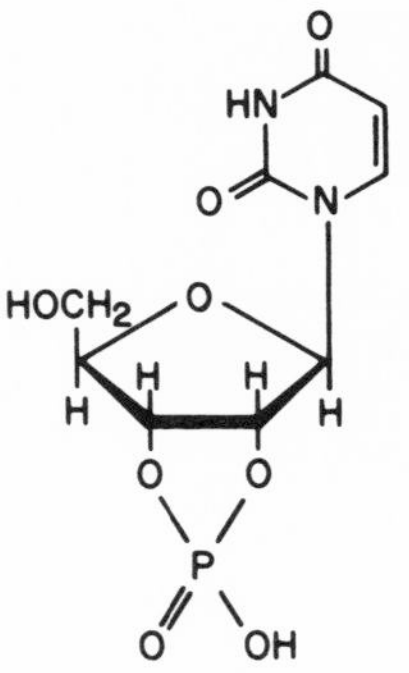

Abbreviations: Urd-2′:3′-P; 2′:3′-cyclic UMP
Formula: $C_9H_{11}N_2O_8P$
Mol. Wt.: 306.17
Calc. %: C, 35.31
H, 3.62
N, 9.15
O, 41.81
P, 10.11
pK$_a$:[31] 9.5

Paper Chromatography: Dissolve the Li$^+$, Na$^+$, K$^+$, or NH$_4^+$ salt in H$_2$O.

Solvent	R_f	R_M	R_M
A	0.32	0.69	
B	0.38	0.25	
C	0.65	0.57	0.86*.**
D	0.94		

Spectral Constants:

pH	λ_{max}	ϵ_{max}	Ref.
7	258–259	9.6	32

N-77
Uridine 5′-Diphosphate
(Uridine 5′-Pyrophosphate)

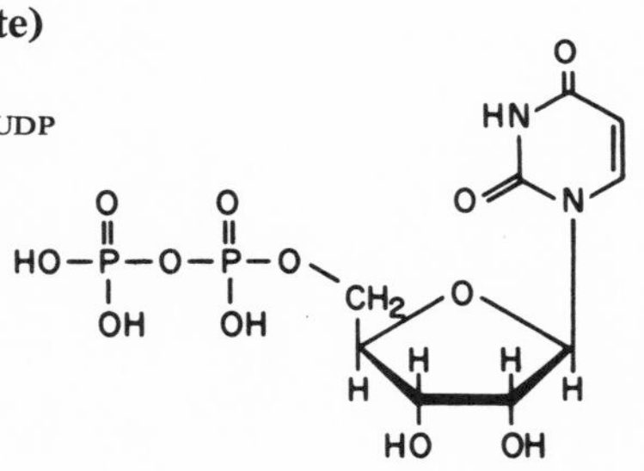

Abbreviations: Urd-5′-P$_2$, 5′-UDP
Formula: $C_9H_{14}N_2O_{12}P_2$
Mol. Wt.: 404.17
Calc. %: C, 26.74
H, 3.49
N, 6.93
O, 47.50
P, 15.33
pK$_a$:[15] 6.5, 9.4

Paper Chromatography: Dissolve the Li$^+$, Na$^+$, K$^+$, or NH$_4^+$ salt in H$_2$O.

Solvent	R_f	R_M	R_M
A	0.14	2.5	
B	0.02		
C	0.16	0.65	1.5
D	0.89		

Spectral Constants:

pH	λ_{max}	ϵ_{max}	$\dfrac{A_{250}}{A_{260}}$	$\dfrac{A_{280}}{A_{260}}$	Ref.
2	262	10.0	0.73	0.39	15
7	262	10.0	0.73	0.39	
11	261	7.9	0.80	0.32	

N-78
Uridine 2′-Phosphate
(2′-Uridylic Acid)

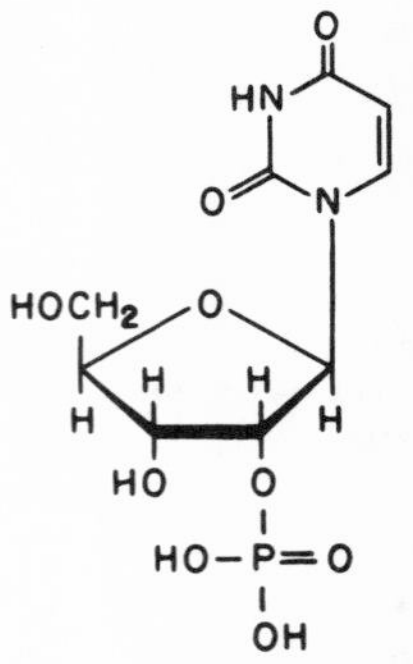

Abbreviations: Urd-2′-P, 2′-UMP
Formula: $C_9H_{13}N_2O_9P$
Mol. Wt.: 324.19
Calc. %: C, 33.34
H, 4.04
N, 8.64
O, 44.42
P, 9.55

Paper Chromatography: Dissolve the Li$^+$ salt in H$_2$O.

Solvent	R_f	R_M	R_M
A	0.32	0.47	0.69
B	0.22	3.8	
C	0.33	0.67*	
D	0.94		

Gas Chromatography: Trimethylsilyl derivative.

MU value	Rel. Peak Area
26.09	100

Spectral Constants:

pH	ϵ_{260}	$\dfrac{A_{250}}{A_{260}}$	$\dfrac{A_{280}}{A_{260}}$	Ref.
2	10.0	0.80	0.28	12
7	10.0	0.78	0.30	
12	7.4	0.85	0.25	

N-79
Uridine 2′(3′)-Phosphate, Mixed Anomers

Abbreviations: Urd-2′(3′)-P, 2′(3′)-UMP
Formula: $C_9H_{13}N_2O_9P$
Mol. Wt.: 324.19
Calc. %: C, 33.34   N, 8.64   P, 9.55
H, 4.04   O, 44.42

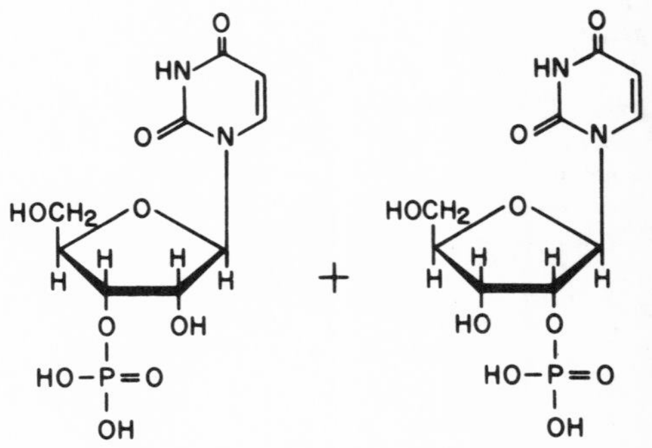

Paper Chromatography: Dissolve the free acid in 1 M NH₄OH; the Li⁺, Na⁺, or K⁺ salt, in H₂O.

Solvent	R_f	R_M
A	0.28	
B	0.18	5.1
C	0.30	
D	0.87	

Spectral Constants:

pH	λ_{max}	ϵ_{max}	Ref.
2	262	9.9	36
7	262	10.0	
12	261	7.3	

N-80
Uridine 3′-Phosphate
(3′-Uridylic Acid)

Abbreviations: Urd-3′-P, 3′-UMP
Formula: C₉H₁₃N₂O₉P
Mol. Wt.: 324.19
Calc. %: C, 33.34
H, 4.04
N, 8.64
O, 44.42
P, 9.55

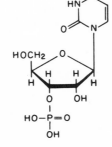

Paper Chromatography: Dissolve the Li⁺ salt in H₂O.

Solvent	R_f
A	0.31
B	0.23
C	0.33
D	0.96

Gas Chromatography: Trimethylsilyl derivative.

MU value	Rel. Peak Area
26.08	100

Water Analysis: 10.9% (calculated for the dihydrate, 9.7%).

Spectral Constants:

pH	λ_{max}	ϵ_{max}	$\dfrac{A_{250}}{A_{260}}$	$\dfrac{A_{280}}{A_{260}}$	Ref.
2	260	10.0	0.76	0.32	12 (ratios)
7	261	9.9	0.73	0.35	NRC sample (λ and ε
12	261	7.3	0.83	0.28	values)

Remarks: Met NRC specifications. Anion-exchange chromatography showed no 2′-isomer or other impurity.

N-81
Uridine 5′-Phosphate
(5′-Uridylic Acid)

Abbreviations: Urd-5′-P, 5′-UMP
Formula: C₉H₁₃N₂O₉P
Mol. Wt.: 324.19
Calc. %: C, 33.34
H, 4.04
N, 8.64
O, 44.42
P, 9.55
Specific Rotation: $[\alpha]_D^{28}$ +3.44° (1.02 g/100 ml, 10% HCl)[87]
pK_a:[15] 6.4, 9.5

Paper Chromatography: Dissolve the free acid in 1 M NH₄OH; the Li⁺, Na⁺, or K⁺ salt, in H₂O.

Solvent	R_f
A	0.22
B	0.11
C	0.28
D	0.90

Gas Chromatography: Trimethylsilyl derivative.

MU value	Rel. Peak Area
28.96	100

Water Analysis: 11.0% (calculated for the dihydrate, 8.9%).

Spectral Constants:

pH	λ_{max}	ϵ_{max}	$\dfrac{A_{250}}{A_{260}}$	$\dfrac{A_{280}}{A_{260}}$	Ref.
2	262	10.0	0.73	0.39	15
7	262	10.0	0.73	0.39	
11	261	7.8	0.80	0.31	

Remarks: Met NRC specifications. Anion-exchange chromatography showed no impurities.

N-82
Uridine 5′-Triphosphate

Abbreviations: Urd-5′-P₃, 5′-UTP
Formula: C₉H₁₅N₂O₁₅P₃
Mol. Wt.: 484.15
Calc. %: C, 22.32
H, 3.12
N, 5.79
O, 49.57
P, 19.19
pK_a:[15] 6.6, 9.5

Paper Chromatography: Dissolve the Li⁺, Na⁺, K⁺, or NH₄⁺ salt in H₂O.

Solvent	R_f	R_M
A	0.20	1.3
B	0.05	
C	0.19	2.2
D	0.94	

Spectral Constants:

pH	λ_{max}	ϵ_{max}	$\dfrac{A_{250}}{A_{260}}$	$\dfrac{A_{280}}{A_{260}}$	Ref.
2	262	10.0	0.75	0.38	15
7	262	10.0	0.75	0.38	
11	261	8.1	0.81	0.31	

N-83
Xanthine

Abbreviation: Xan
Formula: $C_5H_4N_4O_2$
Mol. Wt.: 152.11
Calc. %: C, 39.48
      H, 2.65
      N, 36.84
      O, 21.04
pK_a:[88] 7.7, 11.9

Paper Chromatography: A solution containing 10 A_{260} units/5 μl could not be made. The data reported here were obtained by dissolving 3.5 mg in 0.2 ml of 2 M NH₄OH (3.3 A_{260} units/5 μl) and applying three separate aliquots to the paper, drying after each application.

Solvent	R_f
A	0.54
B	0.32
C	0.35

Gas Chromatography: Trimethylsilyl derivative.

MU value	Rel. Peak Area
20.05	100

Spectral Constants:

pH	λ_{max}	ϵ_{max}	$\dfrac{A_{250}}{A_{260}}$	$\dfrac{A_{280}}{A_{260}}$	Ref.
5	267	10.3	0.57	0.61	12
10	277	9.3	1.29	1.71	
14	284	9.4	1.11	2.39	

Remarks: Met NRC specifications.

N-84
Xanthosine
(9-β-D-Ribofuranosylxanthine)

Abbreviation: Xao
Formula: $C_{10}H_{12}N_4O_6$
Mol. Wt.: 284.23
Calc. %: C, 42.26
      H, 4.26
      N, 19.71
      O, 33.77
Specific Rotation: $[\alpha]_D^{30}$ −51.2° (0.8 g/100 ml, 0.3 M NaOH)[89]
pK_a:[90] 5.7

Paper Chromatography: Dissolve the sample in 1 M NH₄OH.

Solvent	R_f	R_M	R_M
A	0.37	1.3	0.22*,**
B	0.35	0.06*,**	
C	0.41	1.7	0.22*,**
D	0.71	0.52**	

Gas Chromatography: Trimethylsilyl derivative.

MU value	Rel. Peak Area
27.05	100

Water Analysis: 11.7% (calculated for the dihydrate, 11.3%).

Spectral Constants:

pH	λ_{max}	ϵ_{max}	$\dfrac{A_{250}}{A_{260}}$	$\dfrac{A_{280}}{A_{260}}$	Ref.
3	263	9.0	0.75	0.28	12
8	278	8.9	1.30	1.13	
14	276	9.3	1.12	1.16	

Remarks: Although this compound was labeled anhydrous, it was a stable dihydrate. The water was lost only at temperatures above 120 °C.

N-85
Xanthosine 5'-Phosphate
(5'-Xanthylic Acid)

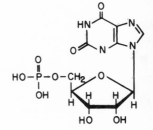

Abbreviations: Xao-5'-P, 5'-XMP
Formula: $C_{10}H_{13}N_4O_9P$
Mol. Wt.: 364.21
Calc. %: C, 32.98
      H, 3.60
      N, 15.38
      O, 39.54
      P, 8.50

Paper Chromatography: Dissolve the Li^+, Na^+, K^+, or NH_4^+ salt in H_2O.

Solvent	R_f	R_M
A	0.16	0.50*
B	0.03	0.33*
C	0.15	
D	0.91	

Gas Chromatography: Trimethylsilyl derivative.

MU value	Rel. Peak Area
31.34	100

Spectral Constants: No spectral data in the literature; use the data for xanthosine (N-84).

REFERENCES

1. *Specifications and Criteria for Biochemical Compounds*, Second Edition, NAS–NRC Publ. 1344, Washington, D. C. (1967), p. 379.
2. D. B. Dunn and R. H. Hall, in *Handbook of Biochemistry. Selected Data for Molecular Biology*, H. A. Sober, ed., Chemical Rubber Company, Cleveland, Ohio (1968), p. G-3.
3. W. C. Butts, *J. Chromatogr. Sci.*, 8, 474 (1970).
4. C. E. Dalgleish, E. C. Horning, M. G. Horning, K. L. Knox, and K. Yarger, *Biochem. J.*, 101, 792 (1966).
5. C. D. Scott, R. L. Jolley, W. W. Pitt, and W. F. Johnson, *Am. J. Clin. Pathol.*, 53, 701 (1970).
6. N. G. Anderson, J. G. Green, M. L. Barber, and F. C. Ladd, *Anal. Biochem.*, 6, 153 (1963).
7. M. Uziel, C. K. Koh, and W. E. Cohn, *Anal. Biochem.*, 25, 77 (1968).
8. M. T. Kelley, R. W. Stelzner, W. R. Laing, and D. J. Fisher, *Anal. Chem.*, 31, 220 (1959).
9. IUPAC–IUB Commission on Biochemical Nomenclature, "Abbreviations and Symbols for Nucleic Acids, Polynucleotides, and Constituents. Recommendations 1970," published in *Biochemistry*, 9, 4022 (1970); *J. Biol. Chem.*, 245, 5171 (1970); and elsewhere.
10. R. A. Alberty, R. M. Smith, and R. M. Bock, *J. Biol. Chem.*, 193, 425 (1951).
11. R. M. Izatt and J. J. Christensen, *J. Phys. Chem.*, 66, 359 (1962).
12. G. H. Beaven, E. R. Holiday, and E. A. Johnson, in *The Nucleic Acids: Chemistry and Biology*, Vol. I, E. Chargaff and J. N. Davidson, eds., Academic Press, New York (1955), p. 493.
13. J. J. Davoll, B. Lythgoe, and A. R. Todd, *J. Chem. Soc.*, 967 (1948).
14. R. Phillips, *Chem. Rev.*, 66, 501 (1966).
15. R. M. Bock, N. S. Ling, S. A. Morell, and S. H. Lipton, *Arch. Biochem. Biophys.*, 62, 253 (1956).
16. M. Smith, G. I. Drummond, and H. G. Khorana, *J. Am. Chem. Soc.*, 83, 698 (1961).
17. P. Reichard, Y. Takenaka, and H. S. Loring, *J. Biol. Chem.*, 198, 599 (1952).
18. R. A. Alberty, R. M. Smith, and R. M. Bock, *J. Biol. Chem.*, 193, 425 (1951).
19. E. Volkin and W. E. Cohn, *Methods Biochem. Anal.*, 1, 304 (1954).
20. G. Embden and G. Schmidt, *Hoppe-Seyler's Z. Physiol. Chem.*, 181, 130 (1929).
21. R. M. Izatt, J. H. Rytting, L. D. Hansen, and J. J. Christensen, *J. Am. Chem. Soc.*, 88, 2641 (1966).
22. M. Prystaš, J. Gut, and F. Šorm, *Chem. Ind.* (London), 947 (1961).
23. J. Jonás and J. Gut, *Collect. Czech. Chem. Commun.*, 27, 716 (1962).
24. A. R. Restivo and F. A. Dondzila, *J. Org. Chem.*, 27, 2281 (1962).
25. D. M. Frisch and D. W. Visser, *J. Am. Chem. Soc.*, 81, 1756 (1959).
26. S. Y. Yang, *Photochem. Photobiol.*, 1, 37 (1962).
27. D. W. Visser, K. Dittmer, and I. Goodman, *J. Biol. Chem.*, 171, 377 (1947).
28. K. Berens and D. Shugar, *Acta Biochim. Polon.*, 10, 25 (1963).
29. D. T. Elmore, *J. Chem. Soc.*, 2084 (1950).
30. J. J. Christensen, J. H. Rytting, and R. M. Izatt, *J. Phys. Chem.*, 71, 2700 (1967).
31. J. Clauwaert and J. Stockx, *Z. Naturforsch.*, B 23, 25 (1968).
32. D. M. Brown, D. I. Magrath, and A. R. Todd, *J. Chem. Soc.*, 2708 (1952).
33. H. S. Loring and N. G. Luthy, *J. Am. Chem. Soc.*, 73, 4215 (1951).
34. L. F. Cavalieri, *J. Am. Chem. Soc.*, 75, 5268 (1953).
35. R. J. C. Harris, S. F. D. Orr, E. M. F. Roe, and J. F. Thomas, *J. Chem. Soc.*, 489 (1953).
36. J. M. Ploeser and H. S. Loring, *J. Biol. Chem.*, 178, 431 (1949).
37. A. M. Michelson and A. R. Todd, *J. Chem. Soc.*, 2476 (1949).
38. D. Shugar and J. J. Fox, *Biochim. Biophys. Acta*, 9, 199 (1952).
39. H. Venner, *Chem. Ber.*, 93, 140 (1960).
40. C. D. Anderson, L. Goodman, and B. R. Baker, *J. Am. Chem. Soc.*, 81, 3967 (1959).
41. W. Klein and S. J. Thannhauser, *Hoppe-Seyler's Z. Physiol. Chem.*, 224, 252 (1934).
42. R. O. Hurst, A. M. Marko, and G. C. Butler, *J. Biol. Chem.*, 204, 847 (1953).
43. W. S. MacNutt, *Biochem. J.*, 50, 384 (1952).
44. J. J. Fox and D. Shugar, *Biochim. Biophys. Acta*, 9, 369 (1952).
45. A. M. Michelson and A. R. Todd, *J. Chem. Soc.*, 34 (1954).
46. L. Grossman, S. S. Levine, and W. S. Allison, *J. Mol. Biol.*, 3, 47 (1961).
47. W. Klein and S. J. Thannhauser, *Hoppe-Seyler's Z. Physiol. Chem.*, 218, 173 (1933).
48. L. A. Manson and J. O. Lampen, *J. Biol. Chem.*, 191, 87 (1951).
49. E. Scarano, *Biochim. Biophys. Acta*, 29, 459 (1958).
50. J. F. Gerster and R. K. Robins, *J. Am. Chem. Soc.*, 87, 3752 (1965).
51. W. F. Hemmens, *Biochim. Biophys. Acta*, 68, 284 (1963).
52. A. Albert and D. J. Brown, *J. Chem. Soc.*, 2060 (1954).
53. S. F. Mason, *J. Chem. Soc.*, 2071 (1954).
54. W. Pfleiderer, *Ann. Chem.*, 647, 167 (1961).
55. J. Davoll and B. A. Lowy, *J. Am. Chem. Soc.*, 73, 1650 (1951).
56. L. G. Bunville and S. J. Schwalbe, *Biochemistry*, 5, 3521 (1966).
57. D. O. Jordan, in *The Nucleic Acids: Chemistry and Biology*, Vol. I, E. Chargaff and J. N. Davidson, eds., Academic Press, New York (1955), p. 447.
58. J. N. Toal, G. W. Rushizky, A. W. Pratt, and H. A. Sober, *Anal. Biochem.*, 23 60 (1968).
59. P. A. Levene, H. S. Simms, and L. W. Bass, *J. Biol. Chem.*, 70, 243 (1926).
60. R. V. Wolfenden, *J. Mol. Biol.*, 40, 307 (1969).
61. P. A. Levene and R. S. Tipson, *J. Biol. Chem.*, 111, 313 (1935).
62. N. Adler, G. J. Litt, and R. G. Johl, *Anal. Biochem.*, 39, 249 (1967).
63. H. Sigel, *Eur. J. Biochem.*, 3, 530 (1968).
64. P. Chang and A. D. Welch, *J. Med. Chem.*, 6, 428 (1963).
65. N. J. Leonard and T. Fujii, *Proc. Nat. Acad. Sci. U.S.*, 51, 73 (1964).
66. M. J. Robins, R. H. Hall, and R. Thedford, *Biochemistry*, 6, 1837 (1967).
67. R. H. Hall, M. J. Robins, L. Stasiuk, and R. Thedford, *J. Am. Chem. Soc.*, 88, 2614 (1966).
68. N. J. Leonard, S. Achmatowicz, R. N. Loeppky, K. L. Carraway, W. A. H. Grimm, A. Szweykowska, H. Q. Hamzi, and F. Skoog, *Proc. Nat. Acad. Sci. U.S.*, 56, 709 (1966).
69. D. M. G. Martin and C. B. Reese, *J. Chem. Soc.* (C), 1731 (1968).
70. S. R. Breshears, S. S. Wang, S. G. Bechtolt, and B. E. Christensen, *J. Am. Chem. Soc.*, 81, 3789 (1959).
71. C. O. Miller, F. Skoog, F. S. Okumura, M. H. Von Saltza, and F. M. Strong, *J. Am. Chem. Soc.*, 78, 1375 (1956).
72. A. D. Broom, L. B. Townsend, J. W. Jones, and R. K. Robins, *Biochemistry*, 3, 494 (1964).
73. J. J. Fox, D. Van Praag, I. Wempen, I. L. Doerr, L. Cheong, J. E. Knoll, M. L. Eidinoff, A. Bendich, and G. B. Brown, *J. Am. Chem. Soc.*, 81, 178 (1959).
74. W. Pfleiderer, *Ann. Chem.*, 647, 167 (1961).
75. P. Brookes and P. D. Lawly, *J. Chem. Soc.*, 3923 (1961).
76. J. W. Jones and R. K. Robins, *J. Am. Chem. Soc.*, 85, 193 (1963).
77. H. T. Miles, *J. Org. Chem.*, 26, 4761 (1961).
78. J. J. Fox, N. Yung, J. Davoll, and G. B. Brown, *J. Am. Chem. Soc.*, 78, 2117 (1956).
79. R. Shapiro and R. W. Chambers, *J. Am. Chem. Soc.*, 83, 3920 (1961).
80. C. T. Yu and F. W. Allen, *Biochim. Biophys. Acta*, 32, 393 (1959).
81. A. Dugaiczyk, *Biochemistry*, 9, 1557 (1970).
82. D. M. Brown, M. G. Burdon, and R. P. Slatcher, *J. Chem. Soc.* (C), 1051 (1968).
83. A. Hampton, J. J. Biesele, A. E. Moore, and G. B. Brown, *J. Am. Chem. Soc.*, 78, 5695 (1956).
84. H. M. Kissman and M. J. Weiss, *J. Org. Chem.*, 21, 1053 (1956).
85. K. Burton and G. B. Petersen, *Biochem. J.*, 75, 17 (1960).
86. W. Klein and S. J. Thannhauser, *Hoppe-Seyler's Z. Physiol. Chem.*, 231, 96 (1935).
87. P. A. Levene and R. S. Tipson, *J. Biol. Chem.*, 106, 113 (1934).
88. W. Pfleiderer and G. Nübel, *Ann. Chem.*, 647, 155 (1961).
89. P. A. Levene and W. A. Jacobs, *Chem. Ber.*, 43, 3150 (1911).
90. A. Albert, *Biochem. J.*, 54, 646 (1953).

Porphyrins and Related Compounds

GENERAL REMARKS

A most useful and convenient criterion of purity and stability for many of these compounds is still their visible and near-ultraviolet-absorption spectra. However, other forms of physical and analytical measurement, for example, nuclear magnetic resonance spectroscopy, infrared spectroscopy, mass spectroscopy, electron spin resonance spectroscopy, and various forms of chromatography, are rapidly becoming conveniently useful and, moreover, provide deeper insight into the problems of establishing meaningful criteria of purity. The material in this section supplements, in a general way, that provided in the following tables and further acts as a guide to desirable future considerations and developments in characterizing these compounds.

Absorption Spectra

Near-ultraviolet and visible-absorption spectra of porphyrins have traditionally been measured in 1 M HCl as a standard solvent because (a) the acid increases the solubility of sparingly soluble porphyrins by the formation of their diprotonic salts and (b) at this acidity various salt and aggregative effects are minimized. In aqueous media the sodium salts of many common dicarboxylic acid porphyrins are relatively insoluble and are thus easily aggregated. Potassium or ammonium salts are more soluble. Di- and trivalent cations also cause aggregation of polycarboxylic acid porphyrins.

Spectroscopic grade pyridine is also to be recommended as a useful solvent because (a) most porphyrinic materials are reasonably soluble and stable in it and (b) the spectral resolution of trace materials from major components in terms of sharper and stronger absorption peaks is often enhanced in this solvent. For the less ionic porphyrins, e.g., esterified derivatives, p-dioxane (peroxide free) and chloroform (phosgene and acid free) are also useful solvents. Tetrahydrofuran and N,N'-dimethylformamide have also proven to be very useful solvents for general spectral studies on these materials. In detecting and evaluating trace impurities in porphyrinic materials, it is recommended that spectra taken in an acidic, neutral, and basic solvent be compared with the appropriate standards. Information on the purification and stability of such solvents is available.[1]

The wavelengths and absorption coefficients of both minima and maxima in the absorption spectra of porphyrins are important criteria of purity. Shoulders are also often observed and should be noted. Bandwidths at half-maximal absorption are also reported, as they are a good measure of purity and usually indicate the occurrence of aggregation or complexation (which is also of theoretical and frequently practical interest). The changes in such properties with dilution or concentration can therefore provide a very useful measure of purity.

Changes of substituents on the periphery at either *beta* position or at *meso* (or bridge) positions modify and shift the porphyrin spectrum in predictable ways. Rules originally worked out by A. Stern have been summarized.[2,3] A further classification for electronic spectra of some porphyrins and metalloporphyrins has been reported.[4,5,6] The spectra of both bridge-reduced (e.g., phlorins)[7] and ring-reduced (e.g., chlorins)[8,9] materials are also recorded. Spectra for several synthetic porphyrins, chlorins, and their metalloderivatives can be found in the literature.[10,11,12] Useful reviews on the theory of these spectra are also available.[13,14]

The ultraviolet and visible absorption spectra of bile pigments are useful in that changes in substituent groups or in chromophore length can alter the spectrum in a predictable way.[15,16] Solvents most frequently used are methanol–3 M HCl (100:1, v/v) and chloroform. Since isomeric bile pigments have nearly identical spectra, the differences found in such spectra are frequently too small to be useful.[17] Care must be taken to account for mixed species, such as the neutral and hydrochloride forms, which are easily interconvertible and exhibit significantly different spectra.[18] Metal complexes also form readily. Thus, characteristic absorption and fluorescence spectra of zinc complexes have proven helpful in determinations of identity and purity.[15,16,19]

Nuclear Magnetic Resonance Spectra

Nuclear magnetic resonance spectra are particularly useful for proof of structure and for detection of impurities.[20–25] In addition to detection of substituent groups and their relative locations with iron porphyrins, information on the spin state, oxidation state, and iron-bound ligands are also determined.[26–30] Amounts from 5 to 30 mg in 0.4 to 0.5 ml of suitable solvents (e.g., deuteriochloroform, pyridine-d_5, trifluoroacetic acid) may be conveniently studied. Even smaller quantities can be handled by microtechniques or accumulation methods. Chemical shifts may vary markedly with concentration due to variations in the extent of association; this sensitivity of chemical shift to concentration is dependent upon porphyrin structure.[24,31] Thus, chemical-shift values computed for infinite dilution become useful.[32] However, in metal-free porphyrins, an acidic solvent (e.g., deuteriochloroform with 2.5% trifluoroacetic acid) gives a protonated species with chemical shifts relatively insensitive to concentration changes.[32]

Mass Spectroscopy

Many porphyrins have sufficient volatility to give mass spectra of value in structure determination and purity analysis.[33,34] Moreover, although ionic and other porphyrins of low volatility often thermally decompose before mass analysis, structural information may still be obtained by detailed analysis of the data from de-

composition products. In many cases, volatility is increased by chemical modification, e.g., by esterification. Very small amounts of impurities may be detected, but some impurities, e.g., metalloderivatives, can actually arise from reactions within the source.

Mass-spectral analysis is also an important aid for identification and evaluation of purity of bile pigments.[35,36] Molecular ions as well as many fragments are detected.[35–38] However, due to their low volatility and general instability, bile pigments undergo drastic changes in the mass spectrometer; secondary products are always detected with otherwise pure samples.[36,39]

Chromatography

Early work on the chromatography of porphyrins[40] and chloroplast pigments[41] has been reviewed. A bibliography on paper and thin-layer chromatography of these systems for 1961–1965 inclusive can also be consulted.[42] Absorption or liquid–liquid partition column chromatography is also an important step in the preparation of large amounts of compounds of high purity. For most of the porphyrins discussed here, paper chromatography with 2,6-lutidine-NH_3-H_2O is a recommended system. This system leads to the separation of porphyrins differing in their number of acidic groups, the distance moved being inversely proportional to the number of acidic functions. Metalloderivatives may prove either slower or faster than the free bases, depending on the structure and the metal. In many cases, the method may be quantified by determination of the fluorescence intensity of the spots. Many workers prefer thin-layer chromatography (TLC) to the paper methods, as it permits more rapid separation and development, particularly for isomer separations.[43] TLC methods separate porphyrins according to the number of carboxylic acids[44] or esters[44–47] and variously substituted dicarboxylic esters[48] and their metal complexes.[49] Free-acid or esterified bile pigments are separated and minor impurities removed by TLC methods.[37,50,51] Thus, isomeric bilirubins and biliverdins can be separated and prepared in pure form.[17,52,53]

Differentiation of Urobilinoids

Urobilinoids, which include the b-bilenes[15] (urobilins, half-stercobilin, and stercobilins), are separated into three classes according to the degree of saturation of the end rings: dipyrrolinones, pyrrolinone–pyrrolidone, or dipyrrolidones.[54] Several methods are available for differentiation among these classes: ferric chloride oxidation,[54] mass spectroscopy,[36] optical rotatory dispersion and circular dichroism,[54] and chromic acid oxidation.[55,57] Natural 1-stercobilin can be specifically identified by its molecular complex with ferric chloride.[56]

Symbols and Terms Used

For the system used here for numbering rings, substituents, and the like of the porphyrins, see the structure of chloroprotoporphyrin IX iron(III) (Po-6). Note that this nomenclature does not follow the most recent IUPAC recommended rules [*cf. J. Am. Chem. Soc.* **82**, 5582 (1960)]. The symbol ϵ_{mM} is used for the millimolar absorption coefficient (unit: mmol^{-1}l cm^{-1}). The symbol nm (nanometer) is used in preference to its equivalent the mμ (millimicron). The symbol 0.5W (nm) is the width of the spectral absorption band at half the maximum absorption; min = position of the minimum in an absorption spectrum; and sh = position of a shoulder on a band in the spectrum. The HCl number is the percent of aqueous HCl saturated with ether that will extract two thirds of a particular porphyrin from water-saturated ether into an equal volume of the aqueous HCl layer (this number can vary considerably in the presence of specific salts, with small amounts of other solvents, and with concentration).

REFERENCES

1. A. Weissberger, E. S. Proskauer, J. A. Riddick, and E. E. Toops, *Organic Solvents,* 2nd edition, Interscience Publishers, New York, (1955).
2. S. Granick and H. Gilder, *Advan. Enzymol.,* **7**, 305 (1947).
3. J. E. Falk, *Porphrins and Metalloporphyrins,* Elsevier Publishing Co., Amsterdam (1964).
4. W. S. Caughey, R. M. Deal, C. Weiss, and M. Gouterman, *J. Mol. Spectry.,* **16**, 451 (1965).
5. W. S. Caughey, W. Y. Fujimoto, and B. P. Johnson, *Biochemistry,* **5**, 3830 (1966).
6. A. Treibs, *Ann. Chem.,* **728**, 115 (1969).
7. D. Mauzerall, *J. Am. Chem. Soc.,* **84**, 2437 (1962).
8. U. Eisner, *J. Chem. Soc.,*   uiG
9. H. W. Whitlock, R. Hanauer, M. Y. Oester, and B. K. Bower, *J. Am. Chem. Soc.,* **73**, 4315 (1951).
10. G. D. Dorough, J. R. Miller, and F. M. Huennekens, *J. Am. Chem. Soc.,* **74**, 4315 (1951).
11. G. D. Dorough, and F. M. Huennekens, *J. Am. Chem. Soc.,* **74**, 3974, (1952).
12. A. Treibs and N. Haeberle, *Ann. Chem.,* **718**, 183 (1968).
13. M. Gouterman, *J. Mol. Spectry.,* **6**, 138 (1961).
14. G. P. Gurinovich, A. N. Sevchenko, and K. N. Solov'ev, *Soviet Physics Uspekhi,* **6**, 67 (1963).
15. C. H. Gray, *The Bile Pigments,* John Wiley & Sons, New York (1953).
16. H. Fischer and H. Orth, *Die Chemie des Pyrrols,* Akademische-Verlagsgesellschaft, Leipzig (1937).
17. R. Bonnett and A. F. McDonagh, *J. Chem. Soc.* (**D**), 237, 238 (1970).
18. C. J. Watson and P. T. Lowry, *J. Biol. Chem.,* **218**, 633 (1956).
19. T. K. With, *Bile Pigments,* Academic Press, New York (1968).
20. NMR *Spectra Catalog,* Varian Associates, Palo Alto, Calif. (1962).
21. R. J. Abraham, A. H. Jackson, and G. W. Kenner, *J. Chem. Soc.,* 3468 (1961).
22. E. D. Becker, R. B. Bradley, and C. J. Watson, *J. Am. Chem. Soc.,* **83**, 3743 (1961).
23. W. S. Caughey and W. S. Koski, *Biochemistry,* **1**, 923 (1962).
24. W. S. Caughey, J. L. York, and P. K. Iber, in *Magnetic Resonance in Biological Systems,* A. Ehrenberg, B. G. Malmstrom, and T. Vanngard, eds., Pergamon Press, Oxford (1967), p. 25 ff.
25. J. J. Katz, R. C. Dougherty, and L. J. Boucher, in *The Chlorophylls,* L. P. Vernon and G. R. Seely, eds., Academic Press, New York (1966), p. 186 ff.
26. G. A. Smythe and W. S. Caughey, *J. Chem. Soc.* (**D**), 809 (1970).
27. W. S. Caughey and L. F. Johnson, *J. Chem. Soc.* (**D**), 1362 (1969).
28. K. Wuthrich, R. G. Shulman, B. J. Wyluda, and W. S. Caughey, *Proc. Nat. Acad. Sci., U.S.* **62**, 636 (1969).
29. R. J. Kurland, R. G. Little, D. G. Davis, and C. Ho., *Biochemistry,* **10**, 2237 (1971).
30. W. S. Caughey, *Adv. Chem. Ser.,* **100**, 248 (1971).
31. R. J. Abraham, P. A. Burbidge, A. H. Jackson, and G. W. Kenner, *Proc. Chem. Soc.,* 134 (1963).
32. W. S. Caughey, J. O. Alben, W. Y. Fujimoto, and J. L. York, *J. Org. Chem.,* **31**, 2631 (1966).
33. A. H. Jackson, G. W. Kenner, K. M. Smith, R. T. Aplin, H. Budzikiewicz, and C. Djerassi, *Tetrahedron Lett.,* **21**, 2913 (1965).
34. A. D. Adler, J. H. Green, and M. Mautner, *Org. Mass. Spectrosc.,* **3**, 955 (1970).
35. A. H. Jackson, G. W. Kenner, H. Budzikiewicz, C. Djerassi, and I. M. Wilson, *Tetrahedron Lett.,* **23**, 603 (1967).
36. D. A. Lightner, A. Moscowitz, Z. J. Petryka, S. Jones, M. Weiner, E. Davis, N. A. Beach, and C. J. Watson, *Arch. Biochem. Biophys.,* **131**, 566 (1969).
37. M. S. Stoll and C. H. Gray, *Biochem. J.,* **117**, 271 (1970).
38. A. W. Nichol and D. B. Morell, *Biochim. Biophys. Acta,* **184**, 173 (1969).
39. H. Budzikiewicz and S. E. Drewes, *Ann. Chem.,* **716**, 222 (1968).
40. J. E. Falk, *J. Chromatogr.,* **5**, 277 (1961).
41. Z. Sestak, *J. Chromatogr.,* **1**, 293 (1958).
42. K. Macek, I. M. Hais, V. Rabek, J. Kopecky, J. Gasparis, eds., *J. Chromatogr.* (Suppl.), 6624, 6652, 8208 (1968).
43. J. Jensen, *J. Chromatogr.* **10**, 236 (1963).
44. T. K. With, *J. Chromatogr.,* **42**, 389 (1969).
45. R. A. Cardinal, I. Bossenmaier, Z. J. Petryka, L. Johnson, and C. J. Watson, *J. Chromatogr.,* **38**, 100 (1968).
46. T. C. Chu and E. J. H. Chu, *J. Chromatogr.,* **28**, 475 (1967).
47. M. Doss, *Z. Klin. Chem. Klin. Biochem.,* **8**, 197 (1970).
48. R. W. Henderson and T. C. Morton, *J. Chromatogr.,* **27**, 180 (1967).
49. M. Doss, *Z. Klin. Chem. Klin. Biochem.,* **8**, 208 (1970).
50. Z. J. Petryka and C. J. Watson, *J. Chromatogr.,* **37**, 76 (1968).
51. Z. J. Petryka, *J. Chromatogr.,* **50**, 447 (1970).
52. P. O'Carra and E. Colleran, *FEBS Lett.,* **5**, 295 (1969).
53. A. F. McDonagh and F. Assisi, *FEBS Lett.,* **18**, 315 (1971).
54. C. J. Watson, M. Weiner, Z. J. Petryka, D. A. Lightner, A. Moscowitz, E. Davis, and N. A. Beach, *Arch. Biochem. Biophys.,* **131**, 414 (1969).

55. C. J. Watson, A. Moscowitz, D. A. Lightner, Z. J. Petryka, E. Davis and M. Weiner, *Proc. Nat. Acad. Sci., U.S.,* **58,** 1957 (1967).

56. C. J. Watson and Z. J. Petryka, *Anal. Biochem.,* **30,** 156 (1969).

57. W. Rvediger, *Z. Physiol. Chem.,* **350,** 1291 (1969).

BIBLIOGRAPHY

B. F. Burnham, "Metabolism of Porphyrins and Corrinoids," in *Metabolic Pathways,* Vol III, 3rd Edition, D. M. Greenberg, ed., Academic Press, New York (1969), p. 403.

W. S. Caughey, "Porphyrin Proteins and Enzymes," *Ann. Rev. Biochem.,* **36,** 611 (1967).

B. Chance, R. W. Estabrook, T. Yonetani, *Hemes and Hemoproteins,* Academic Press, New York (1966).

J. E. Falk, *Porphyrins and Metalloporphyrins,* Elsevier Publishing Co., Amsterdam (1964).

H. Fischer and H. Orth, *Die Chemie des Pyrrols;* Vol. I and IIA; H. Fischer and A. Stern, *Die Chemie des Pyrrols.* Vol. IIB, Akademische Verlagsgesellschaft, Leipzig (1934).

[Now available from the Johnson Reprint Corp., New York New York (1966).]

T. W. Goodwin, *Porphyrins and Related Compounds,* Academic Press, New York (1968).

S. Granick and D. Mauzerall, "Metabolism of Heme and Chlorophyll," in *Metabolic Pathways,* Vol. 2, D. M. Greenberg, ed., Academic Press, New York (1961), p. 525.

J. Lascelles, *Tetrapyrrole Biosynthesis and its Regulation,* W. A. Benjamin Inc., New York (1964).

R. Lemberg and J. W. Legge, *Hematin Compounds and Bile Pigments,* Interscience Publishers, New York (1949).

G. S. Marks, *Heme and Chlorophyll,* D. Van Nostrand Co., London (1969).

W. Siedel, "Pyrrolfarbstoffe," in *Handbuch der Physiologisch und Pathologisch–Chemischen Analyse fur Arzte, Biologen und Chemiker,* Vol. 4, pt. 2, 10th ed., F. Hoppe Seyler and H. Thierfelder, eds., Springer-Verlag, Berlin (1960), p. 845.

L. P. Vernon and G. R. Seely, *The Chlorophylls,* Academic Press, New York (1966).

R. Willstatter and A. Stoll, *Investigation on Chlorophyll,* F. M. Schertz and A. R. Merz (translators), Science Press Printing Co., Lancaster, Pa. (1928).

T. K. With, *The Bile Pigments,* Academic Press, New York (1968).

Po-1
5-Aminolevulinic Acid Hydrochloride

Formula: $C_5H_{10}ClNO_3$
Formula Wt.: 167.59
Calc. %: C, 35.83; H, 6.02; Cl, 21.15; N, 8.36; O, 28.64

$$HO-\overset{O}{\overset{\|}{C}}-CH_2-CH_2-\overset{O}{\overset{\|}{C}}-CH_2-NH_3^+Cl^-$$

Sources: Found in the urine of acute porphyria patients.[1] Synthesis.[2,3,4] Commercially available.

Photometric Determination: Condense with 2,4-pentanedione at pH 4.6 to form 3-acetyl-2-methyl-pyrrole-4-propionic acid. Read optical density 15 min after mixing with an equal volume of modified Ehrlich reagent. ϵ_{mM} at 553 nm is 64, with shoulder ϵ_{mM} of 47 at 525 nm for the colored salt.[5]

Chromatography: For 5-aminolevulinic acid on ascending paper chromatography R_f 0.35 in 1-butanol–acetic acid–water (4:1:5, v/v).

R_f 0.20 in 1-butanol–1.5 M NH_4OH (1:1).

For the 3-acetyl-2-methyl-pyrrole-4-propionic acid:

R_f 0.9 in 1-butanol–acetic acid–H_2O (4:1:5).

R_f 0.2 in 1-butanol–1.5 M NH_4OH (1:1).[5,6]

m.p.: 150–152 °C.

Likely Impurity: NH_4Cl.

Labeled Compounds: Preparations with ^{14}C at positions 1,4,[2] at 2,3,[2] at 4,[3] and at 5[2] have been reported.

Special Reagents: For condensation to form the pyrrole, add 0.1 ml of 2,4-pentanedione to 9.9 ml of solution at pH 4.6 (acetate buffer). Stopper. Place in boiling-water bath for 10 min. For modified Ehrlich reagent, see porphobilinogen (Po-16).

References

1. S. Granick and H. G. Van den Shriek, *Proc. Soc. Exp. Biol. Med.*, **88**, 270 (1955).
2. D. Shemin, *Methods Enzymol.*, **4**, 648 (1957).
3. A. E. A. Mitta, A. M. Ferramola, H. A. Sancovich, and M. Grinstein, *J. Label. Compd.*, **3**, 20 (1967).
4. F. Sparatore and W. Cumming, *Biochem. Prep.*, **10**, 6 (1963).
5. D. Mauzerall and S. Granick, *J. Biol. Chem.*, **219**, 435 (1956).
6. G. Urata and S. Granick, *J. Biol. Chem.*, **238**, 811 (1963).

Po-2
Bilirubin IXα

Formula: $C_{33}H_{36}N_4O_6$
Formula Wt.: 584.67
Calc. %: C, 67.79; H, 6.21; N, 9.58; O, 16.42

Sources: Commercially available as prepared from bile or gall stones of bovine or porcine origin. May be isolated from human or rat bile.[1] Occurs in fresh bile as a mono- or a di-D-glucosiduronic acid.[16]

Spectral Reference Values: In chloroform, λ max at 453–455 nm, $\epsilon_{mM} = 62.6$.[2,*] The colored salt formed with diazo reagent (see below): λ max at 545 mn, $\epsilon_{mM} = 64.1$.[3,4]

m.p.: Decomposes on heating (234–275 °C) without melting.[2]

Mass Spectrum: Reported in Refs. 2, 5, 6.

Other Properties: Readily soluble in 0.1 M NaOH, but unstable. Stabilized by serum albumin and L-ascorbic acid.[7] Isomerization to IIIα and XIIIα compounds is acid catalyzed.[2,8] Hydrogenation (Pd/C catalyst) in ammonia–methanol yields mesobilirubin[1] and in acetic acid yields stercobilinogen.[9] Reduction with Na/Hg yields urobilinogen.[10] Dehydrogenation with benzoquinone[8] or Evelyn–Malloy reagent forms biliveridin. A red fluorescent Zn bilipurpurine (λ max 635–640, 586, 515 nm) is obtained from treatment with alcoholic Zn acetate and iodine.[11]

Chromatography: TLC on silica gel[2] and on polyamide.[12]

Likely Impurities: Biliverdin, mesobilirubin, bilifuscin. Commercial preparations have been found to contain significant amounts (up to 35%) of isomers other than IXα (mainly IIIα and XIIIα).[2,13] In chloroform, IIIα exhibited λ max at 455–458 nm, ϵ_{mM} 65.2 and XIIIα at 449–453 nm, ϵ_{mM} 52.5.[2]

Present evidence suggests that only bilirubin IXα is present in serum; if any of the non-IXα isomers in fact occur, only trace proportions may be expected.[14,17] The presence of the two additional isomers in commercially available preparations of bilirubin IXα does not seriously affect the use of such preparations as standards in the routine clinical measurement of serum bilirubin either by the direct measurement of absorption at 453 nm or the diazo method; however, for use as a standard, a preparation should be the equivalent of those purified by the sodium bicarbonate extraction process[3] or by absorption on sodium sulfate.[15] The National Bureau of Standards issues a purified bilirubin preparation as a standard reference material for clinical use; it is mainly IXα but contains the IIIα and XIIIα isomers as well. The pure IXα isomer may be required for clinical research investigations, but it is presently obtainable only by preparative thin-layer chromatography.[14]

Special Reagents: The diazo reagent involves solution A (1 g sulfanic acid, 15 ml concentrated HCl, and 985 ml H_2O) and solution B (0.5% $NaNO_2$); 10 ml of A and 0.3 ml of B are used. The Evelyn-Malloy reagent consists of 2 ml concentrated HCl, 0.4 ml 30% H_2O_2, and 97.6 ml 95% ethanol.

References

1. H. Fischer and H. Orth, *Die Chemie des Pyrrols*, Vol. II, Part 1, Akademische-Verlag, Leipzig, (1937); U. S. Patents 2,166,073 and 2,386,716.
2. A. F. McDonagh and F. Assisi, *FEBS Lett.*, **18**, 315 (1971).
3. R. Henry, S. L. Jacobs, and N. Chiamori, *Clin. Chem.*, **6**, 529 (1960).
4. J. T. G. Overbeek, C. L. J. Vink, and H. Deenstra, *Rec. Trav. Chim.*, **74**, 81 (1955).
5. B. L. Schram and H. H. Kroes, *Eur. J. Biochem.*, **19**, 581 (1971).
6. R. Bonnett and A. F. McDonagh, *Chem. Ind.*, 107 (1969).
7. T. K. With, unpublished information.
8. R. Bonnett and A. F. McDonagh, *J. Chem. Soc. (D)*, 238 (1970).
9. Z. J. Petryka and C. J. Watson, *Tetrahedron Lett.*, **52**, 5323 (1967).
10. C. J. Watson, *J. Biol. Chem.*, **200**, 691 (1953).
11. C. H. Gray, A. Lichtarowicz-Kulczycka, and D. C. Nicholson, *J. Chem. Soc.*, 2268 (1961).
12. Z. J. Petryka and C. J. Watson, *J. Chromatogr.*, **37**, 76 (1968).
13. Z. J. Petryka, *Proc. Soc. Exp. Biol. Med.*, **123**, 464 (1966).
14. A. F. McDonagh, unpublished information.
15. J. Fog, *Scand. J. Clin. Lab. Invest.*, **16**, 49 (1964).
16. B. H. Billing and F. H. Jansen, *Gastroenterology*, **61**, 258 (1971).
17. Z. J. Petryka, unpublished information.

* The National Bureau of Standards has measured a sample of the same product reported in Ref. 2, and obtained a value of 61.1 (R. Schaffer, unpublished information).

Po-3
Biliverdin IXα

Formula: $C_{33}H_{34}N_4O_6$
Formula Wt.: 582.66
Calc. %: C, 68.02; H, 5.88; N, 9.61; O, 16.49

Sources: From bilirubin IXα by oxidation with H_2O_2,[1] HNO_3,[2] $FeCl_3$ in acid,[2,3] or benzoquinone[4,5] followed by chromatography.[4] From coupled oxidation of hemin.[4]

Spectral Reference Values: In $CHCl_3$, λ max (ϵ_{mM}) 645 nm (13.4); 378 nm (41.7).[3,6] Dimethyl esters in $CHCl_3$, λ max (ϵ_{mM}) 656–664 nm (15.1); 379 nm (51.8).[4]

NMR Spectrum: Dimethyl ester.[4]

m.p.: Dimethyl ester, 208–209 °C.[4]

Other Properties: Color of an alcoholic zinc acetate solution is green, λ max at 385 and 685 nm. After oxidation with iodine, λ max are at 510–515, 588, and 637.5–640 nm.[7]

Likely Impurities: Choletelin, bilifuscin, isomeric biliverdins,[4,8] oxidation products.

Chromatography: TLC on polyamide.[9] Dimethyl esters: TLC on silica gel.[4,5,8,9] Quantitative densitometry is done on the TLC plate.[10]

References

1. R. Lemberg and J. W. Legge, *Aust. J. Exp. Biol. Med. Sci.*, **18**, 95 (1940).
2. R. Lemberg, *Biochem. J.*, **28**, 978 (1934).
3. C. H. Gray, A. Lichtarowicz-Kulczycka, D. C. Nicholson, and Z. Petryka *J. Chem. Soc.*, 2264 (1961).
4. R. Bonnett and A. F. McDonagh, *J. Chem. Soc. (D)*, **4**, 237 (1970).
5. A. F. McDonagh and F. Assisi, *FEBS Lett.*, **18**, 315 (1971).
6. C. J. Watson and I. Bossenmaier, unpublished information.
7. C. H. Gray, A. Lichtarowicz-Kulczycka, and D. C. Nicholson, *J. Chem. Soc.*, 2268 (1961).
8. P. O'Carra and E. Colleran, *FEBS Lett.*, **5**, 295 (1969).
9. Z. J. Petryka and C. J. Watson, *J. Chromatogr.*, **37**, 76 (1968).
10. Z. J. Petryka, *J. Chromatogr.*, **50**, 447 (1970).

Po-4
Chlorophyll a

Formula: $C_{55}H_{72}N_4O_5Mg$
Formula Wt.: 893.51
Calc. %: C, 73.93; H, 8.12; Mg, 2.72; N, 6.27; O, 8.95

Sources: Leaves of higher plants and algae usually as a mixture with chlorophyll b in the approximate ratio of 3:1. Synthesis.[1] Isolation, purification, and properties.[2,3] The method of choice is to chromatograph on powdered polyethylene, and then on sucrose.[4]

Spectral Reference Values:

In ether[2,4]

λ, nm	662	615	578	533.5	430	410	380	326	312	296	282
ϵ_{mM}	90.1	14.6	8.28	3.77	117.5	76.1	49.1	25.9	23.2	20.6	18.0

In 80% acetone[5]

λ, nm	665	618	582	536	433
ϵ_{mM}	81.1	17.5	10.4	4.27	90.7

In benzene (see Bellamy and Lynch)[6]

Quantitative Determination of Chlorophylls a and b in Mixture:

In ether:

chlorophyll a (mg/l) = 10.1 (O.D. 662 nm) − 1.01 (O.D. 664 nm)

chlorophyll b (mg/l) = 16.4 (O.D. 664 nm) − 2.57 (O.D. 662 nm)

In 80% acetone:

chlorophyll a (mg/l) = 11.63 (O.D. 665 nm) − 2.39 (O.D. 649 nm)

chlorophyll b (mg/l) = 20.11 (O.D. 649 nm) − 5.18 (O.D. 665 nm)

Other Properties: NMR and infrared spectra;[7] x-ray spacing in microcrystals;[8,9] density 1.079.

m.p.: 150–153 °C, with decomposition.

Fluorescence: Maxima in ether at 668 and 723 nm, absorbance ratio 1.0/0.17.[10]

Optical Rotation: In acetone $[\alpha]_{270}^{25°}$ −262°.[11]

Solubility: Insoluble in water; soluble in wet petroleum ether; soluble in benzene, ether, acetone, methanol or chloroform.

Distribution Ratio: Petroleum ether–85% methanol, 32:1, petroleum ether–90% methanol, 11:1.

Chromatography: On paper, ascending, with petroleum ether and acetone (2–10%) as developing, gives an approximate order from bottom to top of chlorophyllides (minus phytol), pheophorbide (minus phytol, minus Mg), and oxidized chlorophylls, certain xanthophylls, chlorophyll b, chlorophyll a, pheophytin a, other xanthophylls, carotenes.[2,3,4,] For TLC of plant extracts see Strain.[12]

Phase Test: A negative phase test indicates the oxidation or absence of the cyclopentanone ring. Petroleum ether solutions of chlorophyll a or b underlaid with a 30% solution of KOH in methanol form a colored zone at the interface, red for chlorophyll a, yellow for chlorophyll b, and brown for the mixture of chlorophylls a + b. Colors are transient, and change to green. When shaken, the carotenoids, if present, remain in the upper phase.

Spectrum of Pheophytin a: Wickliff and Aronoff.[13]

Circular Dichroism and Magnetic Circular Dichroism: Houssier and Sauer.[14]

Likely Impurities and Methods of Detection: For chlorophyll b: Chromatography and quantitative determination. For carotenoids: Chromatography, an absorption band at 480 nm; in the phase test, the yellow color is in the upper phase. For pheophytin a (i.e., minus Mg): bands at 505 and 532 nm. For pheophytin b: band at 523 nm. For oxidized cyclopentanone ring: a negative phase test. Chlorophyll derivatives lacking a phytol group are extracted from ether with 0.01 M KOH or with 6 M HCl.

References

1. R. B. Woodward, *Pure Appl. Chem.*, **2**, 383 (1961).
2. J. H. C. Smith and A. Benitez, *Modern Method of Plant Anaylsis*, Vol. IV, Springer-Verlag, Berlin (1955), p. 142.
3. H. H. Strain and W. A. Svec, in *The Chlorophylls*, L. P. Vernon and G. R. Seely, eds., Academic Press, New York (1966), p. 22.
4. A. F. H. Anderson and M. Calvin, *Nature*, **194**, 285 (1962).
5. L. P. Vernon, *Anal. Chem.*, **32**, 1144 (1960).
6. W. D. Bellamy and M. E. Lynch, General Electric Research Laboratory Report 63-RL-3469G (1963).
7. J. J. Katz, R. C. Dougherty, and L. J. Boucher, in *The Chlorophylls*, L. P. Vernon and G. R. Seely, eds., Academic Press, New York (1966), p. 186.
8. G. Donnay, *Arch. Biochem. Biophys.*, **80**, 80 (1959).
9. E. E. Jacobs, A. E. Vatter, and A. S. Holt, *Arch. Biochem. Biophys.*, **53**, 228 (1954).
10. C. S. French, J. H. C. Smith, H. I. Virgin, and R. L. Airth, *Plant Physiol.*, **31**, 369 (1956).
11. A. Stoll and E. Wiedemann, *Helv. Chim. Acta*, **16**, 307 (1933).
12. H. H. Strain, J. Sherma, and M. Grandolfo, *Anal. Biochem.*, **24**, 54 (1968).
13. J. L. Wickliff and S. Aronoff, *Plant Physiol.*, **37**, 584 (1962).
14. C. Houssier and K. Sauer, *J. Am. Chem. Soc.*, **92**, 779 (1970).

Po-5
Chlorophyll b

Formula: $C_{55}H_{70}N_4O_6Mg$
Formula Wt.: 907.49
Calc. %: C, 72.79; H, 7.78; Mg, 2.68; N, 6.17; O, 10.58

Sources: Algae; leaves of higher plants.

Spectral Reference Values:

In ether[2,4]

λ, nm	644	595	549	455	430	375	357	334	325	306	290
ϵ_{mM}	56.2	11.5	6.42	158.6	56.9	19.8	24.0	29.0	27.2	26.3	20.9

In 80% acetone[5]

λ, nm	648.5	600	460
ϵ_{mM}	47.6	13.0	134.3

Other Properties: See Chlorophyll a (Po-4).

m.p.: 183–185 °C, with decomposition.

Fluorescence: Maxima in ether at 649 nm and at 708 nm, ratio 1.0/0.23.[10]

Solubility: See Chlorophyll a (Po-4).

Distribution Ratio: Petroleum ether–85% methanol, 4.5; petroleum ether–90% methanol, 0.67.

Chromatography: See Chlorophyll a (Po-4).

Phase Test: See Chlorophyll a.

Likely Impurities: See Chlorophyll a.

Po-6

Chloroprotoporphyrin IX Iron(III)
(Hemin Chloride or Protohemin Chloride)

Formula: $C_{34}H_{32}N_4O_4FeCl$
Formula Wt.: 651.96
Calc. %: C, 62.36; H, 4.95; N, 8.59; O, 9.82; Fe, 8.57; Cl, 5.44

Source: Blood.[1,2] Esters are obtained from metal-free ester.[3] [The protohemin esters are perhaps most readily characterized as μ-oxo dimers. Thus when the hemin dicarboxylic acid is esterified in 5% H_2SO_4 in methanol at 25 °C, chromatographed on alumina with alcohol-free chloroform, and crystallized from chloroform–isooctane (1:4, v/v), μ-oxo bis-{protoporphyrin IX iron(III)} is obtained in high yield and purity.[4,5] This μ-oxo dimer is readily converted to monomers with any of a number of axial ligands as well as chloro.[5,6]]

PROPERTIES OF THE DICARBOXYLIC ACID

Spectral Reference Values:
In p-dioxane[7]

λ, nm	635.5	538.5	508
ϵ_{mM}	5.4	8.5	9.1

In aqueous NaOH (pH 11.3) with 10% pyridine (where the species is μ-oxo dimer), λ, nm (ϵ_{mM}) for maxima are 599(11.2), 574(12.4), 395(117), and 361(82.3).[8] Other data for species where ligands present are less clear include ϵ_{mM} of 131 at 398 nm in ethanol–0.05 M H_2SO_4 (1:1, v/v)[9] and visible data in several media reported by Shack and Clark.[10]

For purposes of characterization of the iron porphyrin, but not the original axial ligand, the hemin can be reduced to the heme. This iron(II) derivative gives two sharp (hemochromogen) bands in the visible region that are characteristic and useful for both quantitative and qualitative determination of purity. When prepared from 0.1 mM hemin, 6 M pyridine, and 0.2 M aqueous NaOH with reduction by 5 mM $Na_2S_2O_4$, ϵ_{mM} values at 558, 525, and 540 nm (a minimum) are 30.9, 16.2, and 9.06, respectively.[11] This spectrum and the solution are unstable with time.[10,12,13]

NMR Spectrum: As cyano[14] and other[15,16] derivatives.
IR Spectrum: Fe–Cl stretch, 350 cm^{-1}.[17]
X-Ray: See Koenig.[18]
Chromatography: Column ligand–ligand partition.[19] On paper, R_f = 0.34 for ascending, reversed-phase chromatography on Whatman No. 1 paper coated with a silicone; solvent is H_2O:n-

propanol:pyridine (5.5:0.1:0.4, v/v); R_f = 0.45 for the related chloromesoporphyrin IX iron(III).[20] The hemin is best made visible with a benzidine–peroxide spray.[21] Another system uses glacial acetic acid–toluene(saturated with water) (8:100, v/v).[21]

Likely Impurities: Denatured globin, peptides, lipoidal materials, hemotoporphyrin, protoporphyrin, biliverdin, adsorbed solvents.[22,23] Monomers with ligands other than chloro (e.g., acetato) and dimers {e.g., μ-oxo bis-[protoporphyrin IX iron(III)]}. Infrared spectra permit quantitative determination of unmodified vinyl groups[24] and chloro (or other) ligand;[4,17] elemental analyses are, of course, also helpful here.

PROPERTIES OF THE DIMETHYL ESTER

Spectral Reference Values:
In chloroform[25]

λ, nm	916	641	539	512	387
ϵ_{mM}	0.55	5.0	9.9	10.0	100

In benzene[25]

λ, nm	889	637	540	510	385
ϵ_{mM}	0.62	5.5	10.2	10.1	84

In pyridine[25]

λ, nm	928	632	525	409
ϵ_{mM}	0.44	3.3	10.0	121

As the μ-oxo dimer in benzene[5]

λ, nm	601	573	397
ϵ_{mM}	13.0	15.3	138

NMR Spectrum: As cyano[14] and μ-oxo dimer[26] iron(III) derivatives and as iron(II) derivatives.[3,26]
IR Spectrum: Fe–Cl stretch at 345 cm^{-1} in KBr.[17] FeOFe stretch of μ-oxo dimer at 895 cm^{-1} in KBr.[5] Other derivatives.[6]
Far IR Spectra: See Brackett et al.[27]
Mössbauer Spectra: See Moss et al.[28]
Other Properties: Soluble in chloroform, pyridine, benzene, etc. Care must be taken to avoid replacement of the chloro ligand in solution to give monomeric or dimeric species with ligands other than chloro. Characteristic changes in visible, NMR, infrared, far infrared, and Mössbauer spectra accompany such ligand changes.[4,26]

References

1. H. Fischer, Org. Syn., 21, 53 (1941).
2. R. F. Labbe and G. Nishida, Biochim. Biophys. Acta, 26, 437 (1957).
3. J. O. Alben, W. H. Fuchsman, C. A. Beaudreau, and W. S. Caughey, Biochemistry, 7, 624 (1968).
4. N. Sadasivan, H. I. Eberspaecher, W. H. Fuchsman, and W. S. Caughey, Biochemistry, 8, 534 (1969).
5. G. A. Smythe, C. H. Barlow, and W. S. Caughey, unpublished information.
6. S. McCoy and W. S. Caughey, Biochemistry, 9, 2387 (1970).
7. A. Stern and H. Wenderlein, Z. Physik. Chem., 174, 81 (1935).
8. C. H. Barlow, unpublished information.
9. A. C. Maehly and A. Akeson, Acta Chem. Scand., 12, 1259 (1958).
10. J. Shack and W. M. Clark, J. Biol. Chem., 171, 143 (1947).
11. D. L. Drabkin, J. Biol. Chem., 146, 605 (1942).
12. W. A. Gallagher and W. B. Elliott, Biochem. J., 97, 187 (1965).
13. A. D. Adler, in Hemes and Hemoproteins, B. Chance, R. W. Estabrook, and T. Yonetani, eds., Academic Press, New York (1966), p. 255.
14. K. Wuethrich, R. G. Shulman, B. J. Wyluda, and W. S. Caughey, Proc. Nat. Acad. Sci. U.S., 62, 636 (1969).
15. R. J. Kurland, D. G. Davis, and C. Ho, J. Am. Chem. Soc., 90, 2700 (1968).
16. H. A. O. Hill and K. G. Morallee, J. Chem. Soc. (D), 266 (1970).
17. S. McCoy and W. S. Caughey, unpublished information.
18. D. F. Koenig, Acta Cryst., 18, 663 (1965).
19. W. S. Caughey, W. Y. Fujimoto, A. J. Bearden, and T. H. Moss, Biochemistry, 5, 1255 (1966).
20. T. C. Chu and E. J. H. Chu, J. Biol. Chem., 212, 1 (1955).
21. J. L. Connelly, M. Morrison, and E. Stotz, J. Biol. Chem., 233, 743 (1958).
22. K. G. Paul, Acta Chem. Scand., 12, 1611 (1958).

23. A. D. Adler and J. L. Harris, *Anal. Biochem.*, **14**, 472 (1966).
24. W. S. Caughey, J. O. Alben, W. F. Fujimoto, and J. L. York, *J. Org. Chem.*, **31**, 2631 (1966).
25. W. S. Caughey and S. McCoy, in *The Handbook of Biochemistry and Biophysics*, H. C. Damm, ed., The World Publishing Co., Cleveland (1966), p. 446.
26. W. S. Caughey, *Adv. Chem. Sci.*, **100**, 248 (1971).
27. G. C. Brackett, P. L. Richards, and W. S. Caughey, *J. Chem. Phys.*, **54**, 4383 (1971).
28. T. H. Moss, A. J. Bearden, and W. S. Caughey, *J. Chem. Phys.*, **51**, 2624 (1969).

Po-7
Coproporphyrin I

Formula: $C_{36}H_{38}N_4O_8$
Formula Wt.: 654.73 (Tetramethyl ester, 710.83)
Calc. %: C, 66.04; H, 5.85; N, 8.56; O, 19.55

R = CH$_2$—CH$_2$—COOH

Tetracarboxylic acid: R = CH$_2$—CH$_2$—COOH
Tetramethyl ester: R = CH$_2$—CH$_2$—CO$_2$CH$_3$

Sources: From urine and feces of animals and humans having congenital porphyria;[1-5] synthesis of the tetramethyl ester.[6]

PROPERTIES OF TETRACARBOXYLIC ACID

HCl Number: 0.091[7] for all isomers.
Chromatography: Paper: 2,6-lutidine–aqueous ammonia[8,9] separates coproporphyrin I from all other isomers, TLC: 2,6-lutidine–aqueous ammonia gives similar results.[10] Column: 2,6-lutidine–aqueous ammonia and dilute HCl–ether are useful.[11,12,13]
m.p.: Decomposes.
Spectral Reference Values:
In 1 M HCl[14] (identical for all coproporphyrin isomers):

λ, nm	401.3	548	591
ϵ_{mM}	470	17.5	6.5

In 0.1 M HCl:[15] ϵ_{mM} of 489 at 399.5 nm

PROPERTIES OF TETRAMETHYL ESTER

Chromatography: See Chu *et al.*[24,25] and Demole.[26]
m.p.: 251–252 °C.[7,16]
Spectral Reference Values:
In purified dioxane[17]

λ, nm	497	529	567	621
ϵ_{mM}	14.7	9.97	6.72	5.15

Infrared Spectrum: See Falk and Willis,[18] Chu and Chu,[19] Dinsmore and Watson.[20]
NMR Spectrum: See Abraham *et al.*[21,22]
Mass Spectrum: See Jackson *et al.*[23]
Likely Impurities: Other isomers, especially isomer III, porphyrins having 3, 5, or 6 methoxycarbonyl groups (in the esters) of the naturally occurring series; haloporphyrins in the synthetic compound; Cu or Zn chelates.

References

1. H. Fischer and H. Orth, *Die Chemie des Pyrrols*, Vol. II, Part I, Akademische-Verlag, Leipzig (1937), p. 471, 596.
2. H. Fischer and H. Andersag, *Ann. Chem.*, **450**, 201 (1926).
3. C. Rimington, *Acta Med. Scand.*, **143**, 177 (1952).
4. T. K. With, *Biochem. J.*, **68**, 717 (1958).
5. R. A. Cardinal, I. Bossenmaier, Z. J. Petryka, L. Johnson, and C. J. Watson, *J. Chromatogr.*, **38**, 100 (1968).
6. H. Fischer, H. Friedrich, W. Lamatsch, and K. Morgenroth, *Ann. Chem.*, **466**, 147 (1928).
7. H. Fischer and H. Orth, *Die Chemie des Pyrrols*, Vol. II, Part I, Akademische-Verlag, Leipzig (1937), p. 471 ff., 596.
8. L. Eriksen, *Scand. J. Clin. Lab. Invest.*, **10**, 39 (1958).
9. D. Mauzerall, *J. Am. Chem. Soc.*, **82**, 2601 (1960).
10. J. Jensen, *J. Chromatogr.*, **10**, 236 (1963).
11. L. Eriksen, *Scand. J. Clin. Lab. Invest.*, **9**, 97 (1957).
12. G. S. Marks, *Heme and Chlorophyll*, D. Van Nostrand Co., Princeton, New Jersey (1969), p. 71.
13. W. R. Richards and H. Rapoport, *Biochemistry*, **5**, 1079 (1966).
14. D. Mauzerall and S. Granick, *J. Biol. Chem.*, **232**, 1141 (1958).
15. C. Rimington, *Biochem. J.*, **75**, 620 (1960).
16. F. Morsingh and S. F. MacDonald, *J. Am. Chem. Soc.*, **82**, 4377 (1960).
17. J. E. Falk, *Comprehensive Biochemistry*, Vol. 9, M. Florkin and E. H. Stotz, eds., Elsevier, Amsterdam (1963), p. 11.
18. J. E. Falk and J. B. Willis, *Aust. J. Sci. Res. Ser.* A, **4**, 579 (1951).
19. T. C. Chu and E. J. Chu, *J. Biol. Chem.*, **234**, 2751 (1959).
20. H. Dinsmore and C. J. Watson, *J. Lab. Clin. Med.*, **56**, 652 (1960).
21. R. J. Abraham, P. A. Burbidge, A. H. Jackson, and G. W. Kenner, *Proc. Chem. Soc.*, 134 (1963).
22. R. J. Abraham, P. A. Burbidge, A. H. Jackson, and D. B. MacDonald, *J. Chem. Soc.* (B), 620 (1966).
23. A. H. Jackson, G. W. Kenner, K. M. Smith, R. T. Alpin, H. Budzikiewicz, and C. Djerassi, *Tetrahedron Lett.*, **21**, 2913 (1965).
24. T. C. Chu, A. A. Green, and E. J. Chu, *J. Biol. Chem.*, **190**, 643 (1951).
25. T. C. Chu and E. J. Chu, *J. Chromatogr.*, **21**, 46 (1966).
26. E. Demole, *J. Chromatogr.*, **1**, 24 (1958).

Po-8
Coproporphyrin II

Formula: $C_{36}H_{38}N_4O_8$
Formula Wt.: 654.73 (Tetramethyl ester, 710.83)
Calc. %: C, 66.04; H, 5.85; N, 8.56; O, 19.55

R = CH$_2$—CH$_2$—COOH

Tetracarboxylic acid: R = CH₂—CH₂—COOH
Tetramethyl esters: R = CH₂—CH₂—CO₂CH₃

Sources: Chemical synthesis[1,2,3,4,5]

PROPERTIES OF TETRACARBOXYLIC ACID

HCl Number: See coproporphyrin I.
Chromatography: See coproporphyrin I.
Spectral Reference Values: See coproporphyrin I.
m.p.: Decomposes.

PROPERTIES OF TETRAMETHYL ESTER

Chromatography: See coproporphyrin I tetramethyl ester.
m.p.:[6] 288 °C
Spectral Reference Values: See coproporphyrin I, tetramethyl ester.
NMR Spectrum: See Abraham.[7]

References

1. H. Fischer and H. Orth, *Chemie des Pyrrols*, Vol. II, Part I, Akademische-Verlag, Leipzig (1937), p. 471, 596.
2. H. Fischer and H. Andersag, *Ann. Chem.*, **450**, 201 (1926).
3. G. P. Arsenault, E. Bullock, and S. F. MacDonald, *J. Am. Chem. Soc.*, **82**, 4384 (1960).
4. A. H. Jackson and G. W. Kenner, *Nature*, **215**, 1126 (1967).
5. R. L. N. Harris, A. W. Johnson, and I. T. Kay, *J. Chem. Soc.* (C), 22 (1966).
6. F. Morsingh and S. F. MacDonald, *J. Am. Chem. Soc.*, **82**, 4377 (1960).
7. R. J. Abraham, P. A. Burbidge, A. H. Jackson, and D. B. Macdonald, *J. Chem. Soc.*, (B), 620 (1966).

Po-9
Coproporphyrin III

Formula: C₃₆H₃₈N₄O₈
Formula Wt.: 654.73 (Tetramethyl ester, 710.83)
Calc. %: C, 66.04; H, 5.85; N, 8.56; O, 19.55

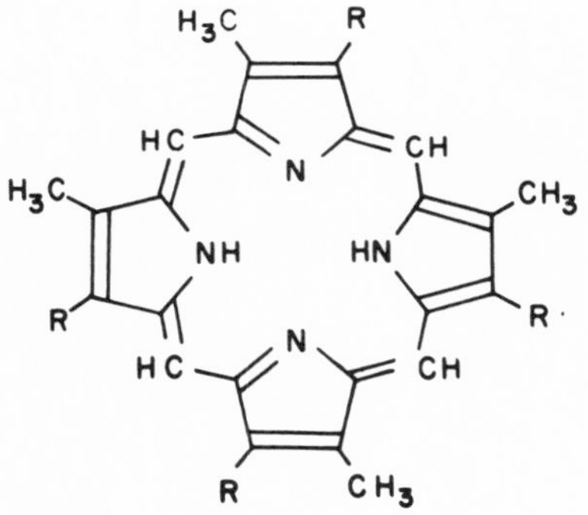

R= CH₂—CH₂—COOH

Tetracarboxylic acid: R = CH₂—CH₂—COOH
Tetramethyl ester: R = CH₂—CH₂—CO₂CH₃

Sources: Bacterial culture medium.[1,2] Synthesis.[3,4,5]

PROPERTIES OF TETRACARBOXYLIC ACID

HCl Number: See coproporphyrin I.
Chromatography: The free acid of the type III isomer is not sepa-

rated from the IV isomer by 2,6-lutidine–aqueous ammonia. See coproporphyrin I.
m.p.: Decomposes.
Spectral Reference Values: See coproporphyrin I.

PROPERTIES OF TETRAMETHYL ESTER

Chromatography: See With[8] and Chu *et al.*[9]
m.p.: 152–154 °C, 165–168 °C, 176–179 °C (polymorphic); Cu complex 216–219 °C.[4]
NMR Spectrum: See Abraham.[6]
Spectral Reference Values: See coproporphyrin I, tetramethyl ester.
Mass Spectrum: See Jackson *et al.*[7]
Other Properties: The solubilities of isomers III and IV are similar, and are higher than those of other isomers. The isomers also form solid solutions (mixed crystals).[4] The Cu chelates of coproporphyrin III and IV methyl esters are not polymorphic and the m.p. can thus be used to distinguish them.[4] Infrared spectra.[4] The HCl number (1.7) is identical with that of other isomers.
Likely Impurities: Other isomers, porphyrins having fewer or more carboxyl groups, haloporphyrins.

References

1. J. B. Neilands and J. A. Garibaldi, *Biochem. Prep.*, **7**, 36 (1960).
2. J. Lascelles, *Tetrapyrrole Biosynthesis and Its Regulation*, Benjamin, New York (1964).
3. A. H. Jackson, G. W. Kenner, G. McGillivray, and K. M. Smith, *J. Chem. Soc.* (B), 294 (1968).
4. F. Morsingh and S. F. MacDonald, *J. Am. Chem. Soc.*, **82**, 4377 (1960).
5. R. L. N. Harris, A. W. Johnson, and I. T. Kay, *J. Chem. Soc.* (C), 22 (1966).
6. R. J. Abraham, P. A. Burbidge, A. H. Jackson, and D. B. MacDonald, *J. Chem. Soc.*, (B), 620 (1966).
7. A. H. Jackson, G. W. Kenner, K. M. Smith, R. T. Alpin, H. Budzikiewicz, and C. Djerassi, *Tetrahedron Lett.*, **21**, 2913 (1965).
8. T. K. With, *J. Chromatogr.*, **42**, 389 (1969).
9. T. C. Chu and E. J. Chu, *J. Chromatogr.*, **21**, 46 (1966).

Po-10
Coproporphyrin IV

Formula: C₃₆H₃₈N₄O₈
Formula Wt.: 654.73 (Tetramethyl ester, 710.83)
Calc. %: C, 66.04; H, 5.85; N, 8.56; O, 19.55

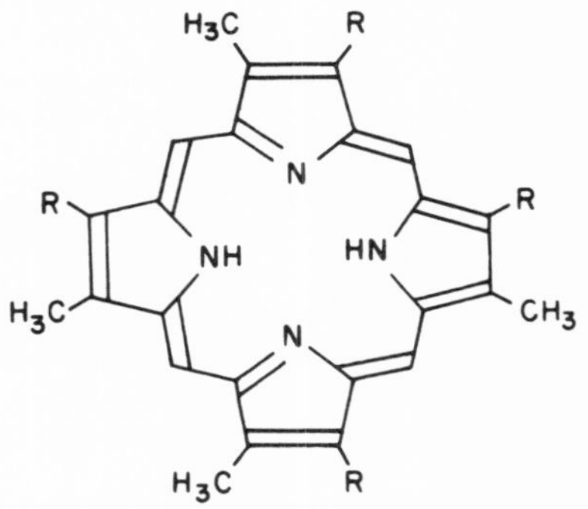

R = CH₂—CH₂—COOH

Tetracarboxylic acid: R = CH₂—CH₂—COOH
Tetramethyl ester: R = CH₂—CH₂—CO₂CH₃

Sources: Synthetic, from the methyl ester.[1,2,3,4]

PROPERTIES OF TETRACARBOXYLIC ACID

HCl Number: See coproporphyrin I.
Chromatography: See coproporphyrin III.
m.p.: Decomposes.
Spectral Reference Values: See coproporphyrin I.

PROPERTIES OF TETRAMETHYL ESTER

Chromatography: See coproporphyrin I tetramethyl ester.
m.p.: 182–186 °C; polymorphic, with solid phase changes at 168–170 °C and 175 °C; Cu complex 230–233 °C.[4]
Spectral Reference Values: See coproporphyrin I tetramethyl ester.
NMR Spectrum: See Abraham.[5]
Other Properties: See coproporphyrin III.

References

1. H. Fischer and H. Orth, *Die Chemie des Pyrrols*, Vol. II, Part I, Akademische-Verlag, Leipzig (1937), p. 471, 596.
2. G. P. Arsenault, E. Bullock, and S. F. MacDonald, *J. Am. Chem. Soc.*, **82**, 4384 (1960).
3. A. H. Jackson, G. W. Kenner, G. McGillivray, and K. M. Smith, *J. Chem. Soc. B*, 294 (1968).
4. F. Morsingh and S. F. MacDonald, *J. Am. Chem. Soc.*, **82**, 4377 (1960).
5. R. J. Abraham, P. A. Burbidge, A. H. Jackson, and D. B. MacDonald, *J. Chem. Soc. B*, 620 (1966).

Po-11
Deuteroporphyrin IX

Formula: $C_{30}H_{30}N_4O_4$
Formula Wt.: 510.59 (Dimethyl ester, 538.65)
Calc. %: C, 70.57; H, 5.92; N, 10.97; O, 12.53

Source: Fusion of protohemin in resorcinol substitutes hydrogens for vinyl groups to give deuterohemin.[1,2,3] Iron is removed in ferrous sulfate–HCl[3] or formic acid–iron powder[1,2] mixtures. Iron removal and acid esterification are conveniently carried out in the same solution.[3]

PROPERTIES OF DICARBOXYLIC ACID

Spectral Reference Value: In 0.1 M HCl, ϵ_{mM} 433 at 398 nm.[4]
Other Properties: HCl No. 0.4.[1] The distribution coefficient between ether and 0.1 M HCl is 0.816 at 23 °C.[5] It is extracted into CHCl₃ from 0.2% HCl. It is crystallized from pyridine–ether or acetic acid–ether in needles. Readily soluble in dilute KOH, acetic acid–ether, HCl, or pyridine.

PROPERTIES OF THE DIMETHYL ESTER

m.p.: After crystallization out of chloroform–methanol: 227 °C,[3] 224.5 °C.[6] (diethyl ester, 204.5 °C.[3])
HCl Number: 2.0.[1]

Spectral Reference Values:
In chloroform,[7] band maxima

λ, nm	619	566	530	497	399	267
ϵ_{mM}	4.1	6.3	8.3	14.4	187	7.4

In dioxane,[8] band maxima

λ, nm	618	593	565	525	495	399
ϵ_{mM}	4.3	1.3	6.8	8.6	15.9	175[4]

with minima

λ, nm	601	590	548	514
ϵ_{mM}	0.68	1.21	1.19	3.08

NMR Spectrum: 2,4-protons at 9.12 δ (infinite dilution value) in CDCl₃ and at 9.43 δ in CDCl₃ with 2.5% F₃CCOOH.[3]
Chromatography: Column: Alumina.[3] Paper: $R_f = 0.45$ ascending, with silicone as stationary phase and water–acetonitrile–propanol–pyridine (3.8:1:2:0.5).[2,9] Also, see References, p. 187.

References

1. H. Fischer and H. Orth, *Die Chemie des Pyrrols*, Vol. II, Part I, Akademische-Verlag, Leipzig (1937), p. 414.
2. T. C. Chu and E. J. H. Chu, *J. Am. Chem. Soc.*, **74**, 6276 (1952).
3. W. S. Caughey, J. O. Alben, W. Y. Fujimoto, and J. L. York, *J. Org. Chem.*, **31**, 2631 (1966).
4. C. Rimington, *Biochem. J.*, **75**, 620 (1960).
5. S. Granick and L. Bogorad, *J. Biol. Chem.*, **202**, 781 (1953).
6. A. H. Corwin and R. H. Krieble, *J. Am. Chem. Soc.*, **63**, 1829 (1941).
7. W. S. Caughey, W. Y. Fujimoto, and B. D. Johnson, *Biochemistry*, **5**, 3830 (1966).
8. A. Stern and H. Wenderlein, *Z. Physik. Chem.*, **A175**, 405 (1936).
9. T. C. Chu and E. J. H. Chu, *J. Biol. Chem.*, **208**, 537 (1954).

Po-12
Hematoporphyrin IX

Formula: $C_{34}H_{38}N_4O_6$
Formula Wt.: 598.70 (Dimethyl ester, 626.76)
Calc. %: C, 68.21
H, 6.40
N, 9.36
O, 16.03

Sources: From protohemin with acetic acid–HBr. Crystallized as the dihydrochloride from 2.5% HCl.[1] The product is a mixture of compounds.[2] The dimethyl ester from esterification with diazomethane can be obtained in high purity following chromatography on alumina.[3]

PROPERTIES OF THE DIMETHYL ESTER

Spectral Reference Values:
In 1 M HCl[4]

λ, nm	402	548	592			
ϵ_{mM}	383	15.6	5.6			
0.5W, nm	13	20	~14			

In chloroform[5]

λ, nm	622	569	534	499	402	269
ϵ_{mM}	4.3	6.9	9.5	15.0	193	8.9

In pyridine[2]

λ, nm	623	569	532	499	401	455(min)	603(min)
ϵ_{mM}	4.3	6.7	9.0	14.8	174	2.1	0.84

NMR **Spectrum:** δ values for the 2,4-(1-hydroxyethyl) groups in CDCl₃ calculated to infinite dilution are CH₃ (2.06) and HO—C—H (6.14), and in CDCl₃ with 2.5% F₃CCOOH are CH₃ (2.22) and HO—C—H (6.59).[3,5,6]
Chromatography: $R_f = 0.08$ in kerosene–CHCl₃ (100:65 v/v) on paper.[7] Alumina column.[3]
Other Properties: m.p. 212–217 °C. HCl No. of unesterified compound 0.1. Aqueous solutions of strong acids cause equilibration between vinyl and 1-hydroxyethyl groups.
Likely Impurities: Protoporphyrin, 2(4)-(1-hydroxyethyl)-4(2)-vinyldeuteroporphyrin, chloro- or bromoethyldeuteroporphyrins. Purification by countercurrent distribution has been achieved.[2] The only recorded optically active sample was one derived from cytochrome c.[8]

References

1. H. Fischer and H. Orth, *Die Chemie des Pyrrols*, Vol. II, Part I, Akademische-Verlag, Leipzig (1937), p. 421.
2. S. Granick, L. Bogorad, and H. Jaffe, *J. Biol. Chem.*, **202**, 801 (1953).
3. W. S. Caughey, J. O. Alben, W. Y. Fujimoto, and J. L. York, *J. Org. Chem.*, **31**, 2631 (1966).
4. D. Mauzerall, unpublished information.
5. W. S. Caughey, W. Y. Fujimoto, and B. D. Johnson, *Biochemistry*, **5**, 3830 (1966).
6. E. D. Becker, R. B. Bradley, and C. J. Watson, *J. Am. Chem. Soc.*, **83**, 3743 (1961).
7. T. C. Chu, A. A. Green, and E. J. Chu, *J. Biol. Chem.*, **190**, 643 (1951).
8. K. G. Paul, *Acta Chem. Scand.*, **5**, 389 (1951).

Po-13
Mesobilirubin IXα

Formula: $C_{33}H_{40}N_4O_6$
Formula Wt.: 588.71
Calc. %: C, 67.33; H, 6.85; N, 9.51; O, 16.31

Source: Hydrogenation of bilirubin (Pd catalyst).[1,2] Natural.[3]
Spectral Reference Values: In chloroform, ϵ_{mM} 54.6 at 434 nm,[4] 54.7 at 429 nm.[5]
m.p.: 315 °C.
Chromatography: Paper,[6] TLC on polyamide.[7]
Other Properties: Oxidation with alcoholic Zn acetate + 0.1 ml of 1% alcoholic iodine yields red fluorescence of Zn chelate of oxidized forms (mesobiliverdin and mesobilivioline, for example) at 626, 597, and 513 nm.
Likely Impurities: Dihydrobilirubin,[1] dihydromesobilirubin,[1] and isomers present in bilirubin employed for hydrogenation (compare Po-2).

References

1. H. Fischer and H. Orth, *Die Chemie des Pyrrols*, Vol. II, Part I, Akademische-Verlag, Leipzig (1937).
2. H. Fischer, H. Plieninger, and A. Weissbrath, *Z. Physiol. Chem.*, **268**, 197 (1941).
3. C. J. Watson, unpublished information.
4. H. Gray, A. Kulczycka, and D. D. Nicholson, *J. Chem. Soc.*, 2268 (1961).
5. C. J. Watson and I. Bossenmaier, unpublished information.
6. L. B. Pearson and C. J. Watson, *Proc. Soc. Exp. Biol. Med.*, **112**, 756 (1963).
7. Z. J. Petryka and C. J. Watson, *J. Chromatogr.*, **37**, 76 (1968).

Po-14
Mesoporphyrin IX

Formula: $C_{34}H_{38}N_4O_4$
Formula Wt.: 566.70 (Dimethyl ester, 594.76)
Calc. %: C, 72.06; H, 6.76; N, 9.89; O, 11.29

Source: Reduction of the vinyl groups of protoporphyrin or its iron complex. Reducing conditions include formic acid with H_2 and PdO_2,[1,2] with H_2 and colloidal Pd,[3] and aqueous KOH with H_2 and PtO_2.[4]

PROPERTIES OF DICARBOXYLIC ACID

Spectral Reference Values:
In 1 M HCl

λ, nm	592	549	401	619	582
ϵ_{mM}	5.8	14.9	390	0.6(min)	3.4(min)

Other Properties: HCl No. 0.5. Distribution coefficient between ether and 0.1 M HCl at 23 °C is 2.93:1.[5] Difficult to dissolve in chloroform, readily soluble in ammonia. Recrystallized from acetic acid–ether. The Na salt is precipitated at concentrations greater than 0.8% of NaOH. Mesoporphyrin I may be separated chromatographically from mesoporphyrin IX.[6]

PROPERTIES OF THE DIMETHYL ESTER

HCl No.: 2.5.
m.p.: 215 °C (Diethyl ester, 207 °C).[2]
Chromatography: See References, p. 187.

Spectral Reference Values:
In chloroform[7]

λ, nm	619	566	532	498	399.5	270
ϵ_{mM}	4.9	6.5	9.9	14.0	168	7.9

In p-dioxane[8]

λ, nm	620	567	528	496	590(min)	549(min)	515(min)
ϵ_{mM}	5.4	16.6	9.8	14.2	1.15	1.19	3.1

NMR Spectrum: For the 2,4-ethyl groups, the CH_2 and CH_3 groups, respectively, appear at δ values of 4.30 and 1.88 in $CDCl_3$, at 4.17 and 1.74 in $CDCl_3$ with 2.5% F_3CCOOH,[2] and at 4.31 and 1.84 in F_3CCOOH.[9]

References

1. J. F. Taylor, *J. Biol. Chem.*, **135**, 570 (1940).
2. W. S. Caughey, J. O. Alben, W. Y. Fujimoto, and J. L. York, *J. Org. Chem.*, **31**, 2631 (1966).
3. S. Granick, *J. Biol. Chem.*, **172**, 717 (1948).
4. E. W. Baker, M. Ruccia, and A. H. Corwin, *Anal. Biochem.*, **8**, 512 (1964).
5. L. Bogorad and S. Granick, *J. Biol. Chem.*, **202**, 781 (1953).
6. C. Rimington and A. Benson, *J. Chromatogr.*, **6**, 350 (1961).
7. W. S. Caughey, W. Y. Fujimoto, and B. P. Johnson, *Biochemistry*, **5**, 3830 (1966).
8. A. Stern and H. Wenderlein, *Z. Physik. Chem.*, **A174**, 81 (1935).
9. R. J. Abraham, A. H. Jackson, and G. W. Kenner, *J. Chem. Soc.*, 3468 (1961).

Po-15
(+)-Phytol

Formula: $C_{20}H_{40}O$
Formula Wt.: 296.54
Calc. %: C, 81.01; H, 13.60; O, 5.39

Source: Hydrolysis of chlorophyll.[1]
Spectral Reference Value: ϵ_{mM} 1.10 at 212 nm in alcohol.
Infrared Spectrum: See Stair and Coblentz[2] and Weigle and Livingston.[3]
Specific Rotation: $[\alpha]_D^{18}$ +0.20°.
Other Properties: Refractive index, n_D^{25} 1.4637; density, d_4^{25} 0.8491; b.p. 132 °C (0.02 mmHg), 150 °C (0.06 mmHg), 203 °C (10.0 mmHg).
Derivatives: Allophanate, m.p. 71 °C;[4] pyruvate semicarbazone, m.p. 72–75 °C.
Chromatography: On paper impregnated with silicone oil, develop with methanol–water (8:2).
   R_f = 0.4–0.5. Paper dried, and treated with iodine in petroleum ether. After the paper becomes white, spray with starch solution, and compare the blue zone with a phytol control; 10–40 μg estimated quantity.[1]
Gravimetric Estimation: 1-Naphthylamine derivative of phytyl 3,5-dinitrobenzoate.[5]
Likely Impurities: Phytadiene, carotenoids, steroids, alcohols of high molecular weight, and other neutral nonsaponifiable substances.[1,5]
NMR Spectrum: See reference 6.

References

1. F. G. Fischer and H. Bonn, *Ann. Chem.*, **611**, 224 (1958).
2. R. Stair and W. W. Coblentz, *J. Res. Nat. Bur. Stand.*, **11**, 703 (1933).
3. J. W. Weigle and R. Livingston, *J. Am. Chem. Soc.*, **75**, 2173 (1953).
4. P. Karrer, H. Simon, and E. Zbinden, *Helv. Chim. Acta.*, **27**, 313 (1944).
5. A. Hrmatka, W. Broll, and L. Stentzel, *Monatsh. Chem.*, **89**, 54, 116, 126, (1958).

Po-16
Porphobilinogen

Formula: $C_{10}H_{14}N_2O_4 \cdot H_2O$
Formula Wt.: 244.25
Calc. %: C, 49.17
       H, 6.60
       N, 11.47
       O, 32.75

Sources: From the urine of acute porphyria patients[1] and from urine of animals treated with such compounds as (2-isopropyl-4-pentenoyl)urea (Sedormid). Enzymically, from 5-aminolevulinic acid.[2,3] By chemical synthesis.[4,5]

Photometric Determination: Condense with modified Ehrlich reagent; the apparent ϵ_{mM} at 553 nm is 61, and there is a shoulder at 525 nm, ϵ_{mM} 47.[6]

Chromatography: $R_f = 0.5$ on ascending paper chromatography with 1-butanol–acetic acid–water (4:1:5, v/v).[6]

m.p.: 174–177 °C with decomposition.[4]

Other Properties: The hydrochloride.[1] The lactam.[1] X-ray powder photograph.[4,7] Infrared spectrum.[5] Quantitatively converted into uroporphyrinogens, either chemically[8] or enzymically.[9] ϵ_{mM} at 212 nm = 6.77.[10] pK_a = 3.70, 4.95, 10.1.[10] The Ehrlich color salt of porphobilinogen in acetate buffer (pH 4–5) is insoluble in butanol, which distinguishes it from the urobilinogen color salt.

Likely Impurities: Not well defined;[4] poly(pyrrylmethanes).

Special Reagents: Modified Ehrlich reagent: Prepare fresh. *p*-Dimethylaminobenzaldehyde (1 g) is dissolved in 30 ml of glacial acetic acid, 8.0 ml of 70% perchloric acid is added and the solution is diluted to 50 ml with acetic acid. Mix with an equal volume of porphobilinogen solution at 23 °C and read at 15 min. The *p*-dimethylaminobenzaldehyde may be recrystallized from hot ethanol–water to give a colorless product.

References

1. G. H. Cookson and C. Rimington, *Biochem. J.*, **57**, 476 (1954).
2. R. Schmid and D. Shemin, *J. Am. Chem. Soc.*, **77**, 506 (1956).
3. H. A. Sancovich, A. M. Ferramola, A. M. del C. Batlle, and M. Grinstein, *Methods Enzymol.*, **17**, 220 (1970).
4. G. P. Arsenault and S. F. MacDonald, *Can. J. Chem.*, **39**, 2043 (1961).
5. B. Frydman, M. E. Despuy, and H. Rapoport, *J. Am. Chem. Soc.*, **87**, 3530 (1965).
6. D. Mauzerall and S. Granick, *J. Biol. Chem.*, **219**, 435 (1965).
7. O. Kennard, *Nature*, **171**, 876 (1953).
8. D. Mauzerall, *J. Am. Chem. Soc.*, **82**, 2605 (1960).
9. L. Bogorad, *J. Biol. Chem.*, **233**, 501, 510 (1958).
10. S. Granick and L. Bogorad, *J. Am. Chem. Soc.*, **75**, 3610 (1953).

Po-17
Protoporphyrin IX

Formula: $C_{34}H_{34}N_4O_4$
Formula Wt.: 562.27 (Dimethyl ester, 590.73)
Calc. %: C, 72.57; H, 6.09; N, 9.97; O, 11.37

Source: Dicarboxylic acid is prepared from hemin chloride by refluxing in 98% formic acid with iron powder to remove iron.[1,2,3] Diester is obtained from hemin chloride by carrying out iron removal and esterification in same solution ($FeSO_4$, HCl, alcohol),[4] as well as by esterification of the metal-free dicarboxylic acid.[4]

PROPERTIES OF THE DICARBOXYLIC ACID

Spectral Reference Values:
In 1 M HCl[5]

λ, nm	600	556	408	520
ϵ_{mM}	5.6	13.5	241	3.2(min)

In 1.37 M HCl, ϵ_{mM} 275 at 408 nm.[6]

Chromatography: On paper in 2,6-lutidine–NH_3,[7] protoporphyrin IX moves as a porphyrin having two carboxyl groups.

Other Properties: Readily soluble in $CHCl_3$ (when acid), mineral acids, or pyridine. Sparingly soluble in 0.01 M KOH in 50% alcohol; colloidal in aqueous alkali. Recrystallized from formic acid–ether in fine needles. HCl No. 2.5.

PROPERTIES OF THE DIMETHYL ESTER

HCl No. = 5.5.

m.p.: Though generally sharp, reports vary from 214 to 232 °C. Diethyl ester, 215 °C.[4]

Spectral Reference Values:
In chloroform[8]

λ, nm	630	576	541	506	407.5	275
ϵ_{mM}	5.0	6.5	11.1	13.8	161	13.0

In *p*-dioxane[9]

λ, nm	630	575	537	503
ϵ_{mM}	5.6	6.8	11.6	14.6

Identification of Vinyl Groups: NMR δ values at 1.64, 3.66, and 3.83 in $CDCl_3$[4] and at 1.71, 3.43, and 3.61 in F_3CCOOH.[10] Infrared absorption of terminal methylenes in $CHCl_3$ are at 1.628 μm (ϵ_{mM}, 0.98) and at 2.116 μm.[4]

Chromatography: Column.[4] Paper.[11] See References, p. 187.
Likely Impurities: Compounds that contain iron, ethyl or 1-hydroxyethyl groups, and oxidation products (650 nm band).

References

1. H. Fischer and H. Orth, *Die Chemie des Pyrrols*, Vol. II, Part 1, Akademische-Verlag, Leipzig (1937), p. 471 ff., 596.
2. V. G. Ramsey, *Biochem. Prep.*, **3**, 39 (1953).
3. W. S. Caughey, W. Y. Fujimoto, A. J. Bearden, and T. H. Moss, *Biochemistry*, **5**, 1255 (1966).
.4. W. S. Caughey, J. O. Alben, W. Y. Fujimoto, and J. L. York, *J. Org. Chem.*, **31**, 2631 (1966).
5. D. Mauzerall, unpublished information.
6. C. Rimington, *Biochem. J.*, **75**, 620 (1960).
7. J. E. Falk, *J. Chromatogr.*, **5**, 277 (1961).
8. W. S. Caughey, W. Y. Fujimoto, and B. P. Johnson, *Biochemistry*, **5**, 3830 (1966).
9. A. Stern and H. Wenderlein, *Z. Physik. Chem.*, **A170**, 337 (1934).
10. R. J. Abraham, A. H. Jackson, and G. W. Kenner, *J. Chem. Soc.*, 3468 (1961).
11. T. C. Chu and E. J. Chu, *J. Biol. Chem.*, **208**, 537 (1954).

Po-18
Stercobilin (Dipyrrolidone)

Formula: $C_{33}H_{46}N_4O_6$
Formula Wt.: 594.76
Calc. %: C, 66.64; H, 7.80; N, 9.42; O, 16.14

Sources: Natural from human feces. (Yields are larger from patients with hemolytic anemia. Absent, or present in small amount, after administration of broad spectrum antibiotics.[1]) Synthesized from bilirubin[2] or from total synthesis.[3]
Spectral Reference Values: Acid form exhibits an ϵ_{mM} of 92.9 at 496 nm in CHCl$_3$[4] and of 90.3 at 492 nm in methanol–3 M HCl (100:1, v/v).[5]
m.p.: 234–236 °C;[6] hydrochloride, 158–160 °C,[1] 140–142 °C.[7]
Specific Rotation:
  $[\alpha]_D^{20} = -4,000°$ as hydrochloride in CHCl$_3$ (natural).[2]
  $[\alpha]_D^{20} = -3,900°$ as hydrochloride in CHCl$_3$ (synthetic).[3]
  $[\alpha]_D^{20} = 0°$ (synthetic).[2,3]
Other Properties: X-ray diffraction powder patterns and infrared spectrum are distinct for dipyrrolidones.[1,3,6] Natural stercobilin is the *trans–trans*-isomer.[8] Both *cis–cis*[9] and *trans–trans*[10] have been synthesized. Stability toward ferric chloride oxidation[10] and ability to form a molecular complex with ferric chloride[11] permit distinction from other urobilinoids. Chromic acid oxidation yields ethyl methyl succinimide.[8,9] Mass spectrum (molecular ion, 594).[12]
Likely Impurities: Pyrrolidone-pyrrolinone, dipyrrolinones, and their oxidation products.[10,12]

References

1. C. J. Watson, P. T. Lowry, V. E. Sborov, W. H. Hollinshead, S. Kohan, and H. O. Matte, *J. Biol. Chem.*, **200**, 697 (1953).

2. I. T. Kay, M. Weiner, and C. J. Watson, *J. Biol. Chem.*, **238**, 1122 (1963).
3. H. Plieninger, K. Ehl, and A. Tapia, *Ann. Chem.*, **736**, 43 (1970); H. Plieninger and U. Lerch, *Ann. Chem.*, **698**, 196 (1966).
4. C. H. Gray, A. Lichtarowicz-Kulczycka, and D. C. Nicholson, *J. Chem. Soc.*, 2276 (1961).
5. C. J. Watson and I. Bossenmaier, unpublished information.
6. C. J. Watson and P. T. Lowry, *J. Biol. Chem.*, **218**, 633 (1956).
7. C. J. Watson, A. Moscowitz, D. A. Lightner, Z. J. Petryka, E. Davis, and M. Weimer, *Proc. Nat. Acad. Sci. U.S.*, **58**, 1957 (1967).
8. C. H. Gray, G. A. Lemmon, and D. C. Nicholson, *J. Chem. Soc. (C)*, 178 (1967).
9. Z. J. Petryka and C. J. Watson, *Tetrahedron Lett.*, **52**, 5323 (1967).
10. C. J. Watson, M. Weimer, Z. J. Petryka, D. A. Lightner, A. Moscowitz, E. Davis, and N. A. Beach, *Arch. Biochem. Biophys.*, **131**, 414 (1969).
11. C. J. Watson and Z. J. Petryka, *Anal. Biochem.*, **30**, 156 (1969).
12. D. A. Lightner, A. Moscowitz, Z. J. Petryka, S. Jones, M. Weimer, E. Davis, N. A. Beach, and C. J. Watson, *Arch. Biochem. Biophys.*, **131**, 566 (1969).

Po-19
Half-Stercobilin
(Pyrrolidone–Pyrrolinone)

Formula: $C_{33}H_{44}N_4O_6$
Formula Wt.: 592.74
Calc. %: C, 66.87; H, 7.48; N, 9.45; O, 16.25

Source: Human feces;[1] bacterial reduction;[2] by-product of catalytic hydrogenation of bilirubin.[3]
Spectral Reference Values: Acid form, absorption maximum between 490.5 and 492.5 nm in chloroform.[1]
Specific Rotation: $[\alpha]_D^{20}$ from $-1,900°$ to $-3,000°$ for acid forms in chloroform.[1]
m.p.:[1] 148–154 °C.
Mass Spectra:[1] Molecular ion 592.
Other Properties: X-ray powder pattern distinct from stercobilin and urobilin.[1] Ferric chloride oxidation[4] and chromic acid oxidation[1] differentiate half-stercobilin from other urobilinoids and indicate presence of impurities.
Likely Impurities: Stercobilin, urobilin, and their oxidation products.

References

1. C. J. Watson, A. Moscowitz, D. A. Lightner, Z. J. Petryka, E. Davis, and M. Weimer, *Proc. Nat. Acad. Sci., U.S.*, **58**, 1957 (1967).
2. C. J. Watson *et al.*, *Biochem. Med.*, **2**, 461 (1969); **2**, 484 (1969); **4**, 149 (1970).
3. Z. J. Petryka, unpublished information.
4. C. J. Watson, M. Weimer, Z. J. Petryka, D. A. Lightner, A. Moscowitz, E. Davis, and N. A. Beach, *Arch. Biochem. Biophys.*, **131**, 414 (1969).

Po-20

MS-Tetraphenylporphin
($\alpha,\beta,\gamma,\delta$-Tetraphenylporphine)

Formula: $C_{44}H_{30}N_4$
Formula Wt.: 614.75
Calc. %: C, 85.90; H, 4.92; N, 9.12

Source: Synthesis—condensation of pyrrole and benzaldehyde in propionic acid.[1] Useful data on preparative procedures and the reaction mechanisms have been reported.[1-5] Metal complexes are readily prepared.[6,7]

Spectral Reference Values:

In benzene,[7-10] band maxima

λ, nm	649	592	548	515	483(sh)	418
ϵ_{mM}	3.8	5.7	8.4	19.8	3.7	480

Mass Spectroscopy: See Adler *et al.*[11]
NMR Spectroscopy: See Badger *et al.*[10]

Chromatography: TLC.[12] A useful column chromatography procedure follows: A solution of crude TPP (1 g) in reagent-grade chloroform (200 ml) is added to a 3 × 12-in. column of *dry* packed neutral or acidic alumina of a chromatographic grade. Elution with chloroform, which develops a slow intense narrow green band and a faster broad diffuse purple-red band, is continued until the green band has reached within 1 in. of the bottom of the column. The material of the purple-red band, after elution and crystallization from chloroform–methanol, is 99.5% pure with adsorbed solvent and a trace of the hydrochloride (acid salt) as the only impurities. Further purification can be accomplished by sublimation.[1]

Likely Impurities: Polypyrrole by-products, MS-tetraphenylchlorin salts, and adsorbed solvents.

References

1. A. D. Adler, F. R. Longo, J. D. Finarelli, J. Goldmacher, J. Assour, and L. Korsakoff, *J. Org. Chem.*, **32**, 476 (1967).
2. A. D. Adler, L. Sklar, F. R. Longo, J. D. Finarelli, and M. G. Finarelli, *J. Heterocycl. Chem.*, **5**, 669 (1968).
3. D. Dolphin, *J. Heterocycl. Chem.*, **7**, 275 (1970).
4. H. W. Whitlock, R. Hanauer, M. Y. Oester, and B. K. Bower, *J. Am. Chem. Soc.*, **91**, 7485 (1969).
5. A. Treibs and N. Haeberle, *Ann. Chem.*, **718**, 183 (1968).
6. D. W. Thomas and A. E. Martell, *J. Am. Chem. Soc.*, **78**, 1338 (1956).
7. A. D. Adler, F. R. Longo, F. Kampas, and J. Kim, *J. Inorg. Nucl. Chem.*, **32**, 2443 (1970).
8. J. A. Mullins, A. D. Adler, and R. M. Hochstrasser, *J. Chem. Phys.*, **43**, 2548 (1965).
9. G. D. Dorough, J. R. Miller, and F. M. Huennekens, *J. Am. Chem. Soc.*, **73**, 4315 (1951).
10. G. M. Badger, R. A. Jones, and R. L. Laslett, *Aust. J. Chem.*, **17**, 1028 (1964).
11. A. D. Adler, J. H. Green, and M. Mautner, *Org. Mass Spectrosc.*, **3**, 955 (1970).
12. R. W. Balek and A. Szutka, *J. Chromatogr.*, **17**, 127 (1965).

Po-21

Urobilin (Dipyrrolinone)

Formula: I. $C_{33}H_{40}N_4O_6$
        II. $C_{33}H_{42}N_4O_6$
Formula Wt.: I.   588.71
           II.   590.73
Calc. %:   I. C, 67.33; H, 6.85; N, 9.52; O, 16.31
           II. C, 67.09; H, 7.17; N, 9.49; O, 16.25

Sources: Infected fistula tube; feces of individuals receiving broad spectrum antibiotics;[1] total synthesis;[2] bilirubin reduction with sodium amalgam[3,4] or catalytic reduction.[4]

Spectral Reference Values: Acids formed in $CHCl_3$ exhibit ϵ_{mM} values of 93.7 for I and 72.1 for II[5] at 499 nm. In methanol–3 M HCl (100:1, v/v), ϵ_{mM} at 494 nm is 90.6 for I and 89.7 for II.[6]

m.p.: I: 172–174 °C.[1] II: 175–177 °C,[3,5] 179–180 °C.[2]

Specific Rotation: $[\alpha]_D^{20}$ for the monohydrochloride in $CHCl_3$ varies from $-4800°$[2] to $+5000°$.[1]

Other Properties: X-ray powder patterns differ among optically active and inactive urobilins, stercobilin, and half-stercobilin. Infrared spectra for urobilins are similar but differ from stercobilin spectra. Urobilins are easily oxidized with ferric chloride[7] or bromic acid[8] and can be separated, as dimethyl esters, on TLC.[9]

Likely Impurities: Mesobiliviolin, glaucobilin, stercobilin, half-stercobilin, mesobilifuscin.

References

1. C. J. Watson and P. T. Lowry, *J. Biol. Chem.*, **218**, 633 (1956).
2. H. Plieninger, K. Ehl, and A. Tapia, *Ann. Chem.*, **736**, 62 (1970).
3. C. J. Watson, *J. Biol. Chem.*, **200**, 691 (1953).
4. H. Fischer and H. Orth, *Die Chemie des Pyrrols*, Vol. II, Part 1, Akademische-Verlag, Leipzig (1934), p. 685.
5. C. H. Gray, A. Kulczycka, and D. C. Nicholson, *J. Chem. Soc.*, 2276 (1961).
6. C. J. Watson and I. Bossenmaier, unpublished information.
7. C. J. Watson, M. Weimer, Z. J. Petryka, D. A. Lightner, A. Moscowitz, E. Davis, and N. A. Beach, *Arch. Biochem. Biophys.*, **131**, 414 (1969).
8. C. J. Watson, A. Moscowitz, D. A. Lightner, Z. J. Petryka, E. Davis, and M. Weimer, *Proc. Nat. Acad. Sci., U. S.*, **58**, 1957 (1967).
9. Z. J. Petryka, unpublished information.

Po-22
Uroporphyrin I

Formula: $C_{40}H_{38}N_4O_{16}$
Formula Wt.: 830.77 (Octamethyl ester, 942.98)
Calc. %: C, 57.83; H, 4.61; N, 6.74; O, 30.82

A = CH₂ COOH

$A = CH_2 COOH$

$R = CH_2CH_2 COOH$

Sources: By hydrolysis of the methyl ester.[1,2] From urine of beef cattle having congenital porphyria,[3] and from urine of humans having congenital porphyria.[4] By chemical synthesis.[5]
Spectral Reference Values: Same as for uroporphyrin III (Po-24) in 1 M HCl.
Chromatography: See uroporphyrin III (Po-24).
Other Properties: See uroporphyrin III. Is decarboxylated to coproporphyrin I, either chemically,[1,6] or enzymically as uroporphyrinogen.[7]

PROPERTIES OF THE OCTAMETHYL ESTER

Spectral Reference Values: Same as for uroporphyrin III ester (Po-25).
NMR **Spectrum:** See Becker *et al.*[8]
Chromatography: $R_f = 0.02$ in kerosene-*p*-dioxane (8:3 by volume); same as for uroporphyrin II ester.[9–11] The R_f is critically dependent on the experimental conditions.
m.p.: 291–292 °C (corrected).[5]
Other Properties: Practically insoluble in ether. May be crystallized as the hydrochloride. X-ray powder photograph.[5,12] Infrared spectrum.[12,13] May also be identified by conversion into coproporphyrin I by quantitative decarboxylation at 180 °C for 4 h in deoxygenated 1 M HCl,[6,14] and by 2,6-lutidine–water chromatography (see coproporphyrin I, Po-7).
Likely Impurities: Isomeric uroporphyrins,[5] hepta- (and lower) carboxylic esters,[9,13,14] haloporphyrins, and Cu or Zn chelates.

References

1. H. Fischer and H. Orth, *Die Chemie des Pyrrols*, Vol. II, Part 1, Akademische-Verlag, Leipzig (1937), p. 504.
2. H. Fischer and J. Hilger, *Z. Physiol. Chem.*, **149**, 65 (1925).
3. T. K. With, *Biochem. J.*, **68**, 717 (1958).
4. C. Rimington and C. A. Miles, *Biochem. J.*, **50**, 202 (1951).
5. S. F. MacDonald and R. J. Stedman, *Can. J. Chem.*, **32**, 896 (1954).
6. P. R. Edmondson and S. Schwartz, *J. Biol. Chem.*, **205**, 605 (1953).
7. D. Mauzerall and S. Granick, *J. Biol. Chem.*, **232**, 1141 (1958).
8. E. D. Becker, R. B. Bradley, and C. J. Watson, *J. Am. Chem. Soc.*, **83**, 3743 (1961).
9. J. E. Falk and A. Benson, *Biochem. J.*, **55**, 101 (1953).
10. J. E. Falk, E. I. B. Dressel, A. Benson, and B. C. Knight, *Biochem. J.*, **63**, 87 (1956).
11. P. A. D. Cornford and A. Benson, *J. Chromatogr.*, **10**, 141 (1963).

12. J. E. Falk and J. B. Willis, *Australian J. Sci. Res., Ser. A*, **4**, 579 (1951).
13. T. C. Chu and E. Chu, *J. Biol. Chem.*, **234**, 2741, 2751 (1959).
14. D. Mauzerall, *J. Am. Chem. Soc.*, **82**, 2601 (1960).

Po-23
Uroporphyrin II

Formula: $C_{40}H_{38}N_4O_{16}$
Formula Wt.: 830.77 (Octamethyl ester, 942.98)
Calc. %: C, 57.83; H, 4.61; N, 6.74; O, 30.82

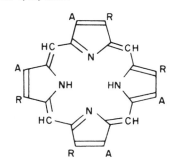

$A = CH_2COOH$

$R = CH_2CH_2COOH$

Source: By hydrolysis of the synthetic methyl ester.[1,2]
Spectral Reference Values: The same as for uroporphyrin III (Po-24).
Chromatography: R_f same as for uroporphyrin III.
Other Properties: Is decarboxylated to coproporphyrin II, either chemically[1,3] or enzymically.[4]

PROPERTIES OF THE OCTAMETHYL ESTER

Spectral Reference Values: The same as those of uroporphyrin III ester (Po-24).
NMR **Spectrum:** See Abraham *et al.*[5]
Chromatography: $R_f = 0.02$ in kerosene-*p*-dioxane, like that of uroporphrin I ester, but unlike those of uroporphyrin III ester or uroporphyrin IV ester.[6,7] R_f is critically dependent on the conditions.[8]
m.p.: 310–313 °C.[1,2]
Other Properties: As the least soluble of the isomers, it may frequently be separated from uroporphyrin mixtures by crystallization from pyridine,[1] or from chloroform–acetone.[1,2] Identify by decarboxylation and 2,6-lutidine–H₂O chromatography of the coproporphyrin II.[7,9] X-ray powder and infrared (mull) spectra.[1,2]
Likely Impurities: Other isomers, haloporphyrins, and Cu and Zn chelates.

References

1. S. F. MacDonald and K. H. Michl, *Can. J. Chem.*, **34**, 1768 (1956).
2. G. P. Arsenault, E. Bullock, and S. F. MacDonald, *J. Am. Chem. Soc.*, **82**, 4384 (1960).
3. P. R. Edmondson and S. Schwartz, *J. Biol. Chem.*, **205**, 605 (1953).
4. D. Mauzerall and S. Granick, *J. Biol. Chem.*, **232**, 1141 (1958).
5. R. J. Abraham, A. H. Jackson, and G. W. Kenner, *J. Chem. Soc.*, 3468 (1961).
6. J. E. Falk and A. Benson, *Biochem. J.*, **55**, 101 (1953).
7. J. E. Falk, E. I. B. Dressel, A. Benson, and B. C. Knight, *Biochem. J.*, **63**, 87 (1956).
8. P. A. D. Cornford and A. Benson, *J. Chromatogr.*, **10**, 141 (1963).
9. D. Mauzerall, *J. Am. Chem. Soc.*, **82**, 2601 (1960).

Po-24
Uroporphyrin III

Formula: $C_{40}H_{38}N_4O_{16}$
Formula Wt.: 830.77
Calc. %: C, 57.83; H, 4.61; N, 6.74; O, 30.82

A = CH$_2$COOH

R = CH$_2$CH$_2$COOH

Sources: From turacin, its Cu chelate.[1,2,*] Enzymically from 5-aminolevulinic acid or porphobilinogen.[3] By synthesis.[4]

Spectral Reference Values:
In 1 M HCl[5]

λ, nm	406	552	593	450(min)	520(sh)	570(sh)	584(min)
ϵ_{mM}	505	17.5	6.15	0.6	2.9	6.1	3.9
0.5W, nm	10.0	20.0	~14				

In 0.5 M HCl, ϵ_{mM} 541 at 406 nm.[6]

Chromatography: Not distinguishable from other uroporphyrin esters by 2,6-lutidine–H$_2$O chromatography
m.p.: Decomposes.
Other Properties: Insoluble in ether, soluble in cyclohexanone, slightly soluble in ethyl acetate–acetic acid. Reversibly reduced to uroporphyrinogen III, but the latter rearranges to a random mixture of isomers in hot 1 M hydrochloric acid.[5] Is decarboxylated to coproporphyrinogen III, either chemically[2,4,7] or, in the form of the uroporphyrinogen, enzymically.[8]
Likely Impurities: Inorganic salts, and Cu and Zn chelates; otherwise, as for the ester (Po-25).

References

1. J. E. Falk, E. I. B. Dressel, A. Benson, and B. C. Knight, *Biochem. J.*, **63**, 87 (1956).
2. R. E. H. Nicholas and C. Rimington, *Biochem. J.*, **50**, 194 (1951).
3. W. H. Lockwood and A. Benson, *Biochem. J.*, **75**, 372 (1960).
4. E. J. Tarlton, S. F. MacDonald, and E. Baltazzi, *J. Am. Chem. Soc.*, **82**, 4389 (1960).
5. D. Mauzerall, *J. Am. Chem. Soc.*, **82**, 2601 (1960).
6. C. Rimington, *Biochem. J.*, **75**, 620 (1960).
7. P. R. Edmondson and S. Schwartz, *J. Biol. Chem.*, **205**, 605 (1953).
8. D. Mauzerall and S. Granick, *J. Biol. Chem.*, **232**, 1141 (1958).

*S. Schwartz and C. J. Watson (unpublished) claim that some samples contain uroporphyrin I.

Po-25
Uroporphyrin III, Octamethyl Ester

Formula: $C_{48}H_{54}N_4O_{16}$
Formula Wt.: 942.98
Calc. %: C, 61.14; H, 5.77; N, 5.94; O, 27.15

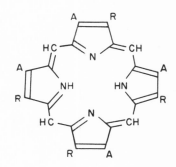

A= CH$_2$COOMe

R=CH$_2$CH$_2$COOMe

Sources: From turacin, the Cu chelate of the porphyrin;[1,2] by synthesis.[3]

Spectral Reference Values:
In chloroform[4]

λ, nm	406	502	536	572	627
ϵ_{mM}	215	15.8	9.35	6.85	4.18
0.5W, nm	22	23.5	16.5	20.5	12

With minima

λ, nm	458	522	554	608
ϵ_{mM}	1.9	3.3	1.4	0.85

Chromatography: $R_f = 0.5$ in kerosine (b.p. 190–250 °C)–*p*-dioxane (8:3 v/v); same as that of uroporphyrin IV ester.[5–7] The R_f is critically dependent on the conditions.[7]
m.p.: 267–269 °C (hot stage);[4] 255–260 °C; polymorphic.[3]
Other Properties: X-ray powder photographs[8,9] and infrared (mull) spectra[10,11] depend on the crystal form;[3] like the infrared spectrum in chloroform, they are worthless as criteria of purity.[3] However, they may suggest the presence of either or both of the uroporphyrin III and IV esters in complex mixtures.[3] The solubilities of the uroporphyrin III and IV esters are similar, and are higher than those of the other isomers. They also form solid solutions ("mixed crystals").[3] Compare also, uroporphyrin IV octamethyl ester. The purity of this isomer can only be assessed by decarboxylation to coproporphyrin followed by 2,6-lutidine–H$_2$O chromatography; this does not, however, distinguish between isomers III and IV.[3,4] The HCl No. is 5.0.[4]
Likely Impurities: Other isomers;[9] porphyrins having fewer methoxycarbonyl groups.[5,12]

References

1. R. E. H. Nicholas and C. Rimington, *Biochem. J.*, **50**, 194 (1951).
2. T. K. With, *Scand. J. Clin. Lab. Invest.*, **9**, 398 (1957).
3. E. J. Tarlton, S. F. MacDonald, and E. Baltazzi, *J. Am. Chem. Soc.*, **82**, 4389 (1960).
4. D. Mauzerall, *J. Am. Chem. Soc.*, **82**, 2601 (1960).
5. J. E. Falk, E. I. B. Dressel, A. Benson, and B. C. Knight, *Biochem. J.*, **63**, 87 (1956).
6. J. E. Falk and A. Benson, *Biochem. J.*, **55**, 101 (1953).
7. P. A. D. Cornford and A. Benson, *J. Chromatogr.*, **10**, 141 (1963).
8. C. J. Watson and M. Berg, *J. Biol. Chem.*, **214**, 537 (1955).
9. O. Kennard and C. Rimington, *Biochem. J.*, **55**, 105 (1953).

10. J. E. Falk and J. B. Willis, *Aust. J. Sci. Res., Ser. A*, **4**, 479 (1951).
11. T. C. Chu and E. J. H. Chu, *J. Biol. Chem.*, **234**, 2751 (1959).
12. T. C. Chu and E. J. H. Chu, *J. Biol. Chem.*, **234**, 2741 (1959).

Po-26
Uroporphyrin IV

Formula: $C_{40}H_{38}N_4O_{16}$
Formula Wt.: 830.77 (Octamethyl ester, 942.98)
Calc. %: C, 57.83; H, 4.61; N, 6.74; O, 30.82

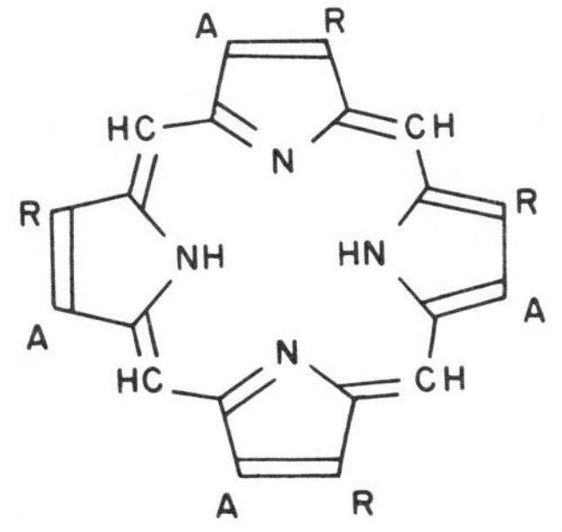

A = CH₂COOH
R = CH₂CH₂COOH

A = CH_2COOH

R = CH_2CH_2COOH

Source: By hydrolysis of the methyl ester[1,2] formed by synthesis.

Spectral Reference Values: Same as those of uroporphyrin III in 1.0 M HCl.

Chromatography: R_f same as that of uroporphyrin III (Po-24).

Other Properties: Is decarboxylated to coproporphyrin IV, either chemically,[1-3] or enzymically[4] (in the form of uroporphyrinogen).

PROPERTIES OF THE OCTAMETHYL ESTER

Spectral Reference Values: Same as for the III isomer (Po-25).

Chromatography: R_f same as for the III isomer (Po-25).

m.p.: 255–258 °C (polymorphic).[1,2]

Other Properties: The x-ray powder photographs and infrared (mull) spectra depend on the crystal form;[1,2] like the infrared spectrum in chloroform, they are worthless as criteria of purity.[5] The purity of this isomer can only be assessed by examining the derived coproporphyrin IV by decarboxylation and 2,6-lutidine–H₂O chromatography; this does not distinguish between isomers III and IV.

Likely Impurities: Other isomers, haloporphyrins.

References

1. S. F. MacDonald and K. H. Michl, *Can. J. Chem.*, **34**, 1768 (1956).
2. G. P. Arsenault, E. Bullock, and S. F. MacDonald, *J. Am. Chem. Soc.*, **82**, 4384 (1960).
3. P. R. Edmondson and S. Schwartz, *J. Biol. Chem.*, **205**, 605 (1953).
4. D. Mauzerall and S. Granick, *J, Biol. Chem.*, **232**, 1141 (1958).
5. E. J. Tarlton, S. F. MacDonald, and E. Baltazzi, *J. Am. Chem. Soc.*, **82**, 4389 (1960).

Radioactive Compounds

BY HORACE S. ISBELL

Much biochemical research is carried on with isotopically labeled compounds. Radioactively labeled compounds, in particular, are now widely available from commercial sources, but specifications for them are neither well understood nor simple to define. The Committee felt unprepared in this edition of *Specifications and Criteria for Biochemical Compounds* to establish specifications for radioactively labeled biochemical compounds, but it did feel the need to outline as follows some of the problems involved. Further attention will be paid in future editions to actual specifications for such compounds.

Nomenclature

Labeled compounds are classified as specifically labeled, uniformly labeled (U.L.), and generally labeled (G.L.). The position of the isotope atom in a specifically labeled compound may be shown by a locant following the name of the compound and preceding the symbol for the radioisotope. The recognized symbol for a radioisotope is the chemical symbol preceded by a superscript specifying the mass number with one exception: For ^{3}H (tritium), T is used alternatively in formulas and *t* alternatively in names, just as, for ^{2}H (deuterium), D is used in formulas and *d* in names. In some publications and catalogs, C^{14} is used for ^{14}C, but this practice is discouraged, because it is not in accord with internationally accepted recommendations for the symbols of isotopes in general. As a word, carbon-14 is acceptable, but C-14 is not, because this abbreviation

designates the position (No. 14) of a carbon atom in a molecule. Where no ambiguity can arise, a labeled atom may be indicated by an asterisk. In naming specifically labeled compounds, the location of the labeled atom may be indicated by the number of the atom involved according to conventional nomenclature rules, but when the labeled atom is attached to the molecule in a unique way, the position may be described by a locant such as *O-t, carboxyl*-^{14}C, *hydroxyl-t, amino-t,* etc., placed in square brackets directly attached to the name of the compound substituted. Typical names and formulas are*

CH_3-$\overset{*}{C}H_2$-OH, [*1*-^{14}C]ethanol; CH_3-CHT-OH, [*1*-^{3}H] ethanol; CH_2T-CH_2-OH, [2-^{3}H]ethanol; CH_3CH_2-OT, [*O*-^{3}H]ethanol; C*H_3S(CH_2)$_2$ CH(NH$_2$) COOH, L-[*methyl*-^{14}C]methionine; $CH_3S(CH_2)_2CH(NH_2)C$*OOH, L-[*carboxyl*-^{14}C]methionine.

The symbol U.L. denotes labeling that is distributed with statistical uniformity throughout the labeled molecule or a specified portion of the molecule. With tritium compounds, the symbol is used to denote uniform labeling in nonlabile positions only. Uniformly labeled carbon compounds are ordinarily prepared from $^{14}CO_2$.

* A complete description of this system for isotopic designation, which has been accepted by the IUB Commission of Editors of Biochemical Journals and by The American Chemical Society and The Chemical Society (London), may be found in the 1972 Instructions to Authors of the Journal of Biological Chemistry and the Biochemical Journal.[5]

Biosynthesis from $^{14}CO_2$ does not necessarily yield a product that truly is uniformly labeled, because of non-labeled endogenous materials that may enter the synthesis from the biological pool. For precise work, degradation of the product is needed to determine the degree of uniformity. For many purposes, a generally labeled (but not uniformly labeled) product is adequate. Compounds are designated G.L. when there is reason to believe that the distribution of the label is not uniform at the various labeled positions.

The specific radioactivity of a labeled compound is defined as the radioactivity percent (by weight). It may be expressed as microcuries per milligram (μCi/mg). The *molar radioactivity* may be expressed in the unit, millicuries per millimole (mCi/mmol), without allowance for the isotopic enhancement of the molecular weight. However, at very high specific activities, an appreciable fraction of the carbon-12 atoms of a compound have been replaced by carbon-14, with a significant increase in molecular weight; hence, the molar radioactivities obtained, if no allowance is made for the isotopic enrichment, are not true values.

The radioisotopes most commonly used in biological research are as tabulated:*

Isotope	Half-Life	β-Particle Energy (MeV)	Specific Activity
Carbon-14	5730 years	0.155	4.4 mCi/mg of ^{14}C
Tritium	12.3 years	0.018	9.6 Ci/mg of ^3H
Sulfur-35	88 days	0.168	42 Ci/mg of ^{35}S
Phosphorus-32	14.3 days	1.71	285 Ci/mg of ^{32}P

The decrease in radioactivity with time for each of the isotopes listed may readily be calculated from the half-life. For the times normally involved in biological research, the change in the radioactivity of a carbon-14 compound is negligible; with the other isotopes listed, the changes are substantial and a correction is necessary to obtain the radioactivity of the compound at any particular time after the measurement was made.

Chemical and Radiochemical Purity[1,2]

The distinctive properties of radioactive products, and the exacting techniques required in their use, make it necessary that labeled compounds meet rigorous requirements for purity. Purity means little unless it defines the suitability of the product for the intended use. Users of radioactive biological compounds are ordinarily interested primarily in radiochemical purity.

Radiochemical purity is the percentage of the radioactivity that arises from the specified isotope, in the specified position, in the specified compound. The

* Data from Ref. 4.

chemical purity is the percentage of the product that is in the stated chemical form. Chemical impurities may be immaterial, or very objectionable, depending on the projected use of the product. Addition to a labeled compound of pure "carrier" consisting of nonlabeled pure compound raises the chemical purity, but leaves the radiochemical purity unchanged unless further purification is carried out. For this reason, chemical purity is not a satisfactory criterion for labeled compounds.

Gas–liquid, paper, and thin-layer chromatography, reverse isotope-dilution analysis, derivative preparation, and spectroscopic methods are employed for testing both the chemical and the radiochemical purity. *Paper and thin-layer chromatography,* in conjunction with radioautography, are the techniques most widely used. The reliability of the methods depends on the use of developing solvents capable of separating the impurities from the pure compound. By searching the literature for specific systems capable of separating suspected impurities, and by examining the product by means of the systems selected, reliable conclusions may be drawn from chromatographic studies, but verification of the purity of the compound by a second method is always necessary.

Isotope-dilution analysis[3] is a very useful technique for determination of radiochemical impurities. The method may be applied to any compound provided that two isotopic forms of it are available. The method depends on (a) the change in the isotopic content (specific radioactivity) that follows dilution of an unknown amount of the labeled compound with a known amount of the pure, nonlabeled compound (reverse-dilution analysis), or (b) the change that follows addition of a known amount of a labeled compound to an unknown amount of the nonlabeled compound (isotope-dilution analysis). For measurement of radiochemical purity, a small sample of the labeled compound is diluted at least 100-fold with the pure, nonlabeled compound. After rigorous purification by recrystallization (or other technique), the specific radioactivity of the purified mixture is determined. The radioactivity of the isotopic compound before dilution, A_1w_1, equals the radioactivity of the compound after dilution, $A_2(w_1 + w_2)$ where A_1 and A_2 are the specific activities and w_1 and w_2 are the weights of the compound and the carrier (unlabeled compound), respectively.

A reverse-dilution analysis of a labeled compound by using an anticipated impurity as the carrier provides a critical test for specific impurity in products. The technique requires use of pure carriers and rigorous purification. When the carrier for the impurity contains some of the product, it will take up some of the radioactivity of the labeled product, and the sum of the activities of the carrier containing the product and of the carrier containing the impurities will exceed 100 percent. This complication is particularly troublesome with compounds that form mixed crystals. For example, samples of D-[*1*-¹⁴C] tagatose and D-[*1*-¹⁴C] sorbose (obtained by rearrangement of D-[*1*-¹⁴C] galactose) gives mixtures that are difficult to separate either by chromatographic methods or by dilution analysis. The chemical and radiochemical purity of a compound depends in large measure on the synthetic route employed in its preparation and the purification steps applied to the product. Hence, the manufacturer's literature accompanying the product should disclose the methods of synthesis and purification, as well as chemical and radiochemical impurities known to be present.

Determination of Isotopic Distribution

Ideally, each batch of a labeled compound should be tested for the location of the isotope, but this would greatly increase the cost of radiochemicals. For this reason, the investigator must make his own evaluation of the distribution of the isotope, taking into account the requirements of his research. In general, compounds (labeled with either carbon-14 or tritium) prepared by chemical or biological methods that have been shown to yield products labeled in specific positions are assumed to have the label in the position stated by the producer, unless evidence to the contrary develops. Methods for determining the distribution of the isotopes at various positions in labeled carbohydrates are discussed in Ref. 3.

RADIATION DECOMPOSITION

Labeled compounds undergo slow decomposition as a result of radiation effects. Only about 0.01 percent of the ¹⁴C atoms per annum are lost by disintegration, but the resulting β-irradiation produces ions and excited molecules that cause far greater degradation of the compound. The rate of decomposition is greatly affected by the physical state of the sample and by the presence of solvents, oxygen, etc. The rate of radiation-induced decomposition increases with increasing specific activity, and varies widely with different compounds. Any labeled compound that has been stored for over 6 months should be checked for decomposition before use.

Recommended storage conditions are (a) low temperature in absence of oxygen, (b) low molar specific activity, (c) dispersal of solids in a dry, inert gas or in vacuum, (d) dilute solution in benzene or similar solvent, and (e) exclusion of light.

Contamination Problems

In the United States, possession and use of radioisotopes are regulated by law, and purchasers of radioisotopes in quantities in excess of certain limited amounts must possess a license issued by the United States Atomic Energy Commission or one of the States, which have assumed local control. Researchers who employ radioisotopes in humans must hold either an AEC-313A License for nonroutine uses, or an AEC Broad License for medical uses. Precautions are needed in the use and handling of radioisotopes, not only to protect health, but also to keep the research laboratory and equipment free from radioactive contamination. For tracer studies, in order to maintain satisfactory working conditions, virtually no contamination can be tolerated.

Recommendations to Producers

The Committee suggests that the following information be made available to users of labeled compounds:

Chemical name, including
1. Type of radioactive label
2. Position or positions of radioactive label
3. Additional descriptive information (configuration, etc.)
Analysis
1. Specific activity by weight
2. Molar radioactivity, expressed on molecular weight of nonlabeled compound
3. Chemical purity
Tests of radiochemical purity
1. Chromatographic scans, with solvent systems used
2. Isotope-dilution analysis
3. Other tests
Distribution of radioactive label
1. How assigned (method of synthesis)
2. Assignment of radioactive label checked by degradation
3. Literature references that show the basis for assignment of the radioactive label to the positions indicated
Method of synthesis or preparation
1. Carriers added
2. Purification steps employed
3. Impurities or by-products removed
Packaging
1. Additives, solvents, and concentration (if in solution)
2. Solids under inert gas, air, or vacuum

REFERENCES

1. E. A. Evans, *Tritium and Its Compounds,* Van Nostrand Company, Inc., Princeton, New Jersey (1965).
2. J. R. Catch, *Carbon-14 Compounds,* Butterworth & Co., London (1961).
3. R. Schaffer, in *The Carbohydrates,* Second Edition, Vol. IIB, W. Pigman and D. Horton, eds., Academic Press, New York (1970), p. 765.
4. *Radioisotopes, Stable Isotopes, Research Materials,* Fifth Revision, July 1967, Isotopes Development Center, Oak Ridge National Laboratory, Oak Ridge, Tennessee.
5. Instructions to Authors, *J. Biol. Chem.,* Jan. 10 issue (1972); *Biochem. J.,* Jan. issue (1972).

Compound Index